DEDICATION

My portion of this book is dedicated to those individuals
with whom I have worked throughout these thirty plus
years, who provided me with an education in the
field. Mostly it is dedicated to my wife for her
patience and support.
—RICHARD KNOTEK

I would also like to acknowledge and thank the people
I have worked with over the years. I would also like to
thank Jane, my wife, for her patience as I worked
on this book.
—JON STENERSON

MECHANICAL PRINCIPLES AND SYSTEMS FOR INDUSTRIAL MAINTENANCE

Richard Knotek

Motion Industries

Jon Stenerson

Fox Valley Technical College

PEARSON

Prentice
Hall

Upper Saddle River, New Jersey
Columbus, Ohio

Library of Congress Cataloging in Publication Data

Knotek, Richard.
 Mechanical principles and systems for industrial maintenance / Richard Knotek, Jon Stenerson.
  p. cm.
 ISBN 0-13-049417-8
  1. Power transmission—Equipment and supplies. 2. Machinery—Maintenance and repair.
 I. Stenerson, Jon. II. Title.

TJ1051.K66 2006
 621.8'16—dc22   2005045886

Editor: Ed Francis
Production Editor: Christine Buckendahl
Production Coordination: Carlisle Publishers Services
Design Coordinator: Diane Ernsberger
Cover Designer: Bryan Huber
Production Manager: Deidra Schwartz
Marketing Manager: Mark Marsden

This book was set in Times Ten Roman by GGS Book Services, Atlantic Highlands. It was printed and bound by R. R. Donnelley & Sons Company. The cover was printed by Phoenix Color Corp.

Pearson Education Ltd.   Pearson Education Australia Pty. Limited
Pearson Education Singapore Pte. Ltd.   Pearson Education North Asia Ltd.
Pearson Education Canada, Ltd.   Pearson Educación de Mexico, S.A. de C.V.
Pearson Education—Japan   Pearson Education Malaysia Pte. Ltd.

PEARSON
Prentice
Hall

10 9 8 7 6 5 4 3 2 1
0-13-049417-8

The transmission of power by electrical and mechanical apparatuses is an integral part of our lives. Everywhere you look, electrical and mechanical devices—both simple and complex—serve people in a variety of ways. These machines accomplish feats of work through mechanical advantage that would be time consuming, difficult, and dangerous without them. We use them to farm, construct, mine, manufacture, produce, transport, protect, communicate, comfort, heal, and even amuse us. The rapid advance of technology has altered our workplace and lives, and in a very short time we have developed increasingly sophisticated and complex equipment. Automated process technology and force transformation are taking us places and doing things for us we only dreamed of a few years ago. Yet we have become increasingly isolated from these electro-mechanical servants that have become an integral part of our society. We need to keep pace with the changes in machine technology as well as have a basic understanding of how they work and how to repair them when they break.

The transmission and control of energy by machines entails the use of various mechanical and electrical power transmission components. This book will introduce you to these devices and give you an understanding of their operation and how they should be maintained. In the ideal world without such realities as friction, gravity, wear, and the unpredictable, machines would operate effortlessly and run forever. In the real world machines run down, wear out, leak, break down, and go clunk in the middle of the night. Because of this, they need to be continuously maintained and mended when they fail. Therefore, a realistic goal with electrical and mechanical systems is to keep them operating continuously at their peak efficiency, with maximum output at minimal cost.

In this book the identification, function, and maintenance of individual components are reviewed, as well as how they work together in a system. It is imperative that the maintenance technician take a systems approach when troubleshooting and repairing electrical and mechanical equipment. If the student reading this book takes only one idea from it and applies it to his or her work, let it be the concept of component connection. Troubleshooting systems is an integral part of any electrical or mechanical maintenance program. Troubleshooting guides are provided in most of the chapters within this book.

Chapter 1 is a general overview of the basic practices and procedures used in the maintenance, recording, troubleshooting, and installation of mechanical power transmission systems. Most manufacturing facilities will employ some form of predictive and preventative maintenance in their organization. This chapter introduces the student to those topics.

Chapter 2 covers industrial safety and lockout/tagout procedures. This is a very important chapter for the reader.

Chapter 3 is an introduction to the fundamentals of mechanical power transmission. Concepts and terms such as power, torque, velocity, mechanical advantage, overhung load, efficiency, and friction are explained.

Chapter 4 covers the basics of various types of lubricants and methods of lubrication. In addition, common terms used in the world of tribology such as viscosity, oxidation, additive, consistency, and compatibility are explained. Emphasis is put on giving the student a basic understanding of the importance of the proper application of the right type of lubricating agent.

Chapter 5 examines rigging. The chapter focuses on typical equipment, terminology, and safe practices for rigging and lifting.

Chapter 6 is a practical overview of the terminology and common types of fasteners. The chapter examines thread forms and specifications, as well as common processes and how to repair damaged threads and parts.

Chapter 7 looks at the issues pertaining to steel shafting on which power transmission components are mounted. Also addressed in this chapter are the various types of tapered locking mechanisms that are used to mount components onto a shaft.

Chapter 8 covers the different types of sealing devices used in machinery. Both static and dynamic forms of seals are detailed along with their installation and maintenance practices.

Chapter 9 takes an in-depth look at the world of bearings. The chapter covers extensively the different types, terminology, application, installation, maintenance, and troubleshooting of bearings. Of particular note is the section on failure analysis within this chapter.

Chapter 10 provides an overview of both V-belt and synchronous belt drives. The operating principles, identification, types, installation, and maintenance of belt drives are reviewed in this chapter.

Chapter 11 covers chain and sprocket drives. The different types of drive chain used in industry, as well as the operating and maintenance of these systems, are addressed.

Chapter 12 is an introduction to clutches and brakes. The basics of brakes, clutches, backstops, hold-backs, and various forms of anti-rotation devices are examined.

Chapter 13 addresses both rigid and flexible types of rotating shaft couplings. Emphasis is on proper identification, installation, removal, and maintenance of shaft couplings. An in-depth look at the precise alignment of shaft couplings is included.

Chapter 14 is a thorough look at gears and gear drives. It is divided into two sections: open and enclosed gears. Besides detailing the various types of gear drives and explaining gearing terminology, this chapter focuses on proper maintenance of gearing. It also includes a failure analysis section for gear troubleshooting.

Chapter 15 introduces the high-tech world of linear motion systems. It covers the various types of components used in a linear system.

Chapter 16 addresses material conveying systems. The different types of material conveyors, along with the components that make up the system, are detailed.

Belted, roller, bucket, drag, and screw conveyors are included in this chapter. A section is devoted to maintaining the proper function of belted conveyors.

Chapter 17 covers the fundamentals of fluid power and typical components. Emphasis is placed on practical examples and applications.

It is our objective that this book will clearly lead the reader through the identification, application, and maintenance of the most common components in a mechanical system. It is designed to be a practical guide that makes use of illustrations and graphics to support the text. Most chapters include troubleshooting concepts that will assist the student in comprehending the necessary maintenance practices associated with the topic. In addition, the student exercises and chapter quizzes should reinforce the objectives outlined at the beginning of each chapter.

Acknowledgments

We would like to acknowledge the numerous companies that provided technical information, pictures, and graphics for this book as listed in the text. A special thank-you goes to those reviewers and professionals who provided valuable comments and suggestions during the development phase: David Baritot, Monroe Community College; Harold Brinkley, Ivy Tech State College; Larry Chastain, Athens Technical College; William G. Frizelle, St. Louis Community College at Florissant Valley; Thomas Kissell, Terra Community College; Mike McCollough, East Mississippi Community College; and Ben Newby, Middle Georgia Technical College.

CONTENTS

CHAPTER 5 Rigging   80

CHAPTER 6 Fasteners   105

CHAPTER 14 Gear Drives   375

CHAPTER 15 Linear Motion Technology   424

Maintenance Practices and Principles

Maintenance of mechanical and electrical systems is necessary to keep them operating at their peak output. The basic organizational structure is discussed along with the importance of continued training of personnel. Concepts such as preventative and predictive maintenance are covered. Emphasis is put on teamwork, record keeping, and the connection that system components have on one another when troubleshooting and maintaining them. The purpose of this chapter is to cover the basics of maintenance terminology, ideas, and methods used in business and industry.

Objectives

Upon completion of this chapter, the student will be able to:
✔ Explain how a maintenance organization should be structured.
✔ Explain preventative maintenance and troubleshooting.
✔ Describe some of the predictive maintenance tools that are available.
✔ Explain why documentation and record keeping are important.
✔ Understand the role that training plays in maintenance.
✔ Explain why the maintenance technician has to look at the entire system when considering repairs.

INTRODUCTION

Machines require maintenance to keep them efficiently producing a quality product. Without friction, gravity, wear, and other unpredictable problems, machines would operate effortlessly and run forever. Unfortunately, machines wear out, leak, break down, and go clunk in the middle of the night. They need to be continuously maintained and repaired when they fail. The goal in any system is to keep it operating at its peak efficiency, with maximum output and minimal cost. There are many maintenance philosophies, programs, and buzzwords describing the process of fixing and maintaining machines. Preventative maintenance (PM), predictive maintenance, and proactive maintenance are just a few of the many ways of thinking and acting to solve maintenance challenges. Many of these programs have merits and an individual can find helpful ideas and solutions in each of them.

The significant idea is that one must be proactive in treating the causes of failure, not simply the symptoms. Being proactive involves many things; chief among them is setting targets and meeting them. Targets such as operation, safety (zero accidents), cleanliness, time, precision, and so forth, must be established and specified. Without targets and goals, it is difficult to measure progress. Precise control, inspection, disciplined monitoring, continuous feedback, and action to resolve problems are all part of being proactive. The maintenance organization and technician must be proactive in striving for failure avoidance and non-events. The reward is reduced costs, safe operation, and continuous production of the plant.

Proper maintenance practices will prevent "downtime" and maintain quality manufacturing standards. *Downtime* is time when the machine is not functioning and represents lost production. Ideally, downtime should not occur unexpectedly. Downtime should be planned and scheduled to replace or repair components that have seen their full service life. Predicting and preventing failure is less costly than fixing catastrophic machine failure. Running until a machine fails should not be the plan, nor is it necessary with modern predictive technology.

In today's competitive world, manufacturing plants cannot afford to produce poor quality product. Proper maintenance of equipment ensures efficient and high quality production. The level of skills required to operate and maintain machines has increased with advancements in technology. Maintenance technicians need to be adequately trained. Most equipment in a modern plant incorporates basic machines integrated with high technology devices. The operating speeds and loads of production equipment are higher now than ever before. Production demands have increased as the available pool of resources and manpower has diminished. The demands on maintenance technicians have also increased. The hammer and wrench are still required, but the toolbox should contain more than the basic tools. Industry needs all the advantages that education and technology can give us, to achieve more "up time" and less "downtime." Maintenance technicians need time, training, and the best tools available to make cost-effective repairs.

In order to succeed, the entire plant staff and organization need to be thought of as an integrated system driven by the right attitude. It is not maintenance versus production nor a competition with one another. The organization must function together as a closed-loop system, focusing on the same goals. In a closed-loop system, the actions and attitude of others affect all. The battle to maintain the peak operating condition of plant equipment is not only on the floor, but also in the overall attitude of the mechanic and management. It has to be emphasized that frequent repair is not acceptable. Getting it right the first time is the preferred option and involves accuracy and precision. Precision maintenance involves many practices such as accurate measuring, precise alignment, correct tensioning, record keeping, and the elimination of sloppy workmanship.

ORGANIZATION

To properly plan, schedule, communicate, and optimize the available resources, a maintenance department must be organized efficiently. There needs to be a chain of command as well as effective communication between workers, technicians, and management. Good communication is imperative to complete the assigned tasks and motivate the staff. All progress, statistics, graphs, results, and achievements should be posted for all to read.

Every employee involved in maintenance has a role to play. Maintenance engineers and management will have an outlook based on the realities of production and management goals, with a big-picture perspective. This is needed to set benchmarks and guide the team toward desired goals. Engineers use their knowledge to calculate, predict, and determine results. At the same time, the technician or mechanic in the trenches has a different perspective and direct exposure to the challenges in maintenance that the management team does not have. The technician's experience and suggestions are reality-based and must be carefully considered. Working together, they make up the team.

The maintenance organization generally will vary with the size of the operation. Small facilities might have only a single maintenance person per shift who is required to be a "jack of all trades." Larger plants will have many technicians grouped by their primary function. The duties and responsibilities of the maintenance staff will be determined by the organizational structure of the maintenance department. This structure can be based on plant operation or the skills of the maintenance personnel (see Figure 1–1).

Some maintenance personnel are hired and trained to perform a specific task such as welding or HVAC repair. These specialists are required to perform functions that the general maintenance mechanic is not qualified to do. In certain instances, licensing and certification is required to ensure that the individual is qualified to successfully perform tasks to industry-recognized standards. For example, a qualified specialist might be needed to weld high-pressure steam pipes to satisfy AWS (American Welding Society) standards.

Increasingly, the maintenance technician is required to be a "multi-craft" or "super-craft" employee. Super-craft technicians are required to be knowledgeable and skilled in multiple functions. The duties and responsibilities of most mechanics have increased due to economics and skilled manpower shortages. Welding, rigging, machine repair, and understanding electrical concepts are all broad-based skills that the average maintenance person needs. Regardless of title, the more knowledgeable the individual is, the more valuable he or she is to the organization. Maintenance technicians with multiple skills are required to economically maintain a modern plant.

Multi-skilled maintenance technicians are valuable because they are flexible, cost effective, and a good source of feedback. A flexible maintenance staff that is capable of handling most machine failures is an asset to the organization. When a system malfunctions, multi-craft technicians are able to troubleshoot, weld, repair, align, or perform whatever function is needed to get the equipment up and running,

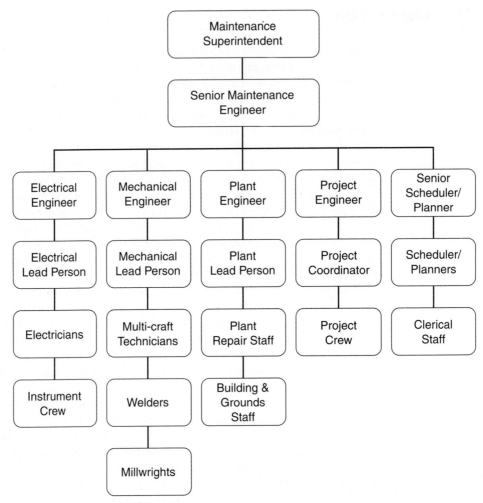

FIGURE 1–1
Typical organizational chart for a large maintenance department.

in many cases, without calling in a specialist. Although the multi-skill technician has broad-based skills, he or she will still have special expertise. Maintenance crews will develop a sub-organization within themselves and assign those most skilled for the task at hand to perform the needed repairs. Multi-craft technicians who are qualified to identify and repair the root cause, as well as communicate the plan of action to management, are an important part of the solution. Successful maintenance technicians must adapt and be proactive as industry and technology advances.

TRAINING AND EDUCATION

Multi-skilled maintenance technicians are not born—they are made. It is true that certain individuals, due to their genetic wiring, are more capable than others at comprehending and servicing an electrical or mechanical problem. These individuals with the right training, experience, and attitude will make exceptionally qualified specialists. Whether the individual is of average or exceptional skill, many things are needed to optimize job performance. Besides the right attitude, proper tools, management direction, and a safe working environment, training and experience are key elements in effective maintenance. Experience comes with time, and hopefully with minimal mistakes. Training needs to be provided for all employees.

Training of maintenance technicians has to be systematic. New machinery and technology require updated skills. On the job training is useful, but unless monitored and planned it can be haphazard or even disastrous. A scheduled training program should be developed by the management, employees, and training specialists and implemented prior to and during the technicians' work experience. The goals of training need to be established based on the requirements of the job and the present skill level of the employee. The objectives need to be made clear and communicated to all involved.

Feedback is essential. Testing prior to, during, and after each training session should be part of the program. This will ensure that the instructor is effectively teaching and the student is learning. The results need to be measured by written and performance-based tests. These results should be used only to provide feedback to the employee and to allow for adjustment to the program. Changes in the program need to be based on feedback, performance testing, and student/teacher evaluations. The program should not stagnate; it should be dynamic and circular in nature. Changes in staff and technology require a continuing educational process.

Training should be a combination of "soft-skill" and "hard-skill" training. Soft-skill training includes such things as computer software, OSHA, communications, and sensitivity. Hard-skill training includes such things as maintenance and machine repair, welding certification, alignment training, and much more.

The location for training can affect the outcome and results of the session. Training can be conducted at an established neutral facility (off-site) or at the plant (on-site). Facilities should be well established with adequate classrooms and lab facilities. The classroom must be clean, quiet, and comfortable and should not allow for interruptions, except in the case of emergencies. The student and teacher must have access to up-to-date manuals and working training fixtures. The lab should be well equipped and well lit. If training is of the hard-skill variety, the lab should be equipped with proper tools and adequate fixtures and mock-ups. Safety during class time and on the job has to be emphasized. It is imperative that a safe, clean, friendly atmosphere is provided for all training to maximize the effort, cost, and end results.

Time is critical to everyone. When the technician is in the classroom, he or she is not on the floor servicing machines. This creates additional expenses for the company in wages, training staff, and shortages of manpower during the training period.

Because of this, the time spent on training must be productive. Objectives must be clear to the instructor and the student. The trainer should have a course syllabus and workbooks that are easily understood by the student. The training should be condensed, but not at the cost of quality. The instructor must be concise in his presentation, capable of verbally communicating the topic, and have the experience to demonstrate the task. What is taught must be important, and what is learned must be useful on the job.

The maintenance technician should be hired from a pool of prospective educated employees. Most technical colleges offer an associate or bachelor degree in industrial maintenance technology and various engineering subjects. These individuals have hopefully received a good educational foundation that can be expanded throughout their employment. The training of maintenance personnel should never stop—it is a lifelong obligation.

Maintenance technicians must not allow their skills to stagnate. It is imperative that their skills and competence increase over time. This can be accomplished by gaining experience, reading journals, participating in seminars, and attending formal classes. Ongoing training and continuing education are available from a variety of institutions and organizations.

PREVENTATIVE MAINTENANCE

Preventative maintenance (PM) is usually a combination of scheduled and unscheduled work on any of the electrical or mechanical equipment in a plant. The goal is to minimize machine failures and optimize up time. PM serves the purpose of keeping the equipment operating at its peak efficiency beyond its normal service life and under safe conditions. The PM tasks and frequency are determined by the original equipment manufacturer (OEM), operating specialists, engineers, and experience. Most large industrial plants will use a computerized maintenance system that details, directs, tracks, and records the maintenance tasks that are needed and completed. Maintenance software systems are excellent tools for record keeping and eliminating guesswork.

The triggering mechanism for PM work is the work order, which lists the specific tasks to be performed. In addition to detailing the necessary steps, it includes information such as time, names, equipment numbers, required tools, and safety specifications. Work orders can be generated by the computerized PM management system.

The priority of tasks is based on their level of importance. The highest level of priority or the most important work would be situations that could compromise the safety of plant staff. Broken railings, loose grates, chemical leaks, hazardous spills, fire hazards, or electrical shorts are all typical examples of this type. The next level is any situation that immediately impacts production. Seized motor bearings, broken V-belts, stuck valves, and cracked housings are examples of this level of priority. The next level is that which affects efficiency and productivity. Vibration, cavitating pumps, small lubrication leaks, and high operating temperatures will cause a loss in efficiency or eventual downtime. The final level of priority is long-term projects. Obviously there needs to be flexibility within any system and priorities will be based on the present circumstances.

The cycle of maintenance tasks can be based on days, weeks, months, or the amount of operating time. Typically the periodic maintenance tasks will include routine inspection of the electrical, hydraulic, and mechanical systems. Scheduled

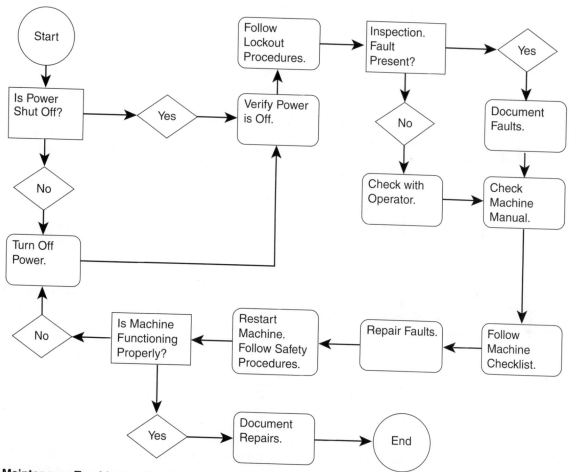

Maintenance Troubleshooting Report Number 1234
Equipment Number 48P   Date: 2-13-05 Report Author: Jim Jilbert
Equipment Description: Sludge Pump   Department: Pulp mill
Description of Problem: V-belt drive on pump is noisy and not functioning properly.
Probable Cause: V-belts are off of the sheaves or slipping.
Repair Procedure: Install, tension, and align new V-belts & sheaves.
Future Recommendations: Have operators routinely check pump drive for proper functioning of system & train staff on V-belt maintenance.

FIGURE 1–2
An example of a troubleshooting chart with documentation.

maintenance tasks at regular intervals will help prevent breakdowns. Unexpected repairs are costly and very inefficient.

Troubleshooting is an integral part of PM. Troubleshooting is the systematic checking of an electrical or mechanical system or circuit to determine the cause of a current or potential problem (see Figure 1–2). PM and troubleshooting are usually performed by the technician using a combination of checklist, experience, expertise, and creative effort to isolate and correct the malfunction. When the technician troubleshoots an electrical or mechanical problem, a recommended solution can be written into a work order. If he/she is qualified, an immediate remedy can save time, effort, and cost by preventing a breakdown.

It is imperative that the technician takes a systems approach when troubleshooting and repairing equipment. A systems approach involves looking at the big picture. Prior to making any adjustments or changes on a machine, the entire system must be considered. Most industrial systems are closed-loop systems. This means that the machine must be viewed as a system. Changes in performance or adjustments made to any component can affect the performance and operation of other components within the system. This connection can be termed "connectivity." Even simple mechanical systems consist of a series of parts connected to another component. Operating or positional changes that may be done to correct an immediate problem can cause a significant failure, over time, in other components. Small changes made to sensitive, complex electrical systems can adversely affect various inputs, sensors, and controls that can result in decreased performance or errors in the entire system. Changes made to one part in the system can cause a chain reaction to other components, which can then work back to the original component and compound the existing problem. For this reason, the direct or indirect connectivity of the various separate components should always be considered prior to repairs and adjustments. The importance of thinking of all electrical and mechanical components as a system is crucial to effective maintenance.

DOCUMENTATION AND SURVEYS

Adequate and accurate documentation of maintenance procedures, plans, prints, inventory, safety sheets, and schedules is crucial in the operation of an efficient maintenance department. Work orders, survey sheets, schedules, logbooks, and other documentation should be up-to-date, accurate, and in a database. This database must be accessible to those maintenance technicians whose job performance would benefit from this information. Increasingly, handwritten documents are being replaced with real-time computer maintenance systems. Paper is slowly being replaced with electronic documentation. A large amount of important data regarding all aspects of a machine can be stored and retrieved in an electronic form. Machines and operators typically record data about the operation of the machine into the company system. The data in the system is used to assist in scheduling, to monitor job performance, and to provide technical information that ultimately saves time and money.

Work Orders

Work orders are one of the most common documents used by most maintenance departments. A work order is a document that lists the various tasks, parts, dates, times, locations, and details required to accomplish a specific job (see Figure 1–3). Work orders can be triggered by request forms, emergencies, special projects, or as a routine procedure. They can be issued for a series of tasks as part of a project. Priority will be assigned to each task based on the level of importance and necessary sequential order. Work orders are part of computerized maintenance packages and can be viewed electronically or in a printed version.

Plant Work Order

Number: _____ Date: _____ Type: Routine/Scheduled

Priority Class: _____   Emergency

                                                     Project

Plant: _____   Machine Number: _____

Originator: _____   Location: _____

Schedule Date: _____ Shift: _____ Crew Number: _____

Beginning Date/Time: _____ End Date/Time: _____

Required Time to Complete Work: _____ Downtime: _____

Task/Work Description:

Parts Requested:

Comments/Recommendations:

Technician Initials: _____ Supervisor Initials: _____

FIGURE 1–3
Typical maintenance work order.

Plant Survey Form

Name: _____   Date: _____

Company: _____   Plant: _____

Location/Address: _____

Contact: _____ Machine/Conveyor Number:_____

Description: _____

Manufacturer: _____   Vendor: _____

Model Number: _____   Serial Number: _____

Parts List

Quantity	Description	Manufacturer	Part Number
_____	_____	_____	_____
_____	_____	_____	_____
_____	_____	_____	_____
_____	_____	_____	_____
_____	_____	_____	_____
_____	_____	_____	_____
_____	_____	_____	_____
_____	_____	_____	_____

FIGURE 1–4
Typical plant survey form.

Logbooks

Logbooks are a written record or electronic file that details the work performed during various shifts. Logbooks list the accomplished tasks and provide information to the maintenance staff on progress and what is required to accomplish a job. Logbooks should be integrated into the maintenance system and be updated regularly. Supervisors and technicians can review the logbook and issue work orders at the beginning or end of each shift based on the information recorded in the log.

Information recorded in a logbook by operators or technicians should be clear, concise, and complete. Good logbook notes can be very helpful.

Operator Manuals

Operator manuals and manufacturer instruction sheets are a reliable source of information. They should be used by those who are responsible for the installation, operation, and maintenance of a piece of equipment. They should be readily available for those who need them. They contain instructions for installation and operation and often have troubleshooting tips. Sophisticated electronic equipment cannot be safely and properly operated without the use of the equipment manual. Wiring diagrams and flow charts are typically included in the manual and are indispensable.

Plant Surveys

A plant survey is a list of all equipment and inventory. It is an assessment of all the mechanical and electrical components that make up a system or industrial facility. The survey should list the machine number, location, serial numbers, and key components (see Figure 1–4). The information can be gathered by plant staff or by outside consulting firms. Once the data are collected in written form, the details can be put into a computer for storage and easy access. The survey results should be used in conjunction with the computerized maintenance system to create master parts lists and equipment files, which are beneficial for planning and scheduling maintenance work. The survey data are also beneficial in determining the required inventory levels for the storeroom. Plant surveys create an equipment database that can be used to reduce maintenance time and cost.

PREDICTIVE MAINTENANCE

Predictive maintenance is monitoring the condition of operating equipment in an effort to predict malfunction or failure. A baseline of conditions and characteristics of individual machines is used as a reference for determining whether the machine is performing satisfactorily. A variety of data is periodically collected, analyzed, and compared to the baseline information to see if it is within established tolerances. A history of the operating characteristics for each piece of equipment is established and trends in performance are detailed. Any changes or events that occur are of special note because they may reveal a detrimental change in the system. An example of this might be your car. If you kept track of the mileage your car achieved on every tank of gas you could establish its baseline average gas mileage. If something detrimental changes in the car, the mileage will suffer over time and you could detect it before a breakdown.

Measuring the physical characteristics of a machine while it is operating is referred to as *condition monitoring*. Experience with the normal operating characteristics of specific machines and skills in interpreting the data are essential. Normal operating parameters must be established for each machine so that when monitoring data is collected, it can be compared to prior readings. A trained, experienced operator will be able to quickly and accurately interpret the data. The measurements should be collected by a variety of methods and procedures over an extended period. Some of these methods include visual, sound, vibration, temperature, oil analysis, and current draw readings. Using a variety of diagnostic tools to monitor the operating condition of equipment is recommended to give the technician a multi-faceted view of the machine operating condition.

Predictive maintenance programs should be planned and organized. Maintenance in a modern facility must not be haphazard. Quality predictive maintenance programs involve the investment of time, tools, skilled manpower, and training. Technically sophisticated equipment, such as thermal imaging guns or vibration data collectors, requires a trained and knowledgeable operator to accurately collect and interpret the information that is gathered. The data can be gathered and analyzed on a random, scheduled, or continuous basis. Random monitoring is usually triggered by a noticeable event that has been reported by a machine operator. This event can be something such as a sudden loud noise or distinct odor. Scheduled monitoring of operating equipment, which is done at specific time intervals, is preferred. This requires the analyzing technician to have an established scheduled and planned route. The data are collected in the plant using hardware of various types and by noting the time, place, machine number, and other pertinent information. This information is brought back and analyzed by downloading or entering the data into a PC with software that has been written for that specific purpose. The technician, with the aid of the software, interprets the data by comparing it to previously gathered information and established baseline tolerances.

Continuous monitoring of equipment can also be done in real time as the information is relayed electronically to a centralized control room. An operator can observe the read-out of various machine characteristics, such as temperature or vibration. Gauges, meters, and audible and visual alarms all can be wired into the monitoring system.

Early warnings of an impending problem will give the greatest chance for successfully preventing additional problems. Should the levels or data exceed preestablished limits, the operator or technician can notify the maintenance department to correct any potential problems. Work can then be scheduled and completed prior to a catastrophic failure.

Immediately after the machine repairs are completed and the machine is functioning at normal operating loads and speeds, monitoring data should again be obtained. This data—such as current draw, vibration, and temperature—can be used as a baseline for future reference.

VIBRATION ANALYSIS

Vibration analysis is one of the predominant predictive maintenance and condition monitoring tools used on rotating equipment. Vibration analysis is the monitoring of individual components and machines to analyze the vibration characteristics to determine their condition. Vibration can be defined as continuous back and forth motion, from a fixed reference, caused by a force over a given time period. Vibration is one of the best indicators used to judge the overall operating condition of equipment and components. Issues such as poor foundations, misalignment, gear-mesh defects, structural resonance, imbalance, and bearing failures can be detected through vibration analysis. If the source of the vibration can be isolated and the cause determined, the problem can be corrected.

Excessive vibration is detrimental to most machines and responsible for the failure of numerous pieces of plant machinery. The causes of vibration are numerous: unbalanced rotors, misaligned flexible couplings, loose hold-down bolts, and damaged bearings, to name a few. The results of vibration sometimes yield annoying noise. Another result is the eventual failure of the vibrating component and associated parts from being subjected to constant shaking. Fatigue failure, wear, and eventually catastrophic failure are all associated with vibration.

Vibration is inevitable, but control is the key. Every operating machine vibrates minimally even if it is functioning properly. This vibration is inherent within the machine due to minor manufacturing defects, clearances, and tolerances. Each machine will have an overall vibration pattern that is a complex sum of the combined signals of the various internal sources. Each one of the rotating or moving components will have a signature pattern that is its own. The science, or craft, is to isolate, measure, trend, and gauge the various vibration signatures and determine if they are within acceptable limits. Modern vibration analyzers are capable of performing this separating and measuring function. Most plants will have a certain level of ambient vibration present. Ambient vibration is the collective vibration caused by all of the operating machines present within the structure of the facility. This ambient vibration can be filtered or recognized by the current level of vibration monitoring equipment available.

To understand the basics of vibration analysis, we need to review certain terms, characteristics, and methods that pertain to vibration analysis (see Figure 1–5). The following definitions used to define the type, level, amount, various properties, and methods of vibration analysis are basic in nature. They are intended to give the student an overview of some of the significant terminology used in the science of vibration analysis.

Resonance

Resonance is the magnification of vibration and noise. The term resonance is often used interchangeably with natural frequency. Most mechanical metal components, such as a shaft or a pipe, have spring-like capability. Connected mechanical metal components are in essence a spring system. When subjected to a force and moved in one direction it will spring back when the force is removed. The spring is said to oscillate. Each spring system has a natural resonant frequency. When matched with a vibration frequency, it will resonate. The more rigid the part, the higher the natural frequency. Flexible parts will have a lower natural frequency. These high and low frequencies can be measured.

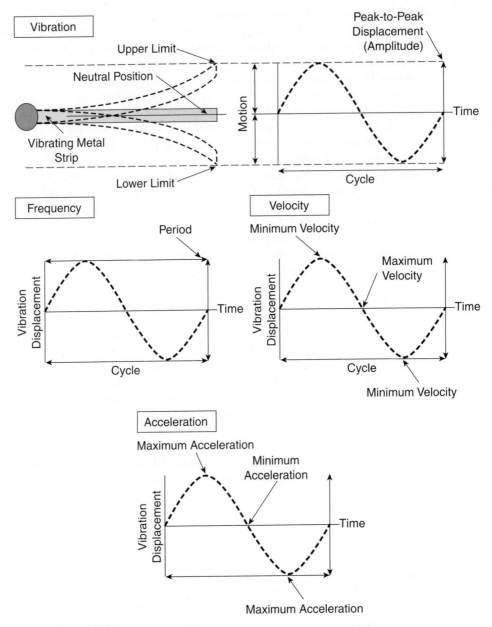

FIGURE 1–5
Vibration analysis terminology.

Rotating mechanisms will have what is known as a "critical speed." The critical speed of a component is the speed at which the part vibrates, or resonates, the greatest. This is sometimes referred to as its resonant frequency. Typically a rotor will pass through a critical speed where the vibration is at its worst. The vibration will lessen as the rotor moves beyond that speed. Rotating components, however, should not be operated at their critical speed. Machines can be dampened or de-tuned to limit the amount of resonance, but this is best left to the experts in the vibration field.

Cycle

Vibration cycle is the upper and lower limits of a motion from a neutral position. Vibration cycles are a frequency in time and are shown typically as a sine waveform in a graphic representation. This waveform graph or spectrum combines the amplitude, cycle, and frequency of the vibration.

Amplitude

Vibration amplitude is the extent or amount of the vibration movement from peak-to-peak. The peaks are the high points and low points from a neutral position. Amplitude is also referred to as the vibration displacement. Vibration displacement is typically measured in mils. A mil equals 1/1000 of an inch (0.001").

Frequency

Vibration frequency is the number of complete vibration cycles within a time period. Common units used to express vibration frequency are cycles per minute (cpm), cycles per second (cps), or multiples of rotational speed (orders). An order is a multiple of the rotating speed in revolutions per minute (rpm) frequency. Orders are generally used in vibration analysis because vibration problems are usually related to the speed (rpm) of the rotating component. Rotating objects are measured in revolutions per minute (rpm) that correspond with the vibrating frequencies expressed in cycles per second (cps). For example, vibration frequency readings for gearing will be the number of teeth multiplied by the speed (rpm), and radial ball bearings will have a high frequency corresponding to the number of balls times the shaft rpm.

Phase

Phase is the position of a vibrating component compared to another vibrating part at a fixed point and time. Phase readings taken on a rotor are expressed in degrees from 0° to 360°. Phase readings are typically used to determine imbalance in electric motor rotors and flexible shaft couplings.

Velocity

Vibration velocity is the rate, speed, and time required to travel from the highest peak of vibration to the neutral position of a vibrating object, which is its maximum velocity. It is expressed in inches per second (ips) or millimeters per second (mm/sec). To measure the velocity, many monitoring machines average the rate of speed using the root mean square (RMS) method. The RMS average is converted to a peak value within the monitoring instrument.

Acceleration

Acceleration is the increasing speed of vibrating movement. It is the change in velocity within a given cycle. Acceleration measurements are usually given in units of gravity (g's). One g is equal to 32.2 feet per second squared.

Transducers

A transducer is an electromechanical device that converts a vibration signal into an electrical signal. It is used in conjunction with analyzers to interpret and meter the vibration signals. Transducers are made in a variety of different types such as velocity transducers, accelerometer transducers, and displacement transducers. They operate in different ways, but all serve the same function to pick up the vibration of a machine or component and send it to an analyzer. Transducers can be the magnetic pick-up type or permanently attached to the machine. The placement of the transducer is critical to receive complete and accurate readings. Deviation in probe placement can produce variations that will not accurately reflect possible changes in vibration. It should be positioned at a consistent position where axial, vertical, and horizontal readings can be taken.

Analyzers

Analyzers are essentiality electronic data collectors and interpreters (see Figure 1–6). Most modern analyzers are portable hand-held microprocessor devices that are capable of transforming complex signals into an easily readable form and display for the operator. The monitor or analyzer is a hardware device that has input capability, display screen, and an attached probe. They have the capacity to scan and record a range of vibration sources, filtering and isolating those of importance. Filtered readings that are isolated are measures of a particular vibration frequency. Some analyzers collect and store the data for downloading into a computer with the appropriate vibration software. This allows the analyst to view and compare the data with past recordings or to standard industry limits.

There are different methods used in vibration analysis. Some of these include peak enveloping, narrow-band, broadband, spectral emitted energy, and signature analysis. Regardless of which method is used, the goal is to collect accurate readings,

FIGURE 1–6
Vibration analyzer used to monitor rotating equipment.
Courtesy of Ludeca Inc.

establish records, and develop a history, which then allows the analyst to compare the data to established limits and watch for trends.

OIL ANALYSIS

Oil analysis is a predictive maintenance tool that detects and analyzes oil samples for the presence of contaminants and to determine current physical properties. Contaminants can be defined as any unwanted substances or particles that enter the fluid. Particles in the lubricating fluids of a machine can be extremely destructive over time. Suspended small particles can be more destructive than large particles because they do not drop out of the oil. They continue in suspension as they are pumped throughout the apparatus, wearing the machine from within. Control of the level of degradation and contamination of the lubricating fluid is key.

Sampling of lubricating and hydraulic fluids is scheduled based on the critical nature of the equipment and if other monitoring methods have indicated a potential problem. For example, a sudden increase in temperature combined with grinding noises within a particular machine would warrant the sampling of the lubricating oil. A sample taken from a machine is analyzed for foreign particles of dirt, metal, acids, water, and other unwanted substances.

A variety of different means is used to count the size, type, and number of particles in a sample. Elemental analysis is one of the most common tests used in oil analysis. Elemental analysis, sometimes referred to as elemental spectroscopy, atomic emission spectroscopy (AES), or simply wear particle analysis, is more than just measuring the concentrations of wear metals such as iron, lead, and copper. Modern sophisticated analyzers are used to determine the concentrations of numerous different elements ranging from wear metals and contaminants to various oil additives. Trace elements and contaminants that are found in the sample are often measured in parts per million (PPM). Those contaminants, along with their size, can indicate the source. On an oil analysis report, the number of contaminants seen typically depends on the component type, oil type, environment, and so forth. They may range from a few parts per million (PPM) or several hundred PPM for wear metals and external contaminants, to several thousand PPM for certain lubricant additives. In addition, the report will indicate the viscosity, rate of oxidation, and general recommendations (see Figure 1–7).

The condition of the sample can be rated based on the results and previous collected data, normal operating environment, and experience. If the sample taken indicates that the oil has reached a critical level where the physical properties are not within acceptable levels and the lubricant is not capable of providing an adequate film, it should be changed.

When analyzing the oil sample data, it is important not only to look at the quantity and composition of each element, but also the trend line. The trend line is the change in element concentrations over consecutive samples, taken at various times from the same machine. This is important because wear rates will be different for each machine, depending on the components, manufacturer, oil type, age, usage, and operating environment, to name a few. Rate-of-change analysis can

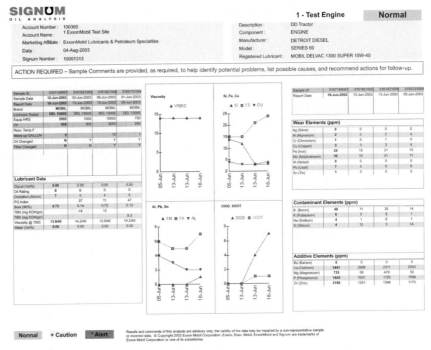

FIGURE 1–7

Sample oil analysis report.

Courtesy of Exxon/Mobil Oil.

be invaluable in finding signs of unexpected wear and contamination-related problems.

Finding the problem is one thing; correcting it is another. Sampling of fluids is the feedback portion of the loop. Potential contaminants need to be initially excluded as well as removed continuously from the system. Simple steps such as replacing filters regularly, effective seals, flushing out old fluids, using clean containers, proper handling and storage, along with common sense housekeeping, will serve the purpose of limiting contamination. Proactive maintenance calls for treating the cause and not the symptom.

The goal is two-fold: 1) Determine if the oil is contaminated and whether it has lost its initial lubricating properties, and 2) change it before it is significantly degraded and yields a loss of fluid film between mating surfaces, both resulting in machine failure.

THERMOGRAPHY

Thermography is the science of monitoring machine temperatures by use of a variety of temperature-indicating equipment. Temperature is one of the most significant indicators of the machine condition. High temperatures are generally detrimental to the optimum performance of any apparatus. Temperature-indicating devices can be portable or a permanent part of the operating equipment. Both portable and permanent types should be used as part of the tools a technician can use to analyze and troubleshoot equipment.

Real-time temperature monitoring devices that are a permanent part of the machine's system instantly relay the temperature to a monitor in a control room. An operator is assigned the task of keeping track of various gauges and screens that indicate the temperatures of numerous key pieces of production machinery. If the temperature becomes excessive an operator can alert the production, maintenance, and engineering departments so corrective action may be taken. Alarms can be triggered and emergency shutoffs can be wired into the system if the temperature reaches critical levels.

Portable devices can be simple tools such as a temperature-indicating crayon, or sophisticated infrared guns that require a trained technician to gather and analyze the data. A temperature crayon or temp-stick will be rated to a particular temperature. When it is touched to an object that is at or above the rating of the crayon, it melts. These devices are crude in comparison to the new technology available to the industrial maintenance technician. Hand-held temperature guns with targeting laser light are used as a tool to safely and quickly ascertain the temperature at various locations on a machine. Because all objects emit radiation, it is important to consider several key factors when using these hand-held devices. The operating environment, atmosphere, angle the gun is held, surface conditions, and color—*all* can affect the readings. The operator should be knowledgeable on the proper use of these instruments.

Modern infrared instruments are capable of pinpoint accuracy, isolating the intended object from interference, recording the data, and viewing the emissions on a screen. This data can be collected and further analyzed by software provided by the manufacturer of the instrument. Current instruments offer a way to simultaneously take temperature measurements and document them with digital photographs (see Figure 1–8). The integration of temperature data with digital photographs elevates

FIGURE 1–8
Hand-held temperature
sensing device.
Courtesy of Raytek.

the process of inspections by creating a real-time image with an object's temperature and its surroundings. These devices are used on electrical equipment, bearings, gearboxes, HVAC equipment, and just about anything that moves or emits heat.

The actual temperature of the monitored equipment and the fluids within are important, but change is the true indicator. As in other monitoring methods, the technician should be trending the data. Records of temperatures must be kept to establish a history and determine if the system has undergone a detrimental change. Portable temperature monitoring devices, in the hands of a skilled and trained operator, can be extremely useful in spotting potential problems.

Questions

1. What is the purpose of machine maintenance?

2. What does being "proactive" in maintenance mean?

3. Why is organizational structure important to maintenance?

4. Define "multi-craft" or "super-craft" maintenance technicians.

5. What role does training and education have in maintenance?

6. Define preventative maintenance.

7. What is a work order?

8. Explain what a "systems approach" is to maintenance work.

9. Define "troubleshooting."

10. Why is record keeping important in maintenance work?

11. What is "predictive" maintenance?

12. List and describe three types of diagnostic tools used in predictive maintenance.

CHAPTER 2

Safety

Safety should be considered the number one concern in the workplace, which has many dangers, including dangerous machinery and electrical hazards. Employees must always be aware of the dangers from which they must protect themselves and others. This chapter examines safe practices and lockout/tagout procedures.

Objectives

Upon completion of this chapter you will be able to:
- ✔ Define what an accident is. Explain why accidents happen.
- ✔ Describe how to prevent accidents.
- ✔ Describe what lockout/tagout is and how it is used in industry.
- ✔ Develop a lockout/tagout procedure.

ACCIDENT DEFINED

An *accident* is an unexpected action that results in injury to people or damage to property. Note that an accident is always unexpected. None of us ever think we will have an accident, but it happens to people just like us. An accident can easily result in the loss of an arm or leg, an eye, or even a life. Accidents cost enterprises huge amounts of money.

Most accidents are minor, but all are important. We should look for ways to prevent even minor accidents from recurring. A minor accident such as a cut could be more serious the next time. Every company is required to have procedures for reporting *all* accidents, even minor ones. Accidents are reported on forms so that the company can gather data in order to improve conditions and reduce the chance of future accidents and injuries. Companies often call these reports *incident report forms*.

CAUSES OF ACCIDENTS

Accidents can be prevented. Some of their causes follow.

1. Carelessness.
2. Use of wrong tools or defective tools or improperly using tools.

3. Unsafe work practices such as lifting heavy objects incorrectly.
4. Horseplay (playing around on the job).
5. Failure to follow safety rules or use safety equipment.
6. Inadequate equipment maintenance.

Items Commonly Sources of Accidents
1. Machines with moving or rotating parts.
2. Electrical machinery.
3. Equipment that uses high-pressure fluids.
4. Chemicals.
5. Sharp objects on machines or tools.

Remember that every piece of equipment can be dangerous!

ACCIDENT PREVENTION

Listed here are several ways to prevent accidents.

1. Design safety into the equipment. Place guards and interlocks on machines to prevent injury from moving parts and electrical shock. Provide lockouts on electrical panels so that an employee can lock the power out to repair equipment. Never remove guards or safety interlocks from machines while the machine is operating.
2. Use proper clothing, and eye and hearing protection including safety glasses. Employees in a noisy environment should wear ear protection. Never wear gloves, loose clothing, or jewelry around moving machinery.
3. Follow warning signs such as high voltage and danger signs. A yellow/black line means danger: Do not cross this line.
4. Follow safety procedures carefully.

SAFE USE OF LAB EQUIPMENT AND HAND TOOLS

Employees must follow procedures and learn good habits that will keep them safe on the job. Following is a list of safety guidelines.

1. Wear proper safety equipment, including eye glasses.
2. Know how to use the equipment. Do not operate equipment you do not understand.
3. Do not hurry.
4. Do not fool around.
5. Keep the work area clean.

6. Keep the floor dry.

7. Always beware of electricity.

8. Use adequate light.

9. Know first aid.

10. Handle tools carefully.

11. Keep tools sharp. A dull tool requires more pressure and is more likely to cause injury.

12. Use the proper tool.

13. Beware of fire or fumes. If you think you smell fire, stop immediately and investigate.

14. Use heavy extension cords. Thin extension cords can overheat, melting the insulation and exposing the wires or causing a fire.

OVERVIEW OF LOCKOUT/TAGOUT

On October 30, 1989, the Lockout/Tagout Standard, 29 CFR 1910.147, went into effect. It was released by the Department of Labor. Titled "The Control of Hazardous Energy Sources (Lockout/Tagout)," the standard was intended to reduce the number of deaths and injuries related to servicing and maintaining machines and equipment. Deaths and injuries resulting in tens of thousands of lost work days are attributable to maintenance and service activities each year.

Lockout is the placement of a lockout device on an energy-isolating device in accordance with an established procedure to ensure that the equipment being controlled cannot be operated until the lockout device is removed. *Tagout* is the placement of a tagout device on an energy-isolating device in accordance with an established procedure to indicate that the equipment being controlled may not be operated until the tagout device is removed. Only authorized employees who are performing the service or maintenance can perform tagout.

The lockout/tagout standard covers the servicing and maintenance of machines and equipment in which the unexpected start-up or energization of the machines or equipment or the release of stored energy could cause injury to employees. Machinery or equipment is considered to be energized if it is connected to an energy source or contains residual or stored energy. Stored energy can be found in pneumatic and hydraulic systems, springs, capacitors, and even gravity. Service and/or maintenance also includes activities such as constructing, installing, setting up, adjusting, inspecting, and modifying machines or equipment. These activities include lubrication, cleaning or unjamming machines or equipment, and making adjustments or tool changes. The standard is intended to cover electrical, mechanical, hydraulic, chemical, nuclear, and thermal energy sources. The standard also establishes minimum standards for the control of such hazardous energy. The standard does not cover normal production operations, cords and plugs under exclusive control, and hot tap operations.

Hot tap operations involve transmission and distribution systems for substances such as gas, steam, water, or petroleum products in the repair, maintenance, and service activities that involve welding on a piece of equipment such as a pipeline, vessel, or tank, under pressure, to install connections or appurtenances. Hot tap procedures are commonly used to replace or add sections of pipeline without the interruption of service for air, gas, water, steam, and petrochemical distribution systems. The standard does not apply to hot taps when they are performed on pressurized pipelines provided that the employer demonstrates that continuity of service is essential, shut down of the system is impractical, documented procedures are followed, and special equipment is used to provide proven, effective protection for employees.

Employers are required to establish a program consisting of an energy control (lockout/tagout) procedure and employee training to ensure that before any employee performs any service or maintenance on a machine or equipment when the unexpected energizing, start-up or release of stored energy could occur and cause

FIGURE 2–1
A typical electrical disconnect that is not locked out.

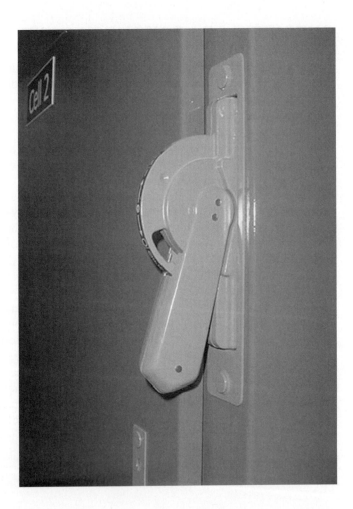

injury, the machine or equipment shall be isolated and rendered inoperative. The employer is also required to conduct periodic inspection of the energy control procedure at least annually to ensure that the procedures that were developed and the requirements of this standard are being followed. Only authorized employees may lock out machines or equipment. An authorized employee is one who has the authority and training to lock or tag out machines or equipment in order to perform service or maintenance on that machine or equipment.

Normal production operations, which include the utilization of a machine or equipment to perform its intended function, are excluded from lockout/tagout restrictions. Any work performed to prepare a machine or equipment to perform its normal production operation is called *setup*.

Another exclusion pertains to an employee working on cord and plug–connected electrical equipment for which exposure to unexpected energization or start-up of the equipment is controlled by unplugging the equipment from the energy source when the plug is under the exclusive control of the employee performing the service or maintenance.

An energy-isolating device is a mechanical device that physically prevents the transmission or release of energy. These devices include manually operated electrical circuit breakers; disconnect switches (see Figure 2–1), manually operated switches to disconnect the conductors of a circuit from all ungrounded supply conductors and to prevent independent operation of any pole; line valves (see Figure 2–2); and locks and any similar device used to block or isolate energy. Push buttons, selector switches, and other control circuit type devices are not energy-isolating devices. An energy-isolating device is capable of being locked out if it has a hasp or other means of attachment to which, or through which, a lock can be affixed or a locking mechanism built into it. Other energy-isolating devices are

FIGURE 2–2
A typical pneumatic disconnect.

capable of being locked out if this can be achieved without dismantling, rebuilding, or replacing the energy-isolating device or permanently altering its energy control capability.

New energy-isolating machines or equipment installed after January 2, 1990, must be designed to accept a lockout device.

LOCKOUT REQUIREMENTS

A lockout device uses a positive means such as a lock to hold an energy-isolating device in the safe position and prevent the energizing of a machine or equipment. A lock may be either a key or combination type. If a device is incapable of being locked out, the employer's energy control program shall utilize a tagout system.

Notification of Employees

The employer or an authorized employee must notify affected employees of the application and removal of lockout or tagout devices. Notification shall be given before the controls are applied to and after they are removed from the machine or equipment. *Affected employees* are defined as employees whose job requires operation or use of a machine or equipment on which service or maintenance is being performed under lockout or tagout, or whose job requires work in an area in which such servicing or maintenance is being performed.

TAGOUT

A tagout device must warn against hazardous conditions if the machine or equipment is energized and must include a clear warning, such as Do Not Start, Do Not Open, Do Not Close, Do Not Energize, Do Not Operate. A tagout device is a prominent warning device, such as a tag and a means to attach it, that can be securely fastened to an energy-isolating device in accordance with an established procedure to indicate that the equipment it controls may not be operated until the tagout device is removed by the employee who applied it. If the authorized employee who applied the lockout or tagout device is not available to remove it, specific procedures, which are given later, must be followed.

When used, tagout devices must be affixed in such a manner to clearly indicate that the operation or movement of energy-isolating devices from the "safe" or "off" position is prohibited. Tagout devices used with energy-isolating devices designed with the capability of being locked must be fastened at the same point at which the lock would have been attached. When a tag cannot be affixed directly to the energy-isolating device, it must be located as close as safely possible to the device in a position immediately obvious to anyone attempting to operate the device.

TRAINING

The employer must provide training to ensure that employees understand the purpose and function of the energy control program and have the knowledge and skills required for safely applying, using, and removing energy controls. Employees should be trained to:

Recognize hazardous energy sources.

Understand the type and magnitude of the energy available in the workplace.

Use methods necessary for energy isolation and control.

Understand the purpose and use of the lockout/tagout procedures.

All other employees whose work operations are or may be in an area where lockout/tagout procedures may be used shall be instructed about the prohibition of attempting to restart or reenergize machines or equipment that are locked out or tagged out.

When tagout procedures are used, employees must be taught the following:

Tags are warning devices and do not provide physical restraint on devices.

An attached tag is not to be removed without authorization of the person responsible for it, and it is never to be bypassed, ignored, or otherwise defeated.

Tags must be legible and understandable by all authorized employees, affected employees, and employees whose work operations are or may be in the area.

Tags may create a false sense of security. Their meaning needs to be understood by all.

Retraining

Retraining shall be provided for all authorized and affected employees whenever there is a change in their job assignments; in machines, equipment, or processes that present a new hazard; or in the energy control procedures. Additional retraining shall be offered when a periodic inspection reveals, or the employer has reason to believe, that an employee does not have adequate knowledge to use the energy control procedures. The retraining shall reestablish employee proficiency and introduce new or revised control methods and procedures as necessary. The employer shall verify that employee training has been accomplished and is being kept up to date. The verification shall contain each employee's name and dates of training.

Requirements for Lockout/Tagout Devices

Lockout and tagout devices must be singularly identified, the only device(s) used for controlling energy, and not used for other purposes. They must be durable, which means they must be capable of withstanding the environment to which they are exposed for the maximum period of time that exposure is expected. Tagout devices must be constructed and printed so that exposure to weather conditions, wet and damp locations, or acids and alkalis will not cause the tag to deteriorate or the message on the tag to become illegible (see Figure 2–3).

Lockout and tagout devices must be standardized within the facility in at least one of the following criteria: color, shape, or size. Print and format must also be standardized for tagout devices; they must be substantial enough to prevent removal without the use of excessive force or unusual techniques, such as the use of bolt cutters or other metal-cutting tools. Tagout devices, including their means of attachment, shall be substantial enough to prevent inadvertent or accidental removal. The devices must have a nonreusable type of attachment that is self-locking and nonreleasable with a minimum unlocking strength of at least 50 pounds. It should have the general design and basic characteristics equivalent to a one-piece, all-environment-tolerant nylon cable tie. They must be attached by hand. Lockout and tagout devices must identify the employee who applied them.

Application of Lockout/Tagout Devices The established procedures for applying lockout or tagout devices cover the following and shall be performed in the sequence indicated.

FIGURE 2–3
Typical tagout tag.

Warning

This machine is locked out

Reason - _____

_____

_____

_____

_____

_____

Name _____
Date _____
Time _____

1. Notify all affected employees that a lockout or tagout system will be used, and ensure that they understand the reason for the lockout. Before turning off a machine or equipment, the authorized employee must understand the type(s) and magnitude(s) of the energy, the hazard(s) of the energy to be controlled, and the method(s) to control the energy for the machine or equipment being serviced or maintained.

2. The machine or equipment shall be turned off or shut down using the procedures established for it. An orderly shutdown must be followed to avoid any additional or increased hazards to employees as a result of the equipment stoppage.

3. All energy-isolating devices needed to control the energy to the machine or equipment shall be physically located and operated in such a manner as to isolate the machine or equipment from the energy sources.

4. Lockout or tagout devices shall be affixed to each energy-isolating device by authorized employees. Lockout devices, when used, shall be affixed in a manner to hold the devices in a "safe" or "off" position.

5. Following the application of lockout or tagout devices to energy-isolating devices, stored energy must be dissipated or restrained. If reaccumulation of stored energy to a hazardous level is possible, verification of isolation shall be continued until the service or maintenance is completed or until the possibility of such accumulation of energy no longer exists.

6. Prior to starting work on machines or equipment that have been locked out or tagged out, the authorized employee shall verify that it has actually been isolated and de-energized. This is done by operating the push button or other normal operating control(s) to guarantee that the equipment will not operate.

The machine is now locked out or tagged out. Caution must be taken to ensure that operating controls are returned to the neutral or off position after the test. Before removing lockout or tagout devices and restoring energy to the machine or equipment, authorized employees shall follow procedures and take actions to secure the following:

The work area has been inspected to ensure that nonessential items have been removed and that machine or equipment components are operationally intact.

The work area has been checked to be sure that all employees have been safely positioned or removed.

Affected employees shall be notified that the lockout or tagout devices have been removed.

Each lockout or tagout device must be removed from each energy-isolating device by the employee who applied it. The only exception to this is when the authorized employee who applied the lockout or tagout device is not available to remove it; the device then may be removed under the direction of the employer provided that specific procedures and training for such removal have been developed, documented, and incorporated into the employer's energy control program. The employer must demonstrate the following elements:

The authorized employee who applied the device was not available.

All reasonable efforts were made to contact the authorized employee to inform him or her that the lockout or tagout device has been removed before he or she resumes work at that facility.

Outside Personnel Working in the Plant When outside service personnel (contractors, etc.) are engaged in activities covered by the lockout/tagout standard, the on-site employer and the outside employer must inform each other of their respective lockout or tagout procedures. The on-site employer must make certain that his employees understand and comply with the restrictions and prohibitions of the outside employer's energy control program.

Group Lockout or Tagout When a group of people perform service and/or maintenance, each person must use a procedure that protects him or her to the same degree that a personal lockout or tagout procedure would. The lockout/tagout standard specifies requirements for group procedures. Primary responsibility is vested in an authorized employee for a set number of employees. These employees work under the protection of a group lockout or tagout device (see Figure 2–4) that guarantees that no one individual can start up or energize the machine or equipment. The authorized employee who is responsible for the group must ascertain the exposure status of individual group members with regard to the lockout or tagout of the machine or equipment. When more than one crew, craft, department, and so on is involved, overall job-associated lockout or tagout control responsibility is assigned to an authorized employee. This employee coordinates affected work forces and ensures continuity of protection. Each authorized employee must affix a personal lockout or tagout device to the group lockout device, group lockbox, hasp (see Figure 2–4), or comparable mechanism when she begins work and shall remove those devices when she stops working on the machine or equipment being serviced or maintained.

Shift or Personnel Changes Specific procedures must be followed during shift or personnel changes to ensure the continuity of lockout or tagout protection between departing and arriving employees and to minimize exposure to hazards from

FIGURE 2–4
Hasps allow multiple personnel to lock out machines or equipment.

the unexpected energization or start-up of the machine or equipment or the release of stored energy.

Tests of Machines, Equipment, or Components In some situations, lockout or tagout devices must be temporarily removed from the energy-isolating device and the machine or equipment energized to test or position the machine, equipment, or component. The following sequence of actions must be followed:

1. Clear the machine or equipment of tools and materials.
2. Remove employees from the machine or equipment area.
3. Remove the lockout or tagout devices as specified in the standard.
4. Energize and proceed with testing or positioning.
5. Deenergize all systems and reapply energy control measures in accordance with the standard to continue the servicing and/or maintenance.

SAMPLE LOCKOUT PROCEDURE

The following is a sample lockout procedure. Tagout procedures may be used when the energy-isolating devices are not lockable, provided that the employer complies with the provisions of the standard that require additional training and more frequent and rigorous periodic inspections. When tagout is used and the energy-isolating devices are lockable, the employer must provide full employee protection and additional training and the more rigorous required periodic inspections. When more complex systems are involved, more comprehensive procedures may need to be developed, documented, and applied.

Lockout Procedure for Machine 37

Note: This normally names the machine when multiple procedures exist. If only one exists, it is normally the company name.

Purpose

This procedure establishes the minimum requirements for the lockout of energy-isolating devices when maintenance or service is performed on machine 37. This procedure must be used to ensure that the machine is stopped, isolated from all potentially hazardous energy sources, and locked out before employees perform any servicing or maintenance where the unexpected energization or start-up of the machine or equipment or release of stored energy could cause injury.

Employee Compliance

All employees are required to comply with the restrictions and limitations imposed upon them during the use of this lockout procedure. Authorized employees are required to perform the lockout in accordance with this procedure. All employees, upon observing a machine or piece of equipment that is locked out for service or maintenance, shall not attempt to start, energize, or use that machine or equipment.

A company may want to list actions to take in the event that an employee violates the procedure.

Lockout Sequence

1. Notify all affected employees that service or maintenance is required on the machine and that the machine must be shut down and locked out to perform the servicing or maintenance.

The procedure should list the names and/or job titles of affected employees and how to notify them.

2. The authorized employee must refer to the company procedure to identify the type and magnitude of the energy that the machine uses, must understand the hazards of the energy, and must know the methods to control the energy.

The types and magnitudes of energy, their hazards, and the methods to control the energy should be detailed here.

3. If the machine or equipment is operating, shut it down by the normal stopping procedure (depress the stop button, open switch, close valve, etc.).

The types and locations of the operating controls of the affected machine or equipment should be detailed here.

4. Deactivate the energy-isolating devices so that the machine or equipment is isolated from the energy sources.

The types and locations of energy-isolating devices should be detailed here.

5. Lock out the energy-isolating devices with assigned individual locks.
6. Stored or residual energy (such as that in capacitors, springs, elevated machine members, rotating flywheels, hydraulic systems, and air, gas, steam, or water pressure) must be dissipated or restrained by methods such as grounding, repositioning, blocking, bleeding down, and so on.

The types of stored energy, as well as methods to dissipate or restrain the stored energy, should be detailed here.

7. Ensure that the equipment is disconnected from the energy sources by first checking that no personnel are exposed; then verify the equipment's isolation by operating the push button or other normal operating controls or by testing to guarantee that the equipment will not operate. *Caution*: Return operating controls to the neutral or "off" position after verifying the isolation of the equipment.

> *The method of verifying the isolation of the equipment should be detailed here.*

8. The machine or equipment is now locked out.

Returning the Machine or Equipment to Service

When the service or maintenance is completed and the machine or equipment is ready to return to normal operating condition, the following steps shall be taken:

1. Check the machine or equipment and the immediate area around it to ensure that nonessential items have been removed and that its components are operationally intact. Check the work area to make sure that all employees have been safely positioned or removed from the area.

2. After all tools have been removed from the machine or equipment, guards have been reinstalled, and employees are in the clear, remove all lockout or tag-out devices. Verify that the controls are in neutral, and reenergize the machine or equipment. (*Note*: The removal of some forms of blocking may require reenergization of the machine before safe removal.) Notify affected employees that the service or maintenance is completed and the machine or equipment is ready for use.

Sample Lockout/Tagout Checklist

Notification I have notified all affected employees that a lockout is required and the reason for the lockout.

Date _____   Time _____   Signature _____

Shutdown I understand the reason the equipment is to be shut down following normal procedures.

Date _____   Time _____   Signature _____

Disconnection of Energy Sources I operated the switches, valves, and other energy-isolating devices so that each energy source has been disconnected or isolated from the machinery or equipment. I have dissipated or restrained all stored energy such as springs, elevated machine members, capacitors, rotating flywheels, pneumatic and hydraulic systems, etc.

Date _____   Time _____   Signature _____

Lockout I have locked out the energy-isolating devices using my assigned individual locks.

Date _____   Time _____   Signature _____

Safety Check After determining that no personnel are exposed to hazards, I have operated the start button and other normal operation controls to ensure that all energy sources have been disconnected and that the equipment will not operate.

Date _____   Time _____   Signature _____

The machine is now locked out.

Questions

1. What is an accident?
2. Explain three ways to prevent accidents.
3. Explain at least five things to do to be safe in a lab.
4. List the sources of energy typically found in an industrial environment.
5. What is an affected employee?
6. What is an authorized employee?
7. Define the term *lockout*.
8. Define the term *tagout*.
9. Describe the typical steps in a lockout/tagout procedure.
10. Write a lockout/tagout procedure for a cell that contains electrical and pneumatic energy.

CHAPTER 3

Mechanical Power Transmission Fundamentals

This chapter examines the basic principles of mechanical power transmission systems. The topics of force, work, power, torque, and the other terms that are imperative for the maintenance technician to comprehend will be addressed in a simple, understandable manner. The common terminology used to describe the actions and reactions of these machines will be explained.

Objectives

Upon completion of this chapter, the student will be able to:
✔ Define the basic function of machines and their components and how they provide a mechanical advantage.
✔ Understand how power and torque are transmitted in a system.
✔ Understand the effects that friction, inertia, and the efficiency of various components have on the transmission of power.
✔ Demonstrate the ability to calculate basic torque, horsepower, and speed changes in a system.
✔ Understand the effect that load classifications and service factors have on the performance and service life of various machine components.
✔ Be able to describe the concepts of overhung load and runout.

INTRODUCTION

The basic laws of physics govern the simplest and most complex machines. In the mechanical world we use the terms "power transmission fundamentals" or "mechanical principles" to describe these laws. It is helpful for the maintenance technician to have a basic understanding of these principles and how they apply to the machines and the terminology used in industry. In this chapter we will address some of the more important terms and aspects of mechanical systems.

Most mechanical industrial machines consist of a prime mover that converts one form of energy to another, the transactional components (mechanical linkages), and the driven equipment. The prime mover most often used in industry is the electric motor or the internal combustion engine. The transactional devices transmit and control the energy to the machine that is doing the work. The various components that transmit the energy are known as force transformers or mechanical power transmission components/devices. Applications such as material handling, pumping, converting, and processing equipment use various force transformers. Almost all of these devices use basic machines like the wedge, inclined plane, screw, and lever. Gear drives, belting, couplings, clutches, and chain and sprockets all transfer energy and in some cases give us a mechanical advantage by multiplying or changing the applied forces.

ENERGY

Energy is the active agent behind any force and has a capacity to do work. If an agent is capable of maintaining force and able to do work it is said to possess energy. Potential energy is stored energy. Kinetic energy is energy in motion. Energy exists in many forms such as heat, light, sound, electrical, chemical, and others. It can be transformed from one form to another but it cannot be created or destroyed. For example, a lump of coal has stored energy that can be released when it is burned.

Prime movers convert one form of energy to another. The electric motor converts electrical energy to mechanical energy. Different types of prime movers have certain operating and starting characteristics. Those characteristics include their efficiency, smoothness of operation, and starting torque capability. The type of prime mover used in a system, and its characteristics, can effect the operation and life of the other mechanical components. For example, a motor runs smoother than an internal combustion engine, resulting in less shock load to the connected devices.

FORCE

Force is defined as a pushing or pulling action on a body. It produces or prevents a change of motion. Force can cause a change of direction and in the velocity of a body. Force has a magnitude, direction, and a point of application. Quantities without direction are called scalars. Temperature and pressure are examples of scalars. In the United States the English measurement system is used, which measures mechanical force in pounds. The Newton is the metric unit used by the System International for computing forces—one pound is equal to 4.45 Newtons. Force and weight are generally interchangeable from a practical standpoint.

Vector analysis is used to calculate various forces acting on a body. A vector is a line that represents the magnitude and direction of the force. Geometry and trigonometry can be utilized to calculate the various forces that do not act in straight lines. When two or more forces are acting on a body (this is usually the case), the resultant force can be determined. If the forces act in the same direction, the resultant force is the sum of those vectors. If they are acting in opposite directions, the resultant force is the difference. Forces that act on a body at any angle can be computed using a graphical method.

The mass or weight of the mechanical component that is being installed, such as a sprocket, causes a force that affects the supporting bearings and shaft. When the chain is wrapped around the sprocket and tightened, additional force in a particular direction is imposed on the shaft and bearings. The maintenance technician must be conscious of the fact that the resultant force direction, magnitude, and point of application will determine the behavior, performance, and ultimately the life of the bearings and power transmission component.

INERTIA

The first law of motion, as defined by Newton, states "all bodies persist in a state of rest, or of uniform motion in a straight line, unless acted upon by an outside force." This means that a body at rest tends to stay at rest, and a body in motion tends to remain in motion unless acted upon by another force. This is the law of inertia. An example would be a flywheel, which stores energy and smoothes out the pulses of the prime mover. If not for the friction from the air and bearing surfaces, it would run forever. Inertia becomes a significant issue when we have to move or stop large, heavy loads quickly. Inertia is that property of an object that requires an outside force to be exerted on it to cause the object to accelerate.

FRICTION

The resistance to mechanical forces is called friction. Friction is the resistance to any effort to slide or roll one body on another. Resistance is inherent whether the surface of the bodies is solid or fluid. In the mechanical power transmission world, friction plays a dual role. It is an asset when we wish to stop or slow an object, as in the functioning of brakes and clutches. Friction between mating surfaces prevents movement. Without friction between mating surfaces, we would not be able to transmit power from one component of a machine to another. In some cases friction is a liability. It works against the free rotation of objects, such as in rolling bearings. Friction also causes a loss of efficiency and contributes to machines not operating at 100 percent of their capacity. The results of friction are energy loss, generation of heat, and component wear.

There are several known causes of friction between surfaces of mechanical systems when they come into contact with each other:

- The intermeshing of small surface irregularities sometimes called aspirates. The relative smoothness of each object can result in more or less friction. This grinding action results in worn surfaces and heat generated.
- Adhesion and cohesion due to a molecular attraction of the two surfaces.

■ The indentation of a hard surface into a softer one.

■ The weight of the object or forces holding the surfaces together.

There are three types of mechanical friction:

1. Static friction is the resistance to movement between two contacting surfaces at rest.
2. Sliding friction is the resistance to continued movement.
3. Rolling friction is between two surfaces that are separated by rollers.

The amount of friction present is dependent upon the force pressing the surfaces together, the type of materials, velocity, and degree of roughness of the surfaces in contact. Lubrication from oils and greases can provide a barrier film between two surfaces. The lubricant barrier aids in reducing the amount of surface contact, thus reducing the friction. Engineers have tested the amount of resistance to movement due to friction of various materials contacting each other and being forced to move. A term known as the "coefficient of friction" (μ) is used to describe the ratio between the force that is required to move an object and its weight. This coefficient of friction number has been established primarily through extensive testing (see Figure 3–1).

FIGURE 3–1

Coefficient of friction table listing the various factors of different mating materials.

Courtesy of Rockwell Automation.

Material	Static		Sliding	
	Dry	Lubricated	Dry	Lubricated
Aluminum on aluminum.	1.35	. . . .	. . . .	. . . .
Canvas belt on rubber lagging	0.30	. . . .	. . . .	. . . .
Canvas belt, stitched, on steel	. . . .	. . . .	0.20	0.10
Canvas belt, woven, on steel	. . . .	. . . .	0.22	0.10
Cast iron on asbestos, fabric brake material	. . . .	. . . .	0.35-0.40	. . . .
Cast iron on brass	. . . .	. . . .	0.30	. . . .
Cast iron on bronze	. . . .	. . . .	0.22	0.07-0.08
Cast iron on cast iron.	1.10	. . . .	0.15	0.06-0.10
Cast iron on copper	1.05	. . . .	0.29	. . . .
Cast iron on lead	. . . .	. . . .	0.43	. . . .
Cast iron on leather	.6	. . . .	. . . .	0.13-0.36
Cast iron on oak (parallel)	. . . .	. . . .	0.30-0.50	0.07-0.20
Cast iron on magnesium	. . . .	. . . .	0.25	. . . .
Cast iron on steel, mild	. . . .	0.18	0.23	0.133
Cast iron on tin.	. . . .	. . . .	0.32	. . . .
Cast iron on zinc	0.85	. . . .	0.21	. . . .
Earth on earth	0.25-1.0	. . . .	. . . .	. . . .
Glass on glass	0.94	. . . .	0.40	. . . .
Hemp rope on wood	0.50-0.80	. . . .	0.40-0.70	. . . .
Nickel on nickel	1.10	. . . .	0.53	0.12
Oak on leather (parallel)	0.50-0.60	. . . .	0.30-0.50	. . . .
Oak on oak (parallel)	0.62	. . . .	0.48	0.16
Oak on oak (perpendicular). . .	0.54	. . . .	0.32	0.07
Rubber tire on pavement.	0.8-0.9	0.6-0.7 *	0.75-0.85	0.5-0.7*
Steel on ice	0.03	. . . .	0.01	. . . .
Steel, hard, on babbit	0.42-0.70	0.08-0.25	0.33-0.35	0.05-0.16
Steel, hard. on steel, hard. . . .	0.78	0.11-0.23	0.42	0.03-0.12
Steel, mild, on aluminum.	0.61	. . . .	0.47	. . . .
Steel, mild, on brass	0.51	. . . .	0.44	. . . .
Steel, mild, on bronze	. . . .	. . . .	0.34	0.17
Steel, mild, on copper	0.53	. . . .	0.36	0.18
Steel, mild, on steel, mild	0.74	. . . .	0.57	0.09-0.19
Stone masonry on concrete . .	0.76	. . . .	. . . .	. . . .
Stone masonry on ground. . . .	0.65	. . . .	. . . .	. . . .
Wrought iron on bronze.	0.19	0.07-0.08	0.18	. . . .
Wrought iron on wrought iron .	. . . .	0.11	0.44	0.08-0.10

*   Wet pavement

ACCELERATION AND DECELERATION

Acceleration and deceleration can be thought of as the change in time a body moves over a given distance. Acceleration is an increase in speed over time.

$$\text{Acceleration} = \frac{\text{Final speed} - \text{Initial speed}}{\text{Time}}$$

Newton's law of acceleration states "The magnitude of the force necessary to produce an acceleration is proportional to the mass of the object being accelerated and the magnitude of the acceleration produced." Force, then, varies directly with mass, and with acceleration, as well as the product of mass times acceleration.

$$F \propto \text{mass} \times \text{acceleration}$$

If a single force is applied to a body, the acceleration produced is inversely proportional to the mass of the object being accelerated. Deceleration is the term used when speed is decreasing.

ANGULAR SPEED, ROTATIONAL DISTANCE, AND VELOCITY

Most prime movers deliver their output power into a rotating shaft. The angular speed is a measurement describing an object rotating around an axis point. The axis point is usually the geometric center, or centerline, of a shaft.

$$\text{Angular speed} = \frac{\text{Angular distance}}{\text{Time}}$$

The angular speed unit usually used in power transmission to measure it is "revolutions per minute" (rpm). The shaft of a typical electric motor might make one complete turn 1,750 times in a minute or 1750 rpm. Another unit used is radians per second (rad/sec). Angular distance can be the number of revolutions that the object turns. Time is usually measured in seconds, minutes, or hours.

To calculate the surface speed of a revolving body, it is necessary to know the diameter of the object and the rpm. The distance moved in one complete revolution (360 degrees) is the circumference of the round body or $\pi \times$ diameter. The constant π is rounded off to 3.1416 (the ratio of the circumference of a circle to its diameter).

In mechanical power transmission the surface velocity of a rotating object is usually measured in feet per minute (FPM). The surface velocity in feet per minute is equal to RPM $\times \pi \times$ diameter in feet or RPM $\times$ .262 $\times$ diameter in inches.

For example: A flat belt pulley has a 20 inch diameter and is rotating at 40 rpm.

$$40 \times .262 \times 20'' = 209.6 \text{ feet per minute (FPM)}$$

RADIUS OF GYRATION AND ROTATIONAL INERTIA (WR²)

WR^2 or WK^2, as it is sometimes referred to, is the radius of gyration. Rotating equipment is a complex sum of various shapes. The machinery designer has usually calculated the rotational inertia of the component. Rotational inertia increases as the square of the distance from the center of rotation to each particle of weight in the body. The radius of gyration is the distance from the axis of rotation to a theoretical point where all of the mass can be considered concentrated. The rotational inertia (WR^2) varies with the weight and the square of the radius of gyration.

WR^2 must be known or calculated for problems involving time or where cycle rates are such that heat calculations must be made. In most mechanical systems the rotating parts, such as gears, drums, rollers, and pulleys, do not necessarily operate at the same speed. In clutch and brake problems, it is common practice to calculate the WR^2 of the parts operating at each speed and reduce them to an equivalent WR^2 at the clutch or brake mounting shaft speed so they can be added together and treated as one unit. WR^2 must be taken into consideration when dealing with applications of large rotating masses or units that start/stop quickly and often.

WORK

Work is accomplished when a force acts on a body and moves it through a given distance. No work is done unless the force produces movement. Work is measured by the product of the force (pounds or Newtons) and the distance (feet or meters) through which motion takes place in a given direction of the resultant force. Linear work is equal to force × distance (see Figure 3–2).

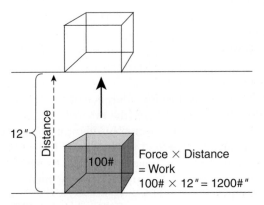

FIGURE 3–2
Linear work example showing the lifting of a 100-pound load 12 inches.

The formula for determining mechanical work is:

$$W = F \times D$$

Where: W = mechanical work in ft.-lb. or N-m

F = force applied in pounds or Newtons

D = distance moved in feet or meters

MECHANICAL ADVANTAGE

When a machine has the ability to increase or multiply the amount of force that is input to an output, it can be expressed as a ratio called the theoretical mechanical advantage. This ratio can be expressed by comparing the force applied to the load moved, resistance to effort, the radius of a wheel to the axle, and in a variety of other ways. This same ratio principle applies to determine mechanical advantage of mechanical power transmission components. One of the fundamental purposes of power transmission mechanisms is to multiply or transform the forces and speed by making use of this principle.

$$\text{Actual mechanical advantage} = \frac{\text{Actual load}}{\text{Actual force}}$$

BASIC MACHINES

A basic or simple machine is a device that effects the work to advantage. That advantage may be measured as the amount of force exerted, direction of movement, or the distance moved. Most complicated machines use the principles and are made up of simple machines. These basic machines are the lever, wheel and axle, pulley, inclined plane, wedge, and the screw.

The lever is one of the most common simple machines used to provide a mechanical advantage. The lever is a rigid bar moving about a fixed point. Practical examples of a lever include a wrench, pry bar, shovel, and wheelbarrow. A lever serves two purposes: It provides mechanical advantage and it can change the direction of the applied forces. There are three classes of levers. A first-class lever has the fulcrum located between the effort force and the resistance force or load (see Figure 3–3a). In a second-class lever, the resistance force is located between the fulcrum and effort force (see Figure 3–3b). With a third-class lever, the effort force is positioned between the fulcrum and the resistance force (see Figure 3–3c).

The wheel and axle combination is one of the oldest forms of a machine. It consists of a wheel rigidly attached to an axle and basically acts as a round second-class lever (see Figure 3–4a).

The pulley consists of a wheel that is mounted in a frame so that it can turn freely on a fixed axis (see Figure 3–4b). Pulleys can be combined to increase mechanical advantage such as in a block and tackle.

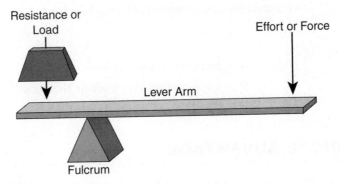

(a) First class lever.

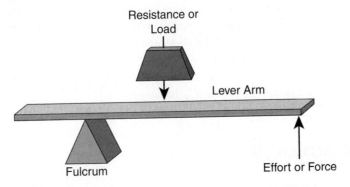

(b) Second class lever.

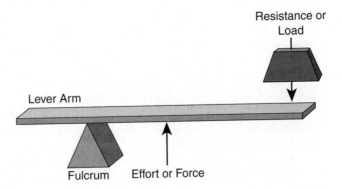

(c) Third class lever.

FIGURE 3–3
Three classes of levers showing relative positions of the applied force, the load, and the position of the fulcrum.

FIGURE 3–4a
Wheel and axle.

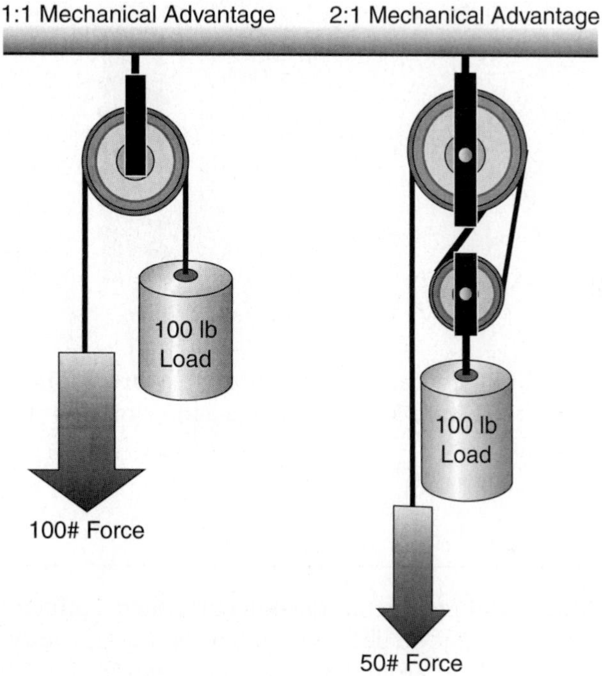

1:1 Mechanical Advantage   2:1 Mechanical Advantage

100 lb
Load

100# Force

100 lb
Load

50# Force

FIGURE 3–4b
Single and double pulley examples.

FIGURE 3–4c
Inclined plane.

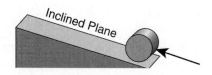

Inclined Plane

The inclined plane reduces the effort required to move a load through a given height. The work done remains constant but the effort (force) required decreases as the distance increases (see Figure 3–4c).

The wedge can be considered two inclined planes put together (see Figure 3–4d). Mechanical power transmission components such as the V-belt and tapered bushing are examples of a wedge in use.

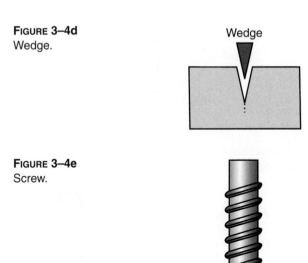

Figure 3–4d
Wedge.

Figure 3–4e
Screw.

The screw can be thought of as a circular inclined plane wrapped around a shaft (see Figure 3–4e). Fasteners and worm gear drives are examples of this sort of basic machine.

TORQUE

Mechanical power transmission components primarily operate with rotary motion. The term used to describe the mechanics of rotary motion is torque. Torque is a twisting effort around an axis (see Figure 3–5). That axis is typically the shaft on which the mechanism is mounted. Torque is defined as the product of the magnitude of an effort (force) by the length of the effort arm. Torque tends to produce the rotation of a lever about an axis. The twisting force (torque) is greater if the force or the length of the lever arm (radius) is increased. A simple formula defining torque is:

Torque = Force × Radius

In the English system, torque is commonly measured in inch-pounds or foot-pounds. The common metric unit is Newton-meter.

Output torque from a drive is needed for two reasons: 1) to accelerate the load to a desired speed and 2) to perform useful output work such as conveying materials, pumping fluids, cutting metals, and so forth. If the driver produces a greater amount of torque than is required by the load, both machines will accelerate. If the load requires more torque than the driver is capable of producing, both will slow down. Many prime movers, such as an electric motor, have 150% load capability for a short time, which may allow the required additional accelerating torque to be obtained without increasing the driver power rating. Higher inertia loads frequently require

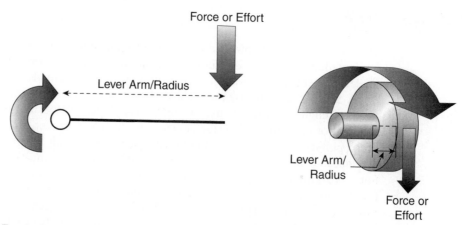

FIGURE 3–5
The above drawings illustrate the concept of torque, or twisting effort, around an axis point.

higher torques for acceleration than is required to maintain a desired running speed. The formula to calculate acceleration torque of a rotating component is:

$$\text{Torque} = \frac{(WR^2)N}{t}$$

Where: WR^2 = The total inertia (ft.²-lb.) that the motor must accelerate

$\quad\quad\quad$ N = Change in speed (RPM)

$\quad\quad\quad$ t = Time to accelerate (seconds)

Manufacturers provide speed-torque curves for both the driver and driven equipment. The curves can be used to select drive machines and/or load equipment. They also can be checked to help predict the performance of the system.

In certain systems, mechanical power transmission components are used to multiply torque and reduce speed, such as with a gear drive. Generally the increase in torque delivered at the output will be inversely proportionate to the decrease in the output speed.

POWER

Power is the product of force times distance divided by time. Work, which is force times distance, needs to be measured against the clock. The time required to accomplish a given amount of work over a given time period, or the rate of doing work, is the definition of power.

$$\text{POWER} = \frac{\text{WORK}}{\text{TIME}}$$

HORSEPOWER

The traditional unit of power adopted in the United States for mechanical engineering has been horsepower. Most prime movers are rated by their horsepower capabilities. In electrical engineering work, the kilowatt is used—which equals 1.34 horsepower.

A Scottish inventor by the name of James Watt patented a more efficient type of steam engine that was in use during the late 1700s. Watt established horsepower as a unit of measure to describe engine capacity. He supposedly determined that a horse, pulling for one minute, could lift about 32,400 pounds to a height of one foot. The number has been rounded off to 33,000. Tests have shown that an average horse working over a period of time works at the rate of about 3/4 horsepower. (Apparently Mr. Watt was also a good salesman.) The average laborer works at the rate of about 1/10 horsepower or less. Man is capable of generating more than a horsepower for only a short duration of time. That is why we find it useful to employ machines to give us a mechanical advantage and increase production in shorter periods of time. The following formulas can be used to calculate horsepower requirements. Formula 1 is used when the time cycle is measured in minutes and Formula 2 is used when the time cycle is in seconds.

HP Formula 1 $\text{Horsepower} = \dfrac{\text{Foot-pounds per minute}}{33,000}$

HP Formula 2 $\text{Horsepower} = \dfrac{\text{Foot-pounds per second}}{550}$

EFFICIENCY

The usable energy output of a machine is always less than the energy supplied to it. All machines are less than 100% efficient due to friction, windage, and misalignment. The efficiency of a machine can be expressed as a percentage and is equal to power output divided by the power input. The efficiency formula is:

% Efficiency = (100) power output/power input

When selecting and sizing power transmission components or determining the power requirements of the prime mover, efficiency must be taken into consideration. Any wear, misuse, poor maintenance, misalignment, or improper installation will contribute to the inefficiency of the device. It is also important to keep in mind that inefficiency multiplies as power is delivered through the system from mechanism to mechanism. In most mechanical power transmission systems, efficiency losses are considered for power requirements but not necessarily for speed and torque. Figure 3–6 shows approximate efficiencies of typical components.

FIGURE 3–6
Power transmission component efficiencies.

MACHINE	TYPICAL EFFICIENCY
Worm gear drives	50–90%
Lead screws	65–85%
V belt drives	95–97%
Helical gearbox	
Single reduction	98%
Double reduction	96%
Flat belt drives	97–98%
Roller chain drives	98%
Synchronous/Timing belt drives	98–99%
Flexible shaft coupling	99%

LOAD CLASSIFICATIONS

Loads imposed by machines can usually be put into three classes: constant torque, constant horsepower, and variable torque.

Constant torque loads are one of the most prevalent in power transmission applications. Figure 3–7 illustrates the load characteristic of machines requiring the same torque at all operating speeds. The horsepower requirements of the load vary directly with speed. Typical constant torque loads are conveyors, positive displacement pumps, piston and screw compressors, hoisting loads, and surface winding machines.

Constant horsepower loads require higher values of torque at lower speeds and lower values of torque at higher speeds. Examples of constant horsepower loads

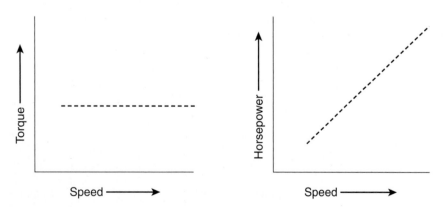

FIGURE 3–7
Constant torque load curves.

are lathes, milling machines, and center winders in the wire and textile industry. Figure 3–8 illustrates constant horsepower load curves.

Variable torque loads or those subject to "fan laws" describe the characteristics of fans and centrifugal pumps operating at various speeds. The torque required varies as the square of its speed. For example, if a fan requires 20 ft.-lb. of torque at 875 rpm, it will require 80 ft.-lb. of torque at 1750 rpm. As the speed doubles, the torque is increased as the square of the speed increases, or by a factor of four. Horsepower requirements increase as the cube of the speed. If the power required at 875 rpm is 3.3 HP, the required power would be 26.7 HP if the speed doubles. As the speed increased two times, its horsepower requirement increased eight times, or by the cube of the speed increase. The drive horsepower is based on the maximum operating speed because power requirements increase rapidly with increasing speed. Fan laws imply that there is no load at zero speed, but all machines have some static friction. That is why the drive selected must produce at least 15% to 30% starting torque at standstill. Examples of variable torque or fan law machines are centrifugal pumps, centrifugal blowers, fans, and centrifugal compressors. Figure 3–9 illustrates fan law.

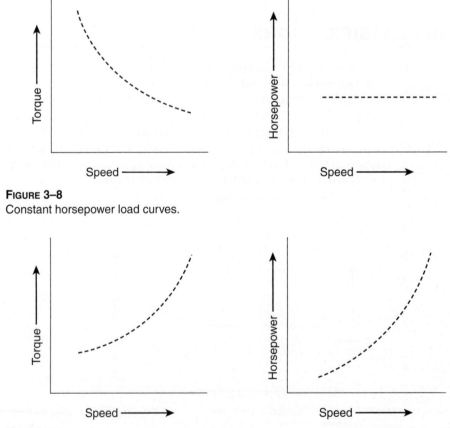

FIGURE 3–8
Constant horsepower load curves.

FIGURE 3–9
Variable torque curves.

SERVICE FACTOR

A service factor is a number applied to the load, power, or torque to reflect the overall operating parameters and conditions the power transmission component will function in. Each application has its own service conditions and requirements. The mode of service that will be applied can classify power sources and the driven components. These service factors reflect the environment, duty cycle, prime mover to be used, shock loads and other conditions. The service factor is applied to redefine the ratings in accordance with actual drive conditions. It is not a safety factor, which relates more to the ultimate strength of the component.

Organizations such as the American Gear Manufacturer's Association (AGMA) and the Rubber Manufacturer's Association (RMA) have made studies of the operating conditions of common industrial equipment. The studies took into account overloads, time periods (duty cycle), torsion, and vibration tendencies as well as the load peculiarities. From this data, machines were grouped into categories and service factors assigned (see Figure 3–10). These service factors are to be

Driven Machine Types Note: Certain machines may require flywheel sheaves or special construction to withstand heavy shock loads. Consult Mfg'r.	Driver: Normal Torque NEMA Des. A or B Motors DC Shunt Wound Motors Multi-Cylinder Engines			Driver: High Torque NEMA Des. C or D Motors DC Series Wound Motors Single Cylinder Engines		
	Service*			Service*		
	Intermit.	Normal	Contin.	Intermit.	Normal	Cont.
Agitators for Liquids Blowers and Exhausters Centrif. Pumps, Compressors Fans up to 10HP Light Duty Conveyors	1.0	1.1	1.2	1.1	1.2	1.3
Belt Conveyors, Bulk Mat'l Dough Mixers Fans over 10 HP Generators Line Shafts Laundry Machinery Machine Tools Punches, Presses, Shears Printing Machinery Positive Displ. Rotary Pumps Revolving & Vibrating Screens	1.1	1.2	1.3	1.2	1.3	1.4
Brick Machinery Bucket Elevators Exciters Piston Compressors Conveyors: Drag, Pan, Screw Paper Mill Beaters Piston Pumps Pos. Displacement Blowers Pulverizers Saw Mill, Woodworking Mach'y Textile Machinery	1.2	1.3	1.4	1.4	1.5	1.6
Crushers: Gyratory, Jaw, Roll Mills: Ball, Rod, Tube Hoists Rubber Calendars, Extruders, Mills	1.3	1.4	1.5	1.6	1.7	1.8
Chokable Equipment, Fire Hazzard	2.0	2.0	2.0	2.0	2.0	2.0

* Note:
Intermittent:
  Up to 6 Hrs/Day
Normal:
  6-16 Hrs/Day
Continuous:
  16-24 Hrs/Day

Adder for Idlers:
Outside on slack
  side 0.1
Inside on tight
  side 0.1
Outside on tight
  side 0.2

FIGURE 3–10

Typical service factor table listing various machines.

Courtesy of Rockwell Automation.

applied to the horsepower or torque rating of the device to yield an acceptable life. The more severe the application, the higher the service factor that should be applied when selecting the component.

> **EXAMPLE**
>
> A power transmission component that is running on a rock crusher, operating 24 hours per day in a poor environment, might have a 1.4 service factor assigned to it. The same component used on a lightly loaded sticky bun machine, operating 8 hours per day, might only require a 1.1 service factor.
>
> These service factors are then applied to the actual prime mover horsepower to give the "design" or "equivalent horsepower" that is used in the component selection. The formula used to determine design horsepower is:
>
> Motor horsepower × Service factor = Design horsepower

RUN-OUT

Run-out is a measure of roundness or eccentricity. Run-out can be measured in the vertical and horizontal planes. Radial run-out is a measure of roundness. Axial run-out is a measure of side-to-side movement or "wobble" (see Figure 3–11).

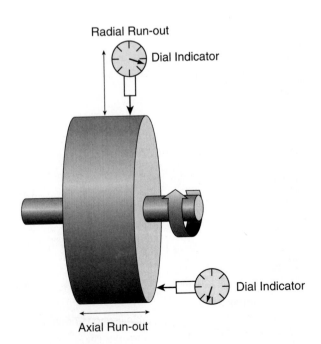

FIGURE 3–11
Measuring radial and axial runout of a rotating round object using dial indicators.

Radial Run-out

Dial Indicator

Dial Indicator

Axial Run-out

Many power transmission components are round and rotate at relatively high speeds. The closer to true roundness that the object is, the less vibration will occur due to imbalance. Excessive vibration is detrimental to the life of the component and the machine in which it resides. A run-out check can be done to analyze machine problems or as a step in an alignment procedure. Magnetic base dial indicators can be used to determine the amount of total indicator run-out (TIR) of the object. Manufacturers generally hold close run-out tolerances on the power transmission equipment they produce. A run-out check of older or worn components on critical or high-speed equipment is recommended. If the piece is not within acceptable run-out tolerances, it should be replaced.

OVERHUNG LOAD

Overhung load (OHL) is a force imposed on a shaft at right angles or perpendicular to its axis, outside of the bearing span (see Figure 3–12). It is the force that occurs when sheaves, pulleys, and gearboxes are mounted on a shaft and used as a power take-off. The positioning, size, and weight of the object, as well as the pull imposed on the shaft, have an effect on the amount of overhung load.

Excessive overhung load is one of the most prevalent causes of premature failure of rotating mechanical equipment. Drive shafts, bearings, seals, housings, flexible couplings, and gearing are all susceptible to damage caused by overhung loads.

Extreme overhung loads can be created in a variety of ways. The actual weight of the machine part can be too great for the capacity of the shafts and bearings. Diameters of rotating sheaves and sprockets that are below industry recommended minimums can cause excessive loading. Thermal growth of machine elements can change the positioning of equipment, imposing the load. Alignment of all shafts plays an important role in reducing overhung loads. Severe misalignment of flexible couplings and improper tensioning of connecting chain or belting will create loads beyond the

FIGURE 3–12
Overhung load on a
shaft and bearings.

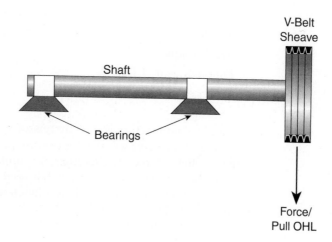

Shaft

V-Belt
Sheave

Bearings

Force/
Pull OHL

capacity of the shafts and bearings. The more tension the maintenance technician places on the belts or chains, the greater the overhung load. The placement of the item on the shaft, relative to the nearest bearing center, will create more or less load. All of these issues need to be taken into consideration when installing and maintaining mechanical power transmission equipment.

Most machinery is designed and produced by manufactures to accept certain reasonable levels of this type of load. Manufacturers provide tables for specifying the proper sizing and placement of the power transmission component onto their machine to prevent failure due to overhung load.

It is possible to calculate overhung load by using the following formula:

$$\text{Overhung load} = \frac{126,000 \times \text{HP} \times \text{Fc} \times \text{Lf}}{\text{PD} \times \text{RPM}}$$

Where: HP = Horsepower

Fc = Load connection factor

Lf = Load location factor

PD = Pitch diameter of the sprocket, sheave, or gear mounted

RPM = Revolutions per minute of the shaft

The load connection factor (Fc) describes the type of power transmission component mounted on the shaft. A flat belt has relatively high tension to transmit the load by friction. A V-belt has significant amounts of tension due to the seating of the belt in the grooves in order to transmit the load by friction. A synchronous belt has slightly less tension than a V-belt. A gear has a separating force related to the pressure angle of the tooth form, which produces moderate amounts of the load. A roller chain, if correctly installed, has very little if any tension. Each type of connecting component is given a different factor to account for the loads. Fc connection factors for various drives are:

Flat belt = 2.5

V-belt = 1.5

Synchronous = 1.3

Pinion or gear = 1.25

Sprocket/roller chain = 1.0

The load connection is directly related to tension or pull on the system. Over-tightening of the drive will result in excessive overhung loads that will cause failure of the machine elements. Therefore, it is imperative that proper tensioning procedures for belts and chain systems be followed.

The load location factor (Lf) is a function of the center of the applied load on the drive shaft. The more distance the load is located on the shaft from the supporting bearing, the greater the moment. Power transmission component manufacturers recommend a location of one shaft diameter from the shaft seal or cover plate. Positioning the sheave, sprocket, and other elements at this distance will keep the loads within acceptable limits. Locating the load as near as possible to the bearing, within alignment constraints, will reduce the moment and decrease the load location factor.

Changing the component to a smaller diameter will increase the overhung load. By increasing the radius at which the net force operates to produce torque, the net force—in proportion to the increase in radius—will be reduced while maintaining the same amount of torque. The result will be that OHL on the shaft will be reduced. Most gearbox and electric motor manufactures list minimum sizes that can be mounted on the appropriate shaft.

Altering the bearing arrangement can be an effective solution to extreme OHL conditions. If practical, a larger bearing with higher load ratings can be installed into the system. If this cannot be done, an outboard bearing supporting the opposite end of the shaft, or setting up a "jack-shaft" assembly, will reduce the OHL. A "jack-shaft" is a separate shaft supported by bearings.

To summarize, there are several different ways to limit the OHL on the shafts and bearings, some or all of which should be considered when applicable:

- Properly tension the connecting drive.
- Increase the diameter of the connecting component within reasonable limits.
- Locate the load and component as close as possible to the supporting bearing.
- Install a larger or another bearing.

The concept of overhung load and the implications it has on the proper functioning of the system cannot be emphasized enough. All maintenance engineers, mechanics, and technicians are routinely involved with OHL. Careful consideration of the causes and effects of overhung load, and implementing proper corrective measures will yield greater machinery life.

Mechanical Power Transmission Formulas

$$WORK = \text{FORCE} \times \text{DISTANCE}$$

$$POWER = \frac{\text{WORK}}{\text{TIME}}$$

$$TORQUE = \text{FORCE} \times \text{RADIUS (lever arm)}$$

$$TORQUE \text{ (inch pounds)} = \frac{\text{HORSEPOWER (HP)} \times 63025}{\text{RPM (revolutions/minute)}}$$

$$VELOCITY \text{ (Feet/Minute)} = .262 \times \text{DIAMETER} \times \text{RPM}$$

$$RPM = \frac{VELOCITY\ (FPM)}{.262 \times DIAMETER}$$

$$HORSEPOWER\ (HP) = \frac{\begin{matrix} DISTANCE & FORCE \\ MOVED\ (feet) \times EXERTED\ (lbs) \div 550 \end{matrix}}{TIME\ (seconds)}$$

$$HORSEPOWER\ (HP) = \frac{FORCE \times VELOCITY\ (FPM)}{33,000}$$

$$HORSEPOWER\ (HP) = \frac{TORQUE\ (inch\ pounds) \times RPM}{63025}$$

$$OVERHUNG\ LOAD = \frac{126,000 \times HP \times Fc \times Lf}{PD \times RPM}$$

Student Exercise

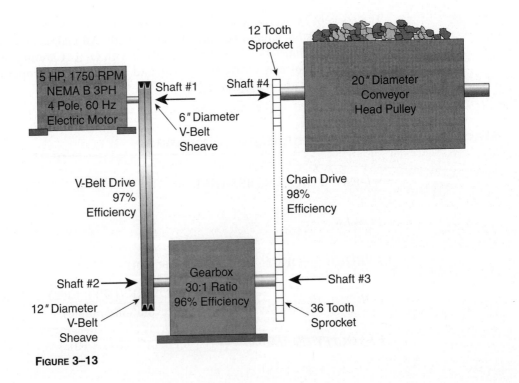

FIGURE 3–13

Student Exercise Data Sheet

Name _____

Date _____

For the application on the exercise sheet, calculate the torque, speeds, horsepower, and ratio at each shaft. Also determine the velocity of the conveyor pulley in feet-per-minute. Fill out the data sheet.

Shaft#	Torque	RPM	Ratio	HP
1				
___	___	___		___
2			___	
___	___	___		___
3			___	
___	___	___		___
4			___	
___	___	___		___

Pulley Velocity (FPM) = _____

Calculate the torque, speeds, horsepower, and ratio at each shaft (Figure 3–13). This is a constant horsepower load. Although the horsepower delivered to the driven load at shaft #4 will be lower than the 5 HP supplied at shaft #1 because of efficiency losses inherent in the power transmission components, the required HP to move the load remains constant. Also determine the velocity of the conveyor pulley in feet-per-minute for the following application. The gravel conveyor operates 24 hours per day. Fill out the data sheet.

Questions

1. What is force?
2. What is a prime mover?
3. What is power?
4. Define linear work.
5. Define inertia.
6. What causes friction between two surfaces?
7. What role does friction play in the transmission of power?
8. How do we measure or calculate the velocity of rotating components?
9. Name the basic machines.
10. Define mechanical advantage.
11. How do power transmission devices yield a mechanical advantage?
12. What is torque?
13. What is the speed/torque relationship?
14. What is the definition of horsepower?
15. Define efficiency.
16. What is a service factor?
17. What are the three load classifications of machinery?
18. Describe radial and axial run-out.
19. What is overhung load?
20. Why is it important to consider overhung load when installing machine components?
21. How can overhung load be minimized?

Lubrication

Bearings, gears, drive chains, and many other types of mechanical systems require lubrication to function properly. The basic types of lubricants and their purpose, the composition of lubricants and their various additives, and the effective application of lubricants are covered in this chapter.

Objectives

Upon completion of this chapter, the student will be able to:
✔ Explain the function of lubrication.
✔ Define the basic important lubrication terms.
✔ Describe the composition of basic lubricating fluids.
✔ Explain the make-up of different types of greases.
✔ Detail the advantages of lubricating with oil or grease.
✔ List basic lubrication systems and describe how they work.
✔ Comprehend and describe safety concerns pertaining to lubricants.

INTRODUCTION

Lubrication is the process of applying lubricants. Lubricants are substances applied to mating surfaces to reduce friction, prevent corrosion and wear, provide a barrier against contaminants, and assist in cooling.

The main function of a lubricant is to form a fluid film between moving machine components. If a lubricating film separates the solid objects, the amount of friction is less and the mating surfaces are not abraded through contact with each other. The study of how lubrication affects friction and moving parts is called "tribology." Tribology is the science of the mechanisms of friction, lubrication, and the wear of contacting surfaces that are in motion.

Friction is the resistance to rolling or sliding movement of one object on another. Friction also causes some of the energy used to produce work to be converted to heat. Heat can damage machine components if it becomes excessive. Lubricants have very low coefficients of friction. Coating metal surfaces that are rubbing against each other with grease or oil lowers the coefficient of friction (see Chapter 3, Mechanical Power Transmission Fundamentals). Bearings and meshing machine elements (gears) function more efficiently and wear less with a proper lubrication film.

Machine breakdown, expensive repairs, and the resulting production downtime are minimized by the proper application of correctly selected quality lubricants.

Large or small particulate contaminants, water, or other undesirable chemicals in the lubricant interrupt the all-important fluid film. Lubricants help keep out most contaminants. The contaminants that do enter are dispelled from the mating parts. Heat generated as a by-product of friction is carried away by circulating lubrication systems. Corrosion is reduced when the surfaces are covered with a lubricant, preventing exposure to air and moisture.

LUBRICATION PRINCIPLES

Formulating, manufacturing, and applying lubricants is complex. Numerous authoritative books and publications have been published on the science of tribology that should be consulted when making lubrication decisions. Complex lubricating compounds are created and refined to produce a product that is designed to work in a particular environment and application, and under very specific circumstances. A change in conditions—or attempting to use them in an application for which they were not intended—will result in problems. Because the operating requirements and environment of the machine determine the selection of lubricating oils and greases, it is recommended that the manufacturer of the machine or component be consulted for any specific lubrication recommendations.

Plant managers and engineers are encouraged to seek out qualified lubricant technicians to assist them in educating maintenance staff and implementing a lubrication program. Surveys should be conducted to determine current usage patterns and the types of lubricants currently on site. Consolidation and elimination of incorrectly applied lubricants might be necessary to prevent confusion and equipment failure. Plant mechanical systems and the environment in which they are operating must be considered as part of the lubricant selection process. Accurate, legible record keeping must take place and be available to staff. The program should establish the types of lubricants to be used and their parameters. The scheduling and frequency of the right lubrication for each machine should be communicated to those charged with the task of lubrication. Periodic review of the adequacy and success of the program should be an ongoing process.

The cost savings realized from properly selected and applied lubricants far exceed any savings generated by the purchase of inferior or misused lubricants.

LUBRICATION TERMINOLOGY

Viscosity

Viscosity is a measure of a fluid's (oil's) resistance to flow, i.e., a measure of the internal friction of lubricating oil that arises between different molecular layers when a liquid is set in motion. The concept of viscosity is the guiding principle in the selection of lubricants. Ideally, the best lubricant is the one with the lowest viscosity that maintains a separating film between metal parts at the various operating temperatures of the equipment. Operating temperatures have an affect on the viscosity of a lubricant. As the temperature rises, the oil will become less viscous; as the temperature falls, it becomes more viscous. Lubricants with a very high viscosity require power to overcome the friction of the oil itself. The friction from this process generates heat that can degrade the oil and result in film barrier loss. Applications with high temperatures, heavy loads, and slow speeds require thicker, heavier, higher viscosity lubricants. Applications with lower temperatures, lighter loads, and higher speeds require thinner, lighter, and lower viscosity lubricants. Thin oils flow freely and are called "light bodied"; thick oils that have high resistance to flow are called "heavy bodied."

Viscosity Index

Viscosity index (VI) is a measure of a lubricant's change in viscosity because of temperature. High VI lubricants have relatively small changes in viscosity across widely changing temperatures. The index is important in applications where oil operating temperatures fluctuate greatly. Hydraulic systems and engines are two examples of where high VI lubrication is required. Oil can be made less temperature dependent by various chemical additives.

Oil Viscosity Classifications

The following groups have established systems for viscosity classifications: American Society for Testing and Materials (ASTM), International Standards Organization (ISO), Society of Automotive Engineers (SAE), and the American Gear Manufacturers Association (AGMA).

The viscosity of industrial lubricants in the United States is measured in Saybolt Universal Seconds (SUS) or Saybolt Seconds Universal (SSU)—the two terms are interchangeable. The American Society for Testing and Materials (ASTM) is the governing body for this system of standards. A viscometer is used to determine the SSU viscosity. Oil is poured from a container at a particular temperature (100°F and 210°F) through an orifice into a "saybolt" flask. The time, in seconds, required to fill the flask to a prescribed amount is used to determine the viscosity of the oil; e.g., if the oil takes 50 seconds to fill the flask to the marked line, it has a viscosity rating of 50 SUS.

The International Standards Organization (ISO) has established lubricant viscosity grades (VG) in an effort to establish standards. The grades are based on kinematic viscosity (absolute viscosity ÷ mass density) rated at temperatures of 40°C and 100°C. Oil flowing through glass capillary tube viscometers at the temperature standards is timed to determine the kinematic viscosity. ISO uses centistokes (cSt) and Celsius metric measures for specifications and classifications (Figure 4–1). ISO 3448 is the accepted world standard.

The Society of Automotive Engineers (SAE) has established classifications for oil viscosity for use in engines. The SAE rating is based on the volume of base oil flowing through a specific orifice at a specified temperature, atmospheric pressure, and time period. The higher the viscosity rating number, the thicker the oil: A 30-weight (30W)

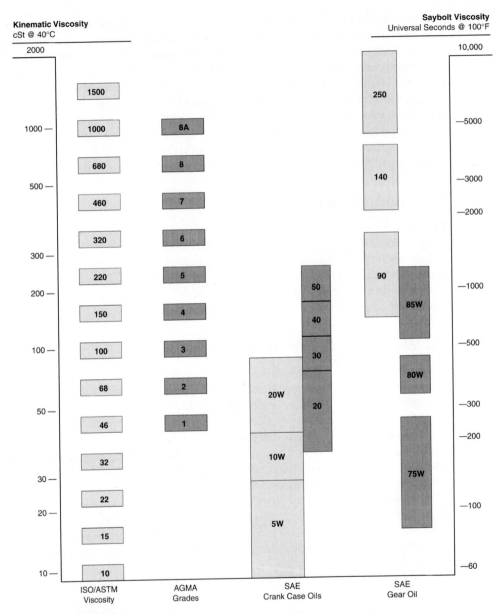

FIGURE 4–1
Viscosity grade comparison chart.
Courtesy of Timken/Torington.

oil is thicker than a 10-weight (10W) oil. Because automobile engines must maintain even viscosity/temperature characteristics at both cold starting and hot running temperatures, they are blended for thermal stability. Most engine oils are formulated to be multigrade by taking low viscosity classification oil and adding viscosity index additives. For example: A 5W-30 oil acts like an SAE 5-weight oil in the cold and an SAE 30-weight oil under the higher running temperatures. SAE also has a classification for transmission/axle applications.

The American Gear Manufacturers Association has a rating system for enclosed gear drives. AGMA 250.03 publication lists specifications that must be met before an oil can be recommended for use in AGMA-rated drives. AGMA numbers are based on ASTM viscosity grades.

API Service Ratings and API Gravity

The American Petroleum Institute (API) has worked with the American Society for Testing and Materials (ASTM) and with the Society of Automotive Engineers (SAE) to establish standards and ratings for SAE engine oils. The petroleum industry, engine manufacturers, and consumers for quality and performance standards use API service classifications.

API gravity scale was established to give weight per unit volume of petroleum oils. It uses specific gravity in its determination. Most oils are lighter than water and have an API number generally in the 20 to 30 range. The higher the assigned number, the lighter the fluid.

Flash and Fire Point

The flash point of a lubricant is the minimum temperature at which sufficient vapors are produced to combine with air to produce a flash when a flame is applied. The fire point is when an ignition source causes vapors to ignite and the sample continues to burn. The fire point temperature is higher than the flash point. Storage areas and operating temperatures must be kept low enough to prevent any possibility of combustion.

Pour Point

Pour point is the lowest temperature at which oil will flow in a controlled laboratory test. It is an indication of the lowest temperature at which oil can be used and still rely on gravity to cause it to flow. Machines running in a low temperature environment require a lubricant with a low pour point. It is recommended that the pour point be 30°F below the starting temperature. Additives that modify the structure of the oil can be added to yield better performance at low temperatures.

Color

Color is determined by the amount of light transmitted by a sample when it is compared with a standard. Exposure to a product such as food or textiles can cause a staining of the finished product. It should also be noted that a change of color from the original shade of the lubricant could be an indication of oxidation.

Oxidation

Petroleum-based lubricants are comprised of complex compounds of hydrocarbons, hydrogen, and oxygen. When they are heated in the presence of air and other contaminants, they undergo changes that include thickening and forming of organic acids. The oil becomes dark, thickens, and can smell burnt. Oil oxidation rates double with each increase of 18°F. Additives that inhibit this process can be added to the lubricant.

Demulsability

Demulsability is the ability of oil to separate from water. Additives can be used to perform this function. Water in the lubricant can interrupt the lube film, allowing metal-to-metal contact.

LUBRICANT ADDITIVES

The purpose of including additives in the lubricating compound is three-fold:

1. To improve the performance properties of the base oil.
2. To impart entirely new performance characteristics to the finished oil compound.
3. To reduce the rate at which undesirable changes take place in the product during its service life.

In lubricating compounds, the role additives play and the determination of their usefulness for the application are an important issue. Additives that improve performance for one application can have a negative effect in another. Care must be taken not to add extra additives to commercial oils that can unbalance the original additive package, resulting in oil with more additives but poorer performance. Certain additives can react adversely with internal pieces of the equipment. There is no substitute for superior base stock oil with a few necessary additives selected for a particular application.

There are a variety of different additives used in oils. The following types are some of the most common:

Oxidation Inhibitors

Oxidation inhibiting additives interrupt the chain reaction of oxidation to prevent or slow oil breakdown. Various acids, varnishes, and sludge will form from exposure to oxygen.

Extreme Pressure (EP)

Extreme pressure additives increase the load-carrying ability of the lubricant film and are commonly found in gear oils and greases. EP additives assist in cushioning shock loads and inhibiting the welding of surfaces. EP agents react chemically with the lubricated surface to form a film. For heavily loaded bearings, it has been customary to recommend the use of greases with EP additives. Sulfur, phosphorus, and chlorine are all EP additives.

Defoamants

Foam is caused by oil agitation and air entrapment. Defoamants promote more rapid breakup of foam bubbles by weakening the oil films between the foam bubbles. Foam interrupts the all-important lubricant film.

Film Strengtheners (Friction Reducers)

Film strengtheners form a film that is strongly attached to the metal surfaces, reducing metal-to-metal contact.

Detergent Dispersants

Detergent dispersants hold tiny dirt and sludge particles in suspension to prevent them from forming engine deposits. These deposits are formed under high temperature conditions.

Anti-Rust Additives

Anti-rust or rust additives create a film on metal surfaces. They displace water and form a film that protects the metal surface from contact with water.

Viscosity Index Improvers

Viscosity index improvers are used to thicken light oil to a higher viscosity without changing the original viscosity–temperature coefficient. This enables a lower viscosity oil to work at higher temperatures and not break down. VI improvers are polymers that function to decrease the slope of the viscosity–temperature relationship. They make the oil more viscous at higher temperatures than it would be without the viscosity improver, while at the same time not making the oil too thick at lower temperatures. VI improvers are used extensively in preparing multigrade engine oils.

Pour Point Depressants

Pour point depressants prevent congealing of the wax molecules found in mineral oils at low temperatures. They are used to lower the temperature at which lubricants will thicken.

Demulsifier Additives

Demulsifier additives enable water to drop out of suspension more easily. Water interrupts the lubricant film.

FIGURE 4–2
Lubricating film separating two surfaces.

Solid additives

Solid additives can improve lubricating properties in a few specific applications. Molybdenum disulfide is one such additive that acts as the lubricant. The particle size should be approximately .2 micron so it will remain in suspension.

LUBRICANT FILM

Lubricant film is needed to separate the metal mating surfaces of machine parts (see Figure 4–2). Under close examination of the surfaces of any metal object, we see peaks and valleys. These irregularities are known as asperities. Interference between opposing asperities in sliding or rolling applications is a source of friction and can lead to metal welding and scoring. The primary function of lubricants is to separate the surfaces, preventing welding, tearing, seizing, and other detrimental conditions from occurring. The film thickness can be of varying quality and depth. The thickness of the generated film depends on the operating conditions such as: velocity, loads, lubricant viscosity, and the pressure/viscosity relationship.

LUBRICATION FILM CONDITIONS

Mixed Film

Mixed film lubrication is the condition in which the film thickness is inadequate to completely separate the surfaces. It is characterized by a combination of solid and fluid friction and is often found in heavily loaded, slow moving and, particularly, shock loaded applications. Because the film is too thin, the surface too rough, or due to shock loads, the high points are forced together allowing metal-to-metal contact. This causes localized welding of peaks of the surface finish and results in high friction wear and surface distress.

Boundary Film

Boundary film lubrication is similar to a mixed film condition. The friction between the two surfaces in motion is determined by the properties of the surfaces. Also, the friction and wear protection properties are determined by the chemical nature of the lubricant rather than its bulk properties. Additives in the lubricant, such as extreme pressure (EP) compounds, coat the surfaces and act as a barrier. There still is some contact between surfaces during starting and stopping, and in periods of severe operation.

Full or Thick Film

This situation is a complete separation of moving surfaces. A full film of lubricant separates the peaks of both surfaces and friction is much less than in a boundary film condition. There are two types of full film lubrication: hydrodynamic and hydrostatic.

In a hydrodynamic state, the shape and relative motion (rotation) of the sliding surfaces, along with the coating ability of the lubricant, cause the formation of a continuous fluid film to prevent any surface contact under sufficient pressure. When we initially look at a shaft supported by a journal or sleeve bearing in a static condition, there is a mixed or boundary lubrication condition. Once the shaft reaches its running speed, a dynamic condition occurs in which the rotating shaft and the sliding velocity of the layers generate hydrodynamic fluid pressure. It is imperative that an adequate supply of oil with the correct viscosity be supplied to the mating area.

Hydrostatic lubrication also is a state where metal-to-metal contact is prevented by a full film. The film is maintained by external pressure from a source such as a lube pump. In some high technology applications—such as spindles, air-turbines, dentist drills, and computers—a pressurized gas lubricant is used to separate the surfaces. The gas can be air or an inert gas that does not react with other elements. The advantages of this system are low friction, high speed, less wear, and the ability to operate in extreme environments of temperature. Disadvantages are cost and the requirement for a clean gas.

Elasto-Hydrodynamic Lube Film

Elasto-hydrodynamic lube film (EHL) is a condition in which surfaces of heavily loaded rolling element bearings are either completely or partially separated by a very thin lubricant film. Pressure rises at the point of contact causing micro deflections in the metal. The deflected surfaces are momentarily pressed together and flattened slightly (elastic deformation). The elastic deformation of the contacting surfaces traps the lubricant, subjecting it to extremely high pressure which increases its viscosity and its load carrying capacity. When the rolling elements roll on, the surfaces return to their original shape and viscosity returns to its original condition.

Surface finishes, lubricant viscosity and properties, speeds, shapes, surface hardness, loads, and pressures all have a role in determining the film condition at the required point of contact. The goal is to minimize the amount of metal-to-metal contact and the resulting friction and component wear.

INDUSTRIAL LUBRICATING OILS

Oil makes an excellent industrial lubricating agent. Because oil is a fluid, it has the ability to be easily pumped, heated, cooled, sprayed, dripped, and cleaned by filtration. Industrial lubricating oils fall into three categories: animal/vegetable, mineral/petroleum, and synthetics.

Animal/Vegetable Oils

From the beginning of recorded history, man has made use of animal and vegetable oils for lubrication. Vegetable oils are derived from various plants that include olive, cotton, linseed, soybean, and many others. Animal oils come from whales, fish, and various types of animal tallow. Animal and vegetable oils tend to be unstable and rapidly form acids. There is some use of animal and vegetable lubricants in the food industry, but their primary application is as an additive to compound oils.

Mineral Oils

Hydrocarbon petroleum fluids are the most prevalent lubricating oil. Mineral oil is refined from crude petroleum oils extracted from the ground. Through the process of refining, many types of petroleum fluids are produced. Gasoline, kerosene, fuel oils, lubricating oils, and many other compounds are extracted during the heating, cooling, and filtering of crude oil. The three types of crude oil stock are paraffinic, napthenic, and asphaltic. Parafinnic oils have a relatively high viscosity index (VI) and contain waxes that need to be removed in the refinement process. Their viscosity changes less with temperature fluctuations than other oils, although typically they have a high pour point which is not suitable for cold operating environments. Napthenic oils have a lower viscosity index (VI). They have a low pour point and flow well at low temperatures. Asphaltic oils, derived from black crudes, are heavy and inexpensive. They are used for paving, roofing, and on large open gear drives. Most modern oils are a blend of several different types of base oils.

Synthetics

Synthetic lubricants are synthesized from pure carbon and hydrogen light gases that are recovered and distilled during the refinement process. The elements are formed in a carefully controlled process to produce a lubricating fluid with predictable properties and ideal performance characteristics. Synthetic-based fluids are tailored to a uniform molecular structure and size, free of unwanted substances and impurities. The result is a lubricant with improved qualities over standard mineral oils. Synthetic lubricants are developed and used for applications where normal petroleum products cannot cope with extreme temperatures. They are very efficient at reducing friction. One major advantage is that lubricant viscosity changes less with temperature variances. The oxidation rate is more stable with synthetics and service life is generally longer than standard mineral oils. Because synthetic lubricants are higher in cost than petroleum oils, they are used selectively. Trained lubricant engineers should determine the replacement of standard mineral oils with synthetics.

INDUSTRIAL LUBRICATING GREASES

Lubricating grease is a mixture of different viscosity oils, thickening agents (base), and a variety of additives. It is defined as a semi-liquid to semi-solid dispersion of a thickening agent in a base oil. It consists of approximately 90 percent mineral or synthetic oil, and 10 percent thickener with a small ratio of additives. A true statement regarding the quality of grease is, "The grease is no better than the oil in it!" By varying the types and amounts of oils, thickeners, and additives, a wide variety of greases can be formulated. The thickener is made up of a matrix of fibers; the cavities of this matrix are filled with oil, like the pores of a sponge. A small amount of oil "bleeds" out of the base to provide a supply of oil to the contacting and sliding surfaces.

Grease has certain advantages over oil lubrication. It acts as a barrier to contaminants, preventing dirt and water from getting into the mechanism. Corrosion is inhibited because the grease coats the stationary and moving metal parts. Grease resists leakage, dripping, or undesirable throw off, and has suitable oxidation and thermal stability. One of the disadvantages of grease is that it retains contaminants and heat because of the absence of flow. It also can separate, soften, or harden.

Classifications of thickening agents are metallic soap, complex soap, and non-soap. The most common metallic soap lubricating greases employ lithium (Li), calcium (Ca), and sodium (Na) as the thickening agents. Complex soap greases contain salt along with a metallic soap. Examples are barium (Ba) and aluminum (Al). Nonsoap greases use bentonite clay, silica, polyurea, and other organic and inorganic compounds. Most manufacturers incorporate additives into grease that can act as thickening agents.

Grease additives are incorporated into the lubricating product to enhance or furnish additional properties. The additives are generally the same as those added to oils.

GREASE TERMINOLOGY

Consistency

Consistency is the degree to which the grease resists deformation under the application of force. It is the measure of stiffness of the grease. The National Lubricating Grease Institute (NLGI) has developed a test and measuring scale to classify consistency. It is based on the degree of penetration achieved by allowing a standard cone to sink into the grease over a specified period of time, after which the depth of penetration is measured in tenths of a millimeter. This gives the "worked penetration index." The softer the grease, the lower the number. For roller bearings, an NLGI number of 1 or 2 is commonly used.

The following chart lists NLGI grades:

NLGI Grade	ASTM Worked Penetration
000	445–475
00	400–430
0	355–385
1	310–340
2	265–295
3	220–250
4	175–205
5	130–160
6	85–115

Dropping Point

The dropping point of grease is determined in a laboratory test by the maximum temperature that the grease withstands before it falls through an orifice. Generally, grease does not have a sharp melting point; instead, upon heating it becomes softer. This process is related to the thickener of the lubricant.

Structural (Mechanical) Stability

Structural or mechanical stability is the ability of grease to resist changes in consistency during mechanical working. Oil producers have developed tests to predict stability under the high shear rates seen in radial anti-friction bearings. Some greases soften as they are worked, which promotes leakage. Bearing applications are especially sensitive to instability.

Water Resistance

Water resistance of a grease is its ability to not break down or dissolve when exposed to water.

Pumpability

The pumpability of grease should be considered if it is delivered through a system of pipes and valves in a centralized system.

Channeling

Channeling is the tendency for grease to form unobstructed paths or channels as the rolling element passes through it.

COMMON GREASE TYPES

Lithium Soap

Lithium grease is one of the most widespread greases used in industry. Many so-called "multipurpose" greases are lithium soap based. Bearing manufacturers often recommend its use in anti-friction bearings because it has excellent heat-resistant qualities. With the proper additives mixed in, lithium grease can be used over a wide temperature range—from −60°F (−51°C) to 300°F (149°C). It adheres well to metal surfaces and is negligibly water soluble.

Calcium Soap

Calcium soap greases are capable of absorbing a large percent of their volume with water. This grease is generally considered water resistant and works well in wet environments, like paper machines and marine applications. Unless specially formulated, it should be used only in applications where the temperature maximum range is 180°F to 200°F; however, calcium complex greases can be effective up to 250°F (120°C).

Sodium Soap

Sodium soap greases absorb water, which decreases their effectiveness, and are washed away under heavy volumes of water. (Sodium-based grease is sometimes referred to as "soda grease.") They can be used over a wider temperature range than normal calcium greases. Some sodium synthetic greases can function in temperatures up to 300°F (149°C). They exhibit good adhesion and sealing properties, though they are not recommended for wet applications.

Complex Soap

Complex soap greases are formulated from various substances and bases to address special applications. They generally can withstand higher temperatures than the corresponding conventional greases.

Synthetic Greases

Synthetic greases make use of highly refined synthetic base oils, such as poly-alfa-olefins (PAO), organic esters, and silicones. The synthetic oil is typically combined with a complex thickener such as lithium soap. These oils do not oxidize as rapidly as mineral oils, contributing to extended service life. A variety of thickening agents are used to provide a high dropping point. Synthetic grease is heat and fire resistant and handles extreme temperature variances. Most of the synthetic greases have

high pumpability at sub-zero temperatures. Certain types of cold weather synthetic greases have low frictional resistance to as low as $-60°F$ ($-51°C$). Some hi-temp synthetic greases have a high end operating temperature of 450°F (232°C). Synthetics were developed for applications in the military and aerospace industries, and for developing robotics and other hi-tech machines. This does not exclude their use in general applications that require a highly effective lubricant.

DRY FILM LUBRICANTS

Dry film or solid lubricants are materials such as graphite, molybdenum disulfide, polytetrafluoroethylene (PTFE), or metal oxides. They are used to provide a dry film between mating surfaces in high load, slow speed, and oscillating conditions. Sometimes these solid lubricants are blended or incorporated as an ingredient in sintered metals or plastics. In some cases it is a coating/lubricant that is bonded to the surface of a substrate.

Graphite, in the form of very small plates or flakes, is a common dry film lubricant. In the right form and size, it has a very low coefficient of friction and slides easily over itself. It adheres well to metal surfaces that run parallel. But it requires moisture, usually from the humidity in the air, to be effective. It is capable of functioning in extreme temperatures up to 930°F (500°C).

Molybdenum disulfide (MoS_2) is a black, lustrous powder that serves as a dry film lubricant. It functions exceptionally well in certain high temperature and high vacuum applications. It does not require the presence of moisture to function properly, which makes it ideal for dry and hot environments. It is also used in the form of a paste, which is applied to parts prior to assembly to help prevent scoring. It is similar to graphite in that it is sometimes used as an additive in oils and greases to impart residual lubrication properties to the lubricant.

Polytetrafluoroethylene (PTFE) is a polymer produced from the synthesis of hydrocarbons. It is combined with other substances and applied to the surface areas of the sliding parts to form a film. It has a very low coefficient of friction and low wear characteristics. PTFE is nontoxic so it is suited for the food processing and pharmaceutical industries.

Solid film coatings are materials with inherent lubricating properties that are bonded to the surface of a substrate by various methods. The lubricating materials prevent the surface-to-surface contact that produces friction and wear. Bonding is achieved with resins, physical vapor deposition, and other complex processes. The material can be a compound mix of graphite, molybdenum disulfide, and other additives. Application is achieved by spraying, dipping, brushing, and tumbling. The film thickness achieves optimum wear properties when applied in uniformly thin multi-layers totaling .0003" to .0007". Proper pretreatment of the surface before application of the film is necessary. Effective lubrication can be accomplished in temperature ranges of $-395°F$ ($-201°C$) to over 2000°F (1093°C) and withstand loads in excess of 250,000 psi when properly selected, bonded, and applied. This type of lubrication is ideal on inaccessible parts, extreme environments, chemical atmospheres, and any application where operating temperatures exceed the range of oils and greases.

LUBRICANT COMPATIBILITY

Mixing lubricants that are incompatible can result in inadequate lubrication at the very least and catastrophe at the worst. The intermixing of complex hydrocarbon molecules subjected to heat, loads, and centrifuging can cause serious problems with lubricant properties and function. The base thickeners for various greases can be incompatible; for example, when lithium grease is mixed with sodium grease, the resulting blend will be softer than if used separately. The mixture will have a lower operating temperature and carrying capacity. When incompatible greases are mixed, they can yield either a softer or stiffer mixture. When oils that are incompatible are mixed, various acids, waxes, and varnishes can be produced. In some cases, different lubricants correctly applied to separate components are close to one another within a machine. Care should be taken not to mix them. For example, a bearing using a standard NLGI #2 lithium-based grease might lie close to an o-ring seal that requires a special synthetic silicone lubricant. In this case, and all others, the supplier of the lubricants should be consulted to check for the compatibility of the two greases.

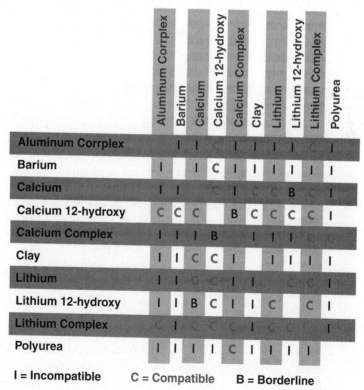

I = Incompatible   C = Compatible   B = Borderline

FIGURE 4–3
Grease compatibility.
Courtesy of NSK.

Pouring purchased special additives into a stock lubricating oil that has been supplied by a reputable manufacturer of lubricants can be a waste of money. Adding non-factory blended additives to degraded oil is not recommended. There is no assurance that the additive will stay in suspension or perform its function.

One of the determining factors in lubricant compatibility is heat. Some lubricants are never compatible and some become incompatible at a determined temperature. There is no one set temperature for every product, but it can be stated that when two greases are not compatible, they will separate at the temperature at which they become incompatible. The separation results in a complete loss of lubricating qualities that either product had initially.

It is inevitable that lubricant substitution or a complete product change will occur in a plant. The previous oil or grease should be cleaned from the reservoirs and purged from the supply lines before the replacement product is introduced into the machine. Gearboxes should be thoroughly flushed of the previous lubricant and contaminants before adding the new replacement. Following this recommendation will help make the transition safe and effective.

Lubricant compatibility is an important issue that should be addressed and monitored by qualified lubricant engineers and technicians. Manufacturers' representatives can provide the technician with lubricant compatibility tables (Figure 4–3). With all blending of products, it is recommended that the lubricants be treated as incompatible unless there is technical data available to prove otherwise.

LUBRICANT DELIVERY SYSTEMS

Lubricants must be delivered and applied to where they are required. They might be dispensed simply by dropping oil from a can or using a complex automated system. Initial lubrication of the mechanism upon assembly or during production is desirable and, in some cases, all that is possible. Closed units such as sealed ball bearings are "lubricated for life." Most mechanical apparatuses require periodic replenishment of the lubricating medium as the machine functions. A variety of systems including bath, wick, drip, splash, circulating, jet, mist, automated pressurized, and manual are used to dispense lubricants.

Bath, Submersion, and Splash Systems

Oil bath systems are used to lubricate chains, bearings, and gear drives. Chains can be run through a bath of oil, contained within the guard, to thoroughly submerge all parts of the chain in the lubricant. Certain types of large babbited bearings have shaft-riding oil slinger rings that pick up oil from the sump and deliver it to the bearing surfaces. The slinger ring, which has a greater diameter than the shaft journal, hangs down into an oil reservoir contained within the bearing. As the shaft turns, oil is picked up by the ring and transferred to the shaft. If systems incorporate the use of rings or splash devices to spread the oil during operation, they can be considered splash systems. Splash systems are common on enclosed gear drives

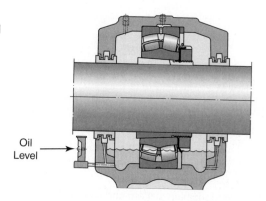

FIGURE 4–4

A splash lubricated bearing inside of a housing showing the oil sump and proper oil level.

Courtesy of SKF.

Oil Level →

where lubrication of the contact surfaces of both bearings and gears is needed. The rotating gears will act as splashing mechanisms within the gear case during operation. Gutters, troughs, or wells are designed into the case on large, enclosed gear drives to feed oil to the bearings supporting the shafts and gears. If the unit has been idle or rebuilt, it is recommended that these channels be primed with oil prior to start-up. This will prevent any metal-to-metal seizing that might occur with start-up loads. Many rolling element bearings that are lubricated with oil rely on the splashing of the oil from a sump (see Figure 4–4).

The level of lubricating fluid in all machinery is critical to provide adequate lubrication where it is required without flooding the system and causing churning and spills. Manufacturers usually provide service manuals that inform the user of the correct fill quantities for the equipment. The oil levels should be high enough to provide thorough splashing to all bearings. Rotating parts, such as gears, should never be completely submerged. If the oil level is too high, the resulting friction from increased resistance and churning can increase operating temperatures. Sight glasses and dipsticks should be used to check oil levels. Oil level checks are best accomplished when the machine is down, unless the operating levels are known and clearly marked. When the machine is turning, running levels will be considerably higher than static levels. If a gear case is turned to a vertical position, care must be taken that the top bearing receives lubrication. A breather pipe extension, based on the manufacturer's recommendations for that particular unit, should be installed to prevent the oil from spilling out. Relocation of various plugs and drains might be required. Routine inspection is necessary to determine the frequency and amounts needed for refilling due to leakage. The housing must have a breather to allow for any pressure differentials. A magnetic drain plug at the bottom of the reservoir is helpful in collecting ferrous contaminants.

Automated or Pressurized Systems

Centralized automated systems deliver the required lubricant, under pressure, to the mechanism by a series of pipes, valves, a pump, and a reservoir (see Figure 4–5). Automated systems are advantageous in many respects. The lubricant is isolated

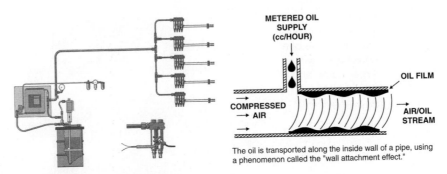

METERED OIL SUPPLY (cc/HOUR)

OIL FILM

COMPRESSED AIR

AIR/OIL STREAM

The oil is transported along the inside wall of a pipe, using a phenomenon called the "wall attachment effect."

FIGURE 4–5
Automated lubricating system and its operation.
Courtesy of SKF.

from contamination and metered out under controlled pressure. Either oil or grease can be used if the system is correctly plumbed. Oil is transported along the inside of the pipes using a phenomenon called "wall attachment effect." There are two types of systems: closed and open loop. In some grease and oil lubricated systems, the lubricant makes a one-way trip (open loop). If the system is a circulating oil type, it will have additional plumbing to return the lubricant to a reservoir (closed loop). Cleaning, cooling, and filtering are required before the lubricant can be recirculated to the load-bearing surfaces.

A pump of some type is used to provide a continuous supply of clean, cooled, and filtered lubricant to all the required points. It must have sufficient pressure and volume to deliver the lubricant through all channels and orifices.

The reservoir can be as simple as a barrel of lubricant with the pump attached to its top, drawing the fluid up through a tube. The reservoir on a closed loop system will be a metal tank that contains the lubricant and allows it to cool through radiation. Some reservoirs that are used on automated systems are transparent for easy lubricant level checks. Other reservoirs use an attached sight glass to indicate the current levels within the tank. The reservoir levels should be checked and maintained to prevent a loss of suction. Low reservoir levels will not allow heat to be dissipated and will cause the flow of lubricant in the system to be interrupted. The reservoir should be drained periodically and cleaned of contaminants, then refilled with the proper lubricant.

The oil temperature within the circulating system must be prevented from becoming too extreme. Cooling will occur naturally by the reservoir acting as a heat sink to dissipate the heat to the surrounding atmosphere. In hot-running applications, cooling tubes with circulating water or water/anti-freeze mix can provide additional cooling capacity. The tubes are submerged in the reservoir and a separate pump with coolers continuously distributes the cooling agent through the reservoir which absorbs the heat and returns the oil at a lower temperature. If breathers are required, filtered breathers that prevent both dirt and moisture from contaminating the system are recommended.

Circulating systems will benefit from having a lubricant filter. Filtering the lubricant prevents contaminating particles from interrupting the lube film and causing wear of the surfaces. Filter elements must be checked and changed out periodically. Screen-type strainers made from wire mesh are capable of removing particles in the system down to approximately 10 microns (0.0004″). The size of the particle removed is determined by the mesh openings. Absorbent filters will provide a finer level of filtration than a strainer. The typical range of particle size removed is from 20 to 0.5 microns (0.0008″ to 0.00002″). Resistance to flow is increased when using a filter in the lubrication system. As the fluid flows through the filter, large particles are deposited on the surface and smaller ones are trapped within the filter medium. Eventually the filter will become saturated and need to be replaced. Replacement intervals are determined by the operating environment, condition of the filter, type of filter, and hours of service recommended by the manufacturer. Materials that filter are made from everything from paper to porcelain. The filter medium material is determined in part by the fluid, environment, types of contaminant, and cost. Most modern quality filters are a composite of various materials. Most imperative is that a filter is used and maintained.

Some lubricating systems use an electrically driven piston pump to provide pressure in the system. Air-operated grease lubricating systems are actuated by regulated and filtered compressed air at predetermined intervals. An air-operated barrel pump delivers lubricant to the injectors. Lubricant injectors are mechanisms that can be adjusted to dispense a precise amount of lubricant directly to a bearing or to a manifold with feed lines. The lines can be directly plumbed to a bearing or will have spray nozzles to deliver the lubricating fluid. Most of these systems are available with a programmable automatic controller. This controller allows the operator to have flexibility in regulating the timing of the lubricant flow. Audible or visual alarm devices often will be incorporated into the controller to warn of low levels, pressure losses, or complete failure.

A type of circulating pressurized oil system frequently used on high speed bearing applications is the oil-jet or oil-mist lubrication system (see Figure 4–6). Bearings that are operating under a combination of high speeds, loads, and temperatures require a constant supply of clean circulating oil. The metered oil is atomized and sprayed directly onto the bearings through narrow jets. The systems must be

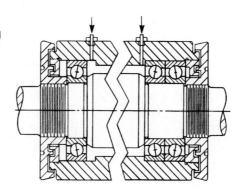

FIGURE 4–6
Oil-mist/jet lubricating system on a machine tool bearing spindle. Arrows reflect pressurized oil entry point.
Courtesy of MRC.

started prior to turning on the equipment to ensure the pre-lubrication or wetting of all bearings. Electrically interlocking the oil pump motor with the primary machine drive will safeguard the oil supply.

Manual Systems

Probably the most common form of a manual system is the grease gun (see Figure 4–7). A grease gun is a hand-operated device that is filled with grease and has a hose with a fitting that is attached to the end. Typically it is attached to a bearing housing fitting so the operator can fill the bearing with fresh grease. The operator manually pumps the handle to force grease into the bearing. The amounts and pressures are difficult to control. More sophisticated manually pumped systems can have a single refillable reservoir that can feed multiple lines through a manifold valve arrangement.

Wick Systems

A wick-feed system is an old, inexpensive method of lubrication (see Figure 4–8). Some sleeve bearings use this method to deliver oil by gravity and capillary action to the bearing surfaces. A porous wick of felt or some similar material determines the rate of flow by its thickness, length, and degree of submersion. The wick also acts as

FIGURE 4–7
Classic manual grease gun.
Courtesy of SKF.

FIGURE 4–8
Wick-feed lubricating system drawing oil from a reservoir.
Courtesy of SKF.

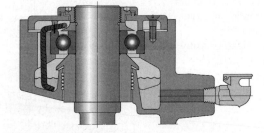

a filter. Napathetic and synthetic oils should be used with wick-feed systems because they do not deposit crystals on the fibers that inhibit flow, as paraffinic oils do.

Drip Systems

Drip systems also rely on gravity to provide oil one drop at a time to where it is needed (see Figure 4–9). A transparent glass or plastic receiver containing the oil is mounted above the intended mechanism. It is sometimes referred to as an oil "cup." The reservoir is visually checked and manually kept full by periodic refilling. The rate of oil flow is controlled by an adjustable needle valve. The oil can be dripped onto the mechanism or it may flow through a contact brush.

Automatic Grease Lubricators

Automatic grease lubricators are small, sealed grease reservoirs that thread directly onto the fitting of a bearing (see Figure 4–10). A measured amount of grease that meets the requirements of the application is dispensed directly into the bearing from a cartridge. Pressure from a spring, gases, or actuator of some sort automatically dispenses small amounts at low pressure on a continuous basis. These lubricators remove much of the guesswork of how much and how often lubrication is required.

FIGURE 4–9
Drip lubricating system with oil reservoirs mounted above the bearings.
Courtesy of Timken/Torington.

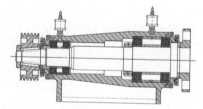

FIGURE 4–10
Replaceable automatic grease lubricating cartridge mounted into the grease entry plug.
Courtesy of SKF.

Applications that are difficult or dangerous to access are ideal locations for this mechanism.

LUBRICATION HANDLING AND SAFETY

Safe and proper handling of lubricating fluids is an integral part of any lubrication program. All plant personnel involved in handling and dispensing lubricants should be educated on proper safety procedures pertaining to their use. Good housekeeping procedures carried out by plant staff and enforced by supervisors will prevent lubrication contamination and any potential hazards. Because lubricants are made with many different materials and chemicals, they are considered hazardous materials. Special precautions must be taken when applying and handling lubricants to prevent injury and illness.

The storage areas should be clean, dry, and organized. Containers must be marked or tagged with the name and number of the fluid they contain. All labels on containers must be clearly legible and positioned for ease of reading.

Storage rooms should have suitable fire protection. They should have water sprinkler systems or CO_2 dispensing systems for fire suppression. Fire extinguishers should be inside and outside of the rooms. Never weld, cut, or burn in a storage room.

Lubricant containers should be used only for their intended purpose. Do not re-use the containers to hold or contain anything else. The containers or drums that held the lubricant must be drained and disposed of in an environmentally safe manner. All lubricants will burn if they are exposed to higher temperatures. Care must be taken not to exceed the flash or fire point of the fluid. Never attempt to modify, drill, or grind on the container or expose it to flames because it may explode.

Clean up spillage immediately and report the leak to the required authorities. All plant and government rules and regulations should be followed when disposing of used lubricants and their containers.

Exposure to petroleum and synthetic lubricants should be kept to a minimum to prevent possible health problems. Both the liquids and vapors should be recognized as potential risks. Hands should be gloved to protect from extended and repeated contact with the fluid. Skin irritation can occur from prolonged exposure and pores may become plugged. Clean hands and exposed skin surfaces by thoroughly washing with soap and warm water or specially formulated cleaners.

Pressurized systems have the potential to forcibly expel lubricant into eyes or to damage soft tissue. Care must be taken to prevent this from occurring.

The Occupational Safety and Health Administration (OSHA) sets standards and requirements for handling hazardous chemicals such as lubricants. OSHA requires that documentation pertaining to the ingredients and safety concerns of the substance be provided upon request. Material Safety Data Sheets (MSDS) and manufacturers' safety publications will list the ingredients, physical data, and necessary precautions for the lubricant. These MSDS documents should be on hand and accessible to all. A list of every type of oil, grease, and hydraulic fluid used in a plant should be on hand and up-to-date.

Questions

1. What is a lubricant?
2. What are the functions of a lubricant?
3. Name three types of lubricating oils used in industry.
4. What is viscosity?
5. How does temperature affect the viscosity of a lubricant?
6. Name four additives sometimes added to oils and state their purpose.
7. What are the advantages of using grease as a lubricant?
8. What are the advantages of using oil as a lubricant?
9. What does NLGI stand for?
10. Name three common thickening agents used in greases.
11. What is a dry film lubricant?
12. Name four methods used in delivering lubricants.

Rigging

R igging is of critical importance to the safety of workers. Many people have been seriously injured and even killed in accidents caused by poor rigging practices.

Objectives

Upon completion of this chapter, the student will be able to:
- ✔ Explain terms such as rigging, balancing, slings, and so on.
- ✔ Calculate the weight of a load.
- ✔ Find the balance point of a load.
- ✔ Describe various types of rigging slings.
- ✔ Describe proper procedures for rigging and lifting various loads.

INTRODUCTION

Rigging is the preparation of a load prior to being moved. This is crucial for the safety of the operator and equipment to be moved. There are many serious injuries and even deaths each year due to improper rigging techniques. Improper rigging techniques can cause a load to shift, a sling to break, or a load to fall and injure or kill. This chapter will concentrate on cable sling rigging, chain sling rigging, and synthetic web slings.

The weight and shape of a load must be known before lifting, as well as the location where it should be lifted. Rigging is securing machinery or equipment prior to lifting. Chains ropes or webbing may be used. The weakest component in a lifting system determines how much may be safely lifted. You must determine the limitations of every component you will use.

SLINGS

Loads may vary in physical dimensions, shape, and weight. Where and how to attach slings is important to a rigger.

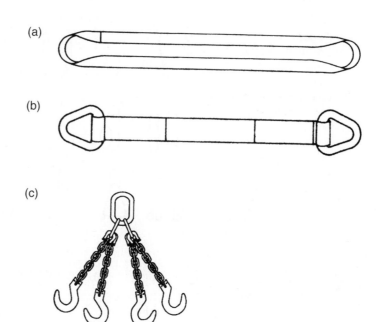

(a)

(b)

(c)

FIGURE 5–1
Three types of slings.
Courtesy of MAZELLA Lifting Technologies

An *endless sling* is a continuous loop of rope or webbing (see Figure 5–1a). Because it is an endless loop, the portion of the web that contacts the material can be varied each time it is used to distribute the wear across the whole loop of the sling.

A *plain sling* is a single web, rope, or chain with a fitting at each end to attach to the load and hoist (see Figure 5–1b). These are also called *single-part slings*.

A *bridle sling* has two or more single-part slings connected together by a common ring (see Figure 5–1c).

Sling Types

Wire Rope/Cable Slings A sling is a synthetic webbing designed into a configuration for hoisting, lifting, and lowering applications (see Figure 5–2).

Cable slings or wire rope slings are made up of individual strands of wire. The number of strands, number of wires, type of material, and nature of the core will depend on the intended purpose of the wire rope. We must always determine the weight of the load before we can begin selecting the sling type.

You must carefully inspect all cable slings before using them. Remove them from service if the following conditions are present:

Corrosion of the rope or end attachments.

Wear or scraping of outside wires reducing the individual wire diameter by 1/3.

FIGURE 5–2
Common configurations
of cable sling hitch
types.

Courtesy of MAZELLA Lifting
Technologies

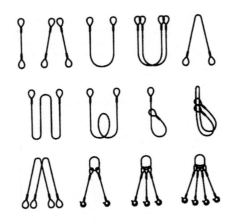

Kinking, crushing, bird caging, or any other damage resulting in distortion of
the wire rope structure.

Evidence of damage due to heat.

Ten randomly distributed broken wires in one rope lay, or five broken wires in
one strand in one rope lay.

End attachments that are cracked, deformed, or worn.

When using a wire rope sling, always follow these safety precautions:

A sharp edge will reduce the life of the sling. Use blocking or padding when the
cable must go around sharp corners (see Figure 5–3).

Use a sling large enough for the load. The smaller the angle of the sling, the
lower the lifting capacity (see Figure 5–4).

Remove kinks in the sling before using.

Do not jerk loads. Jerking may double the stress on your sling. Lift loads
gradually.

Hang the sling up after every use. This will help to keep it clean, undamaged,
and ready for use for the next job.

A radius that is too small for a wire rope sling can result in some of the individual
wires being overloaded. It may also cause a permanent kink in the wire rope. This
can cause permanent damage to the wire rope that may cause a catastrophic failure
later.

Chain Slings Try to avoid using chains when it is possible to use wire rope. The
failure of a single link of a chain results in the complete failure of the chain,
whereas a cable is made up of many wires and strands and they must all fail before
the rope breaks. Chains gives no warning as to when they are going to fail; a wire
rope will show visible signs of impending failure. Chains are better suited for certain

FIGURE 5–3

Protect the sling from sharp corners.

Courtesy of MAZELLA Lifting Technologies

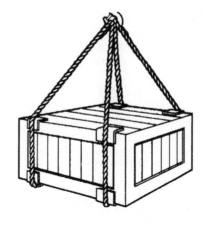

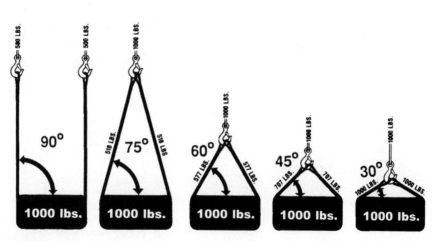

FIGURE 5–4

Estimated rating for cable slings using different sling angles.

Courtesy of MAZELLA Lifting Technologies

FIGURE 5–5

The letter A signifies that this chain is a welded alloy chain and could be used for lifting.

jobs as they will withstand rougher handling and they won't kink. Chains are much more resistant to abrasion and corrosion than wire ropes. Chains are well suited as slings in the machine shop for lifting heavy castings. Only welded alloy chains should be used for lifting. They are easy to identify because they will have a letter A on them (see Figure 5–5).

Follow these Recommendations for Safer Chain Sling Use

1. Visually examine the sling before each use. Look for stretched, gouged, bent or damaged links and components, including hooks, with opened throats, cracks or distortion. If damaged, remove from service.

2. Know the load — determine the weight, center of gravity, angle of lift and select the proper size and type of sling.

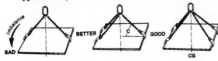

3. Never overload the sling — check the working load limit on the identification tag. Always consider the effect of Angle of Lift — the tension on each leg of the sling is increased as the angle of lift, from horizontal, decreases. Use the charts in this catalog.

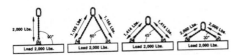

4. Do not point load hooks — load should bear on the bowl of hook.

5. Make sure chain is not twisted, knotted or kinked before lifting the load.

6. Slings should not be shortened with knots, bolts or other makeshift devices.

7. Protect chain with padding when lifting sharp edged loads.

8. Lift and lower loads smoothly, do not jerk.

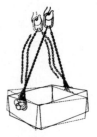

9. Hands and fingers should not be placed between the sling and load while sling is being tightened around the load. When lifted, the load should not be pushed or guided by employee's hands directly on the load.

10. Do not expose A8A alloy chain or slings to temperatures above 500°F.

11. Protect chain slings from corrosion during storage.

12. Store slings properly on an A-Frame.

FIGURE 5–6
Recommended chain sling use.

Courtesy of MAZELLA Lifting Technologies

FIGURE 5–7

Double eye nylon sling.

Courtesy of MAZELLA Lifting
Technologies

Use the following chain safety tips as a guide.

Never exceed the estimated rating capacity of the chain.

Take up slack, then start to load slowly. Chain has no elasticity and does not react well to shock loading.

Always use softeners on the corners of rectangular loads.

Never shorten a chain by tying a knot in it. A chain has its maximum strength with the load running in a straight line.

Do not lift from the point of the hook.

Lift from the center of hooks.

Distribute the load evenly on all legs.

Inspect chains regularly. Look for elongated links.

Don't drop loads onto the chain.

Chain is designed to carry the load in a straight line. Chain that is twisted or knotted cannot develop its full strength and may fail. If you must shorten a chain, use a shortening clutch. Figure 5–6 shows recommendations for the use of chain slings.

Synthetic Web Slings Synthetic web slings are used extensively in the machine shop for rigging purposes (see Figure 5–7) because they are easy to use and won't mar or scratch finished surfaces. Web slings are usually made of nylon or polyester; nylon has a higher strength rating than polyester. Figure 5–8 shows the basic nylon and polyester sling types.

Synthetic web slings, because of their flexibility and elasticity, are a very popular rigging tool. The material with which these synthetic slings are made allows them to be very flexible, but it also makes them susceptible to heat and sharp edges. Synthetic web slings must be well maintained and the user must always be aware of sharp corners, welding sparks, and metal chips.

Slings used in environments where they are subject to continuous exposure to ultraviolet light should be proof tested to two times the rated capacity annually or more frequently depending on severity of exposure.

BASIC PRINCIPLES IN THE USE OF SLINGS

Sling Selection

Some of the considerations when choosing a sling include the temperature of the load, the presence or absence of sharp corners, and the surface of the load. A hot load will probably rule out rope or webbing slings. For sharp corners, a chain may

Type I—Web sling made with a triangle fitting on one end and a slotted triangle choker fitting on the other end. It can be used in a vertical, basket or choker hitch.

Type II—Web sling made with a triangle fitting on both ends. It can be used in a vertical or basket hitch only.

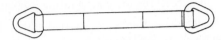

Type III—Web sling made with a flat loop eye on each end with loop eye opening on same plane as sling body. This type of sling is sometimes called a flat eye and eye, eye and eye, or double eye sling.

Type IV—Web Sling made with a both loop eyes formed as in Type III, except that the loop eyes are turned to form a loop eye which is at a right angle to the plane of the sling body. This type of sling is commonly referred to as a twisted eye sling.

Type V—Endless web sling, sometimes referred to as a grommet. It is a continuous loop formed by joining the ends of the webbing together with a load-bearing splice.

Type VI—Return eye (reversed eye) web sling is formed by using multiple widths of webbing held edge to edge. A wear pad is attached on one or both sides of the web sling body and on one or both sides of the loop eyes to form a loop eye at each end which is at a right angle to the plane of the web sling body.

Figure 5–8
Typical nylon and polyester sling types.
Courtesy of MAZELLA Lifting Technologies

be more durable but edge protection must be provided. If the load has a polished surface or is fragile in some manner, fiber webbing or a round sling of fiber rope may be the best choice.

Securing the Sling to the Load

Slings must be firmly secured to loads. Slings must not be kinked, knotted, or twisted. When multi-leg slings are used, two legs will often carry the majority of the weight while the other legs balance the load. Adequately pack sharp corners to prevent damage to the sling. When using slings:

The sling and the method of use must be suitable for the load to be lifted.

The load does not damage the sling.

The sling does not damage the load.

The attachment of the sling to the lift and the sling to the load must be secure.

The method used must ensure that the load is secure and that the load cannot fall from the sling.

No part of a sling should be overloaded by the method, or by the weight of the load.

The load must be balanced so that it will not shift violently when lifted.

If these principles are not followed, it is possible that weight will not be evenly distributed between the legs. If this is the case, the leg or legs that are carrying the extra weight must be de-rated accordingly. If a multi-leg sling is used—with less than the total number of legs attached to the load—the load must be reduced. For example if only two legs of a four-leg sling are used, the safe working load should be reduced by 2/4 or 1/2.

Shackles

A shackle is one type of attachment that can be used to make a connection between rigging and the hoisting device. Shackles are U-shaped devices with the ends drilled to receive a bolt or a pin. The pin or bolt can be removed to enable one or more loop eyes to be attached to complete a sling. A chain shackle can be used as a

FIGURE 5–9
Shackles.

Courtesy of MAZELLA Lifting Technologies

Scew Pin   Bolt Type

connector for a single lifting device such as a one loop eye for a vertical lift. A chain shackle normally uses shackle washers to prevent side shifting. An anchor shackle uses a rounded eye to allow more than one lifting device.

Figure 5–9 shows some examples of shackles. Shackles are made from alloy or high-tensile strength steel.

Hooks

Hooks are designed to support the load in the bowl, not on the tip. Use of the tip may lead to hook deformation or hook failure.

FUNDAMENTAL HITCHES

There are three basic types of hitches: vertical, choker, and basket (see Figure 5–10). Every lift uses one of these. Another very useful hitch is the adjustable hitch.

Vertical Lift

The vertical lift is also called a straight lift. The vertical lift connects a sling between a hook and the load. In a vertical lift, the full-rated lifting capacity can be utilized but must not be exceeded. A tagline should be used to prevent the load from rotating and causing damage to the sling. If two or more slings are attached to the same lifting hook, the load is distributed equally among the individual slings.

Choker Lifts

A choker hitch is a contact sling hitch in which the sling passes entirely around the load. In its simplest form, the sling has a loop or eye on each end and is referred to as a choker sling or choker. One loop passes through the other forming a slip noose.

Choker hitches reduce the lifting capacity of a sling because the method impacts the ability of the wire rope components to adjust during the lift. Choker lifts

FIGURE 5–10
Three basic hitches.
Courtesy of MAZELLA Lifting
Technologies

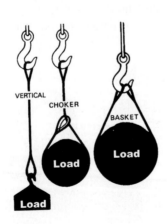

FIGURE 5–11
Capacity adjustment.

Courtesy of MAZELLA Lifting
Technologies

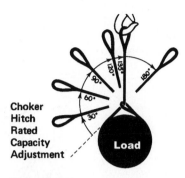

Choker
Hitch
Rated
Capacity
Adjustment

For wire rope slings in choker hitch when angle of
choke is less than 120°.

Angle of choke in Degrees	Rated Capacity Percent*
OVER-120	100
90-120	87
60-89	74
30-59	62
0-29	49

are used when the lift requires the sling to snug up against the load. They should not be used when the load will be damaged by the sling body or the sling damaged by the load. The diameter of the bend where the sling contacts the load should keep the point of the choke against the sling body, never against a splice or the base of an eye. When a choke is used at an angle of less than 120 degrees, the sling rated capacity must be adjusted downward.

A choker hitch should not be pulled tight before the lift is made, nor pulled down during the lift. If a load is hanging free, the choke angle would normally be approximately 135 degrees. If the angle is less than 135 degrees, you must adjust the rated capacity. Figure 5–11 shows the adjustment that must be made for various angles. It is always dangerous to only use one choker hitch on a load that might shift or slide out of the hitch.

Double choker hitches seem to be preferred by many good riggers because they are twice as strong as single choker hitches in the same sling type. When this hitch is made in the right way, both legs will automatically equalize over the crane hook; however, when it is not made correctly, there is no equalization and one of the legs will support most of the load. A double wrapped choker hitch is also acceptable, where overhead space is limited.

Basket Hitch

Basket hitches, whether single or double (see Figure 5–12), may be used successfully in a variety of applications. A basket hitch normally distributes a load equally between the two legs of a sling.

The adjusting hitch is particularly useful when lifting an object that is heavier on one end than the other. The adjusting hitch makes it fairly easy to adjust the length of the legs of the bridle to maintain the load level (see Figure 5–13). This hitch should not be loaded any more than a single basket hitch. The effective length

FIGURE 5–12
Basket hitches.
Courtesy of MAZELLA Lifting Technologies

FIGURE 5–13
Adjusting hitch.
Courtesy of MAZELLA
Lifting Technologies

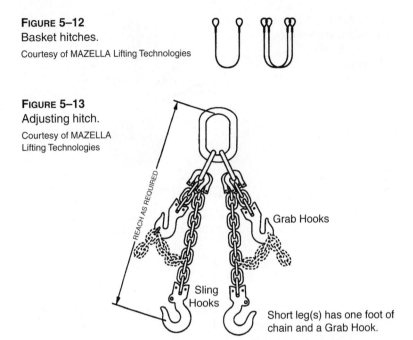

Grab Hooks

Sling
Hooks

Short leg(s) has one foot of
chain and a Grab Hook.

REACH AS REQUIRED

of the adjusting hitch can be easily changed to suit job conditions. Once the weight
of the load comes onto the sling, no further change in length occurs.

CALCULATING THE WEIGHT OF THE LOAD

The most important precaution in rigging and lifting is to determine the weight of
the load before attempting to move or lift it. Its center of gravity also needs to be
determined. Knowing the approximate weight will ensure that you choose rigging
and lifting components that can safely handle the load. To find the weight you must
calculate the volume of the item to be lifted. Figure 5–14 gives the weights of some
of the more common materials. Simple cubic shapes can be easily calculated. More
complex shapes can usually be broken into separate shapes to make the calculation
easier. Figure 5–15 shows a few of the more common formulas for finding volume.

Safe Working Load for Lifting

The factor of safety is the ratio between the minimum breaking load and the safe
working load. The safety factor is recommended to be 4–6:1. Remember that the
safety factor is dependent on the weakest link in the components you are using.
You should check your company's procedures or the manufacturer's information to
find the appropriate safety factor for your components.

Manufacturers supply tables that should always be used to find the safe working
load for slings. The tables normally provide a safe working load for a single vertical

FIGURE 5–14

Table of weights of common materials.

Material	Approximate Weight in Pounds Per Cubic Foot
Aluminium	166
Brass	525
Brick	125
Concrete	140–160
Copper	560
Gold - 24 Carat	1,204
Granite	165–170
Gravel	115–125
Iron (Cast)	450
Lead	712
Limestone	170
Marble	165
Silver	655
Steel	490

Courtesy of JBR.

FIGURE 5–15

Formulas for finding volumes of common shapes.

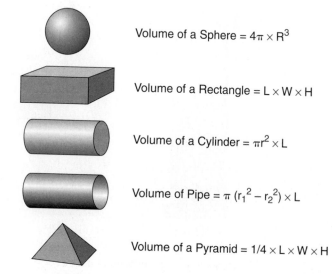

Volume of a Sphere = $4\pi \times R^3$

Volume of a Rectangle = $L \times W \times H$

Volume of a Cylinder = $\pi r^2 \times L$

Volume of Pipe = $\pi \left(r_1^2 - r_2^2\right) \times L$

Volume of a Pyramid = $1/4 \times L \times W \times H$

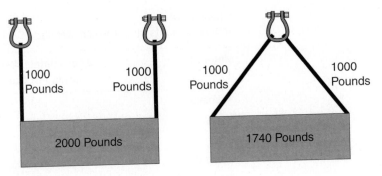

FIGURE 5–16
Examples of a vertical lift and an angular lift.
Courtesy of MAZELLA Lifting Technologies

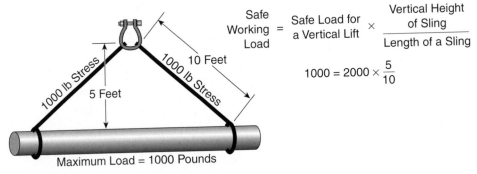

$$\text{Safe Working Load} = \text{Safe Load for a Vertical Lift} \times \frac{\text{Vertical Height of Sling}}{\text{Length of a Sling}}$$

$$1000 = 2000 \times \frac{5}{10}$$

FIGURE 5–17
Example and formula for load derating.
Courtesy of MAZELLA Lifting Technologies

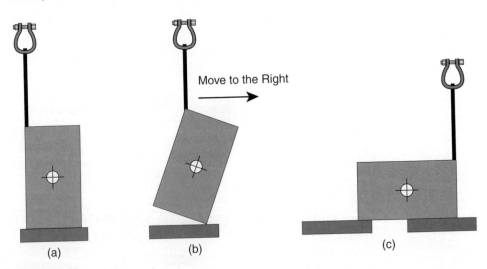

FIGURE 5–18
Changing the orientation of a load.

load. Figure 5–16 shows two examples of lifts. The example on the left shows two 1000 lb slings used vertically to lift 2000 lbs. The example on the right of Figure 5–16 shows the same slings used at an angle, reducing the working load capacity to 1740 lbs. A simple calculation can be used to derate the load for a given angle (see Figure 5–17).

Figure 5–18 is an example of one way to change the orientation of an object with a single hoist. The hoist is connected to the corner of the object. As the hoist is slowly moved the object begins to tilt because of the location of the center of gravity. Once the hoist has moved past the center of gravity, the object can be slowly lowered.

CALCULATING THE LOAD ON EACH LEG OF A SLING

Figure 5–19 shows how the load on each leg increases as the included angle between the legs of a sling increases. Note that when the angle is 90 degrees, the load on each sling supporting the 1000 pound weight is 500 pounds per leg. When the angle is decreased to 75 degrees, the load increases to 518 pounds on each leg. When the angle decreases to 30 degrees, the load on each leg is 1000 pounds!

The table in Figure 5–19 can be used to calculate the load on each leg.

1. Calculate the total load to be lifted.
2. Divide the total load by the number of legs that will be used.
3. Determine the angle between the legs of the sling and horizontal.
4. Multiply the load per leg calculated in step 2 (above) by the load factor from the table. This will give you the actual load on each leg. Remember that the actual load must never exceed the rated sling capacity.

> **EXAMPLE**
>
> Assume a 1000 pound load at 60 degrees with two legs. 1000/2 = 500, 500 × 1.154 = 577 pounds. Each leg will have a load of 577 pounds.

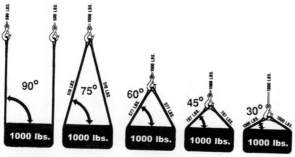

LEG ANGLE (Degrees)	LOAD FACTOR
90	1.000
85	1.003
80	1.015
75	1.035
70	1.064
65	1.103
60	1.154
55	1.220
50	1.305
45	1.414
40	1.555
35	1.743
30	2.000

FIGURE 5–19
Table to calculate load on each leg of a lift.
Courtesy of MAZELLA Lifting Technologies

PACKING THE LOAD

Packing is used to provide an adequate cushion and radius around a load without unacceptable loss of load carrying capacity (see Figure 5–3). When a sling is bent around a corner, its strength is reduced considerably. The packing will help grip the load and prevent a sling from being cut on a sharp corner. The packing will not, however, prevent the loss of strength that is due to any incorrect loading of the sling.

The material that is chosen for packing must be capable of withstanding the crushing forces of the load during the lift. When you position packing you must be sure that it will stay in place during the lift. If packing falls or is forced out during a lift, a shock load on the lifting components may cause a load to shift or fall.

BALANCING THE LOAD

Stability

Stability could be defined as resistance to toppling. A tall object with a small footprint (high center of gravity) will take less force to topple than a low, wide object (low center of gravity).

As the center of gravity increases relative to the width of the base, a point is reached at which an object will topple unless it is supported by some external means. Acceleration, braking, or even wind can potentially cause the load to topple. Loads are always unstable if the center of gravity is above the lifting point (see Figure 5–20).

FIGURE 5–20
This load is unstable because the center of gravity is above the lift point.

Courtesy of MAZELLA Lifting Technologies

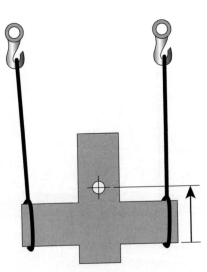

Balance

In most cases when lifting a load, we want the load to remain level as it leaves the ground. To achieve this, the lifting hook must be placed directly above the center of gravity. Legs of slings must be distributed as evenly as possible. The angles of each leg should be similar. If a load shifts while lifting the load on the sling, the legs will be unequal.

A load's shape usually determines its center of gravity. Loads can be symmetrical or asymmetrical. Symmetrical means that each half of the item to be lifted is a mirror image of the other. Some loads have lifting eyes that are supplied by the manufacturer to make the product easy to lift and move. These eyes are usually screwed into the product or in some cases welded to the product.

The center of gravity is the balance point of a load. It is crucial that you determine the center of gravity before lifting a load. Figure 5–21 shows a properly rigged load. Note that the lift is centered directly above the centerpoint of the load. If you do not find the center of gravity before you attach the load, the load may be prone to shifting and/or tipping. Figure 5–22 shows an example of an improperly rigged load. The center of gravity is off center. This load will shift when it is lifted and could cause severe consequences.

FIGURE 5–21
Properly rigged load.
Note that the hoist is
directly above the
center of gravity.

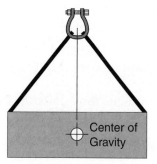

FIGURE 5–22
Improperly rigged load.

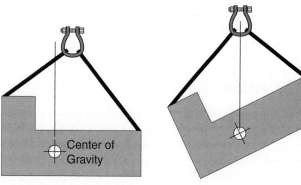

For a simple uniform shape like a cube or a sphere, the center of gravity is in the center of the object. Nonsymmetrical parts are more difficult to judge and should be carefully estimated. Then attach the rigging and slowly begin lifting the object. If the object begins to tilt, reestimate the center of gravity. The lift must be moved toward the side of the object that tilted down (heavier side).

LIFTING THE LOAD

Lifting should be done only by a certified lift or crane operator. Before a lift can be used the operator must know all of the functions of the lift. Although this unit of instruction is not intended to demonstrate how to operate a crane or hoist, some safety tips to follow while around lifting devices include:

Start and stop any lift slowly.

Lift straight up to ensure that the load does not swing.

Do not permit anyone to ride on the hook or the load.

Make certain that no people pass under a load.

Ensure that the load will safely clear nearby equipment.

Make all moves slowly to prevent swinging.

Never use side pulls.

Make sure that the rigging and the hoist will handle the specified load weight.

Signals for crane movement can be given by one person only.

Lifts can be divided into three parts: the lifting device, the hitch, and the load.

Lifting Device

The condition, capacity, and limitations of the lifting device must be known, as well as whether the capacity is adequate for the lift. Make sure the device will be able to lift the load high enough and that the horizontal reach will be adequate. Check the hook to be sure it is the right size so it will not distort the sling eye. Check the hook for cracks and for evidence of point loading or bending to one side of 15% or more.

The Hitch

The type of hitch determines the choice of the sling. Figure 5–8 shows various slings. Before a sling is selected you should choose the type of hitch that will be used. A hitch should be chosen that will protect the load, protect the sling, and do the job. The type of hitch you select may determine the type of sling body that will work best. The type of hitch will also determine the length of sling that will be needed. Lift height, overhead clearance, and hook travel will also affect the choice of hitch and length of sling.

The Load

You must know the weight of the load. You must also protect the load from damage by the slings and protect the slings from damage by the load.

GUIDELINES FOR THE RIGGER

Because rigging is so important to the safety of workers, the information in Figure 5–23 provides quick references and guidelines for the rigger.

Common Hazards

Know the safe working loads of the equipment and components you are using and never exceed the limits.

Examine all of the components you are using before you use them. Defective components should be destroyed to prevent future use.

Beware of unsafe equipment. If you have any reason to believe that equipment is unsafe, you should not use it until you have reported it to a supervisor, its safety has been confirmed, and orders to proceed have been issued by a person in authority who is then responsible for the safety of all involved.

Personnel must also be aware of electrical hazards. One of the most frequent killers of rigging and lifting personnel is electrocution caused by the hoisting device or load contacting power lines.

Factors that Reduce Lift Capacity

A safe working load is based on ideal conditions that are rarely achieved in normal practice. Personnel must be able to recognize and identify factors that may reduce the capacity of equipment.

Crane and Rigging Safety Rules

Check limit switches before rigging the load.

Make sure the load does not exceed rated capacity.

Know the center of gravity of the load.

Attach load above the center of gravity for stability.

Select hitch that will control the load.

Know the rated capacities of rigging and slinging.

Inspect all rigging before use.

Protect the sling from sharp corners.

Allow for increased tension due to sling angle.

Equalize loading on multiple leg slings.

To turn or reposition a load, either one or two lifting devices may be employed. Always use a choker hitch or a single-leg direct attachment. Never attempt to turn a load with a basket, since the load will slide in the hitch, against the sling body—resulting in damage to both the sling and the load, and possibly a dropped load.

One Hook Load Turning

To turn a load with one hook, attach the sling directly to the load ABOVE the Center of Gravity. The lifting hook must be able to move, or travel, in the direction of the turn to prevent sliding of the pivot edge of the load just as the load leaves the ground. It may be necessary to lift the load clear to reposition it after the turn is completed, and irregular shapes sometimes will require blocking for support during and after the turn.

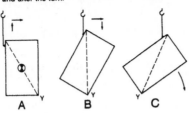

Two-hook turning is employed when it is desired to turn the load freely in the air. Main and auxiliary hoists of a crane can often be used, or two cranes can be used.

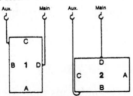

To turn from side (A) to (B) in 1 & 2 above, attach on side (B) above the Center of Gravity and on side (D) at the Center of Gravity, then lift both hoists equally until load is suspended. Lower auxiliary until turn is completed; detach sling at (B) before lowering load completely.

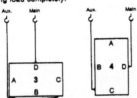

To turn from side (B) to (C) in 3 & 4 above, lift balanced load at (D) directly above the Center of Gravity; then attach auxiliary at (B) and lift to desired position. Lower both hooks simultaneously until side (C) is in desired position.

FIGURE 5–23

Useful guidelines for the rigger.

Courtesy of MAZELLA Lifting Technologies

Turning with double choker gives good control, with weight always applied against a tight sling body and no movement between sling and load. To rig, place both eyes on top of load, pointing opposite direction of turn. Body of sling is then passed under load, through both eyes and over lifting hook. Blocking should be used under load to protect sling and facilitate removal.

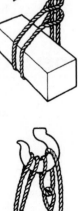

Lifting unbalanced loads when exact length slings are not available can be accomplished by rigging a choke on the heavy end, as right. Length can be adjusted before weight applies, but once the load comes onto the sling, the hitch is locked in position for the lift.

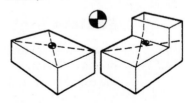

Center of Gravity of a rectangular object with homogenious characteristics will usually be below the junction of lines drawn diagonally from opposite corners. When a rectangular object has weight concentrated at one end, Center of Gravity will be situated toward that end—away from the intersection of diagonal lines. To avoid an unbalanced lift, the lifting hook must be rigged directly above the Center of Gravity.

To locate the approximate Center of Gravity of an irregularly shaped article, visualize it enclosed by a rectangle. Where diagonals from opposite corners intersect will usually provide a lift point near the Center of Gravity.

Some Useful Guidelines For the Rigger

On the following pages are some useful tips to help the rigger do his job more efficiently and safely. Prevailing work rules and government regulations place full responsibility for proper performance upon the rigger, so it is his duty to be familiar with the condition and capability of all tools and equipment used, as well as techniques employed. One basic rule always applies: Always know . . . never guess.

Each lift may be divided into three parts, providing a convenient plan for proceeding:

1. **The Lifting Device** – Know its capacity and limitations, and its condition. When was it last inspected? If in doubt about capacity, check the placard.

2. **The Hitch** – Here is where the rigger can exercise ingenuity . . . but it's also the easiest place to make a mistake. This book can help you decide which sling to use, and how to rig it properly.

3. **The Load** – The weight must be known. But you must also protect the load from possible damage by the slings . . . and protect the slings from damage by the load.

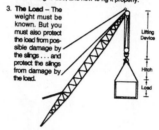

Is the lifting device adequate?

Check the placard on the crane or hoist, and then answer three questions:
1. Is capacity adequate for this lift?
2. Will if lift high enough?
3. Is horizontal reach adequate?

Check the hook and reeving.

1. Are sheaves properly rigged? If multi-part reeving, will if support the load?

2. Is the hook the right size so sling eye won't be distorted when put over the hook?

3. Check for cracks in bowl of the hook, and for evidence of point loading or bending to one side of 15% or more.

FIGURE 5–23 (*Continued*)

Type of Hitch Determines Choice of Sling

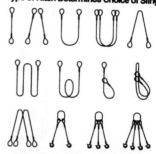

Before you select a sling for a specific lift, determine the most effective hitch to do the job, protect the load, and protect the sling. One of three basic hitches will usually do the job.

The type of hitch you select may determine the type of sling body that will best do the job, as well as the length of sling that will be needed. Lifting height, overhead clearance and hook travel will affect choice of hitch and length of sling.

Choose a sling body type which will best support the load while providing adequate rated capacity. The proper choice will provide:

1. Lifting capacity needed.
2. Proper D/d Ratio.
3. Handling characteristics needed for rigging.
4. Minimal damage to the sling.
5. Minimal damage to the load.

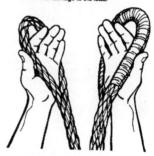

Protect the sling during the lift with blocking or padding at sharp corners or where the sling body would be bent severely.

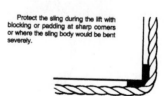

Use a spreader bar between legs of a sling to prevent excessive side pressure on the load by the sling during the lift.

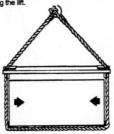

When attaching a sling to eye bolts, always pull on line with the bolt axis. When hitching to bolts screwed into or attached to a load, a side pull may break the bolts.

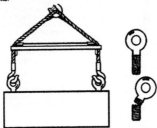

Use a shackle in the sling eye during a choke to protect sling body against excessive distortion. Always put shackle pin through sling eye, rather than against the sling body – since sliding movement of sling body could rotate pin, causing it to come loose.

A sliding hook choker is superior to a shackle or unprotected eye, since it provides a greater bending radius for the sling body.

Use blocking or padding to protect hollow vessels, loose bundles and fragile items from scuffing and bending. Remember that blocking becomes part of the lift, and must be added to total weight of the sling.

When lifting crates or wooden boxes with a basket hitch, be sure load can withstand side pressure as tension is applied to sling. Use spreader bars and corner protectors to prevent damage to contents.

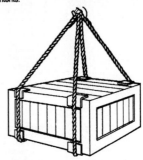

When lifting a bundled load with a single sling near the center of gravity, a choke is more effective than a basket hitch to prevent unbalance and slipping of the load in the sling.

Some riggers will use a double wrap around the load, for 360° gripping of the load, to prevent slippage during the lift.

FIGURE 5–23 (*Continued*)
Useful guidelines for the rigger.

Courtesy of MAZELLA Lifting Technologies

You can reduce the angle of a choke with a wooden block, or blocks, between the hitch and the load. This also increases the angle between the two legs to improve sling efficiency. See page 6 for formula to adjust for capacity loss in chokes at various angles.

When rigging two or more straight slings as a bridle, select identical sling constructions of identical length – with identical previous loading experience. Normal stretch must be the same for paired slings to avoid overloading individual legs and unbalancing the load during the lift.

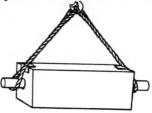

Single-part hand-spliced slings must not be permitted to rotate when rigged in a straight, vertical hitch. Rotation can cause the splice to unlay and pull out, resulting in dropping of the load.

WARNING
Hand-spliced slings should not be used in lifts where the sling may rotate and cause the wire rope to unlay.

Anytime a load is lifted beyond arm's reach with a single-part load line or straight eye-and-eye sling, use a tagline to prevent load rotation. If a wire rope is permitted to rotate, the strands may unlay and the rope's capacity will be reduced.

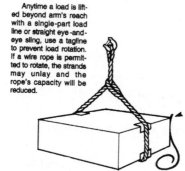

Two basket hitches can be rigged with two slings to provide better balance for long loads. Be sure that slings cannot slide toward one another along the load when the lift is made.

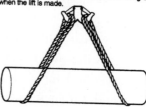

Use an equalizing bar with double basket hitches to reduce tendency of slings to slide together, and to keep loads level. By adjusting the hook point and using a come-along or chain block to support the heavy end, the load can be kept level during the lift.

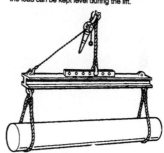

Proper Use of Cribbing

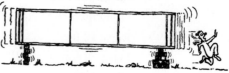

Incorrect

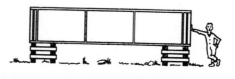

Correct

FIGURE 5–23 (Continued)

Allow for load reductions when using choker hitches.

Attach tag line prior to lift.

Keep personnel clear of lift area.

Wear hard hats when making overhead lifts.

Lift load a few inches and verify rigging.

Check for any loose items.

Know limitations of hoisting device.

Start and stop SLOWLY! Watch for obstructions (not only hook and load but outboard end of the bridge).

Check that pathway is clear before making a lift (use a spotter for blind spots).

Verify that hook completely closes.

Use appropriate hand signals.

Maintain load control at all times.

Report suspected drum wrappings immediately (if drum has fewer than 2.5 wraps remaining).

Never leave load unattended.

Crane Safety - After Use

Spot crane in the approved location.

Lower the load to the ground.

Disconnect the load and slings.

Raise all the hooks to upper limit switch.

Place all controls in the off position.

Visually check for dangerous conditions.

Never leave a load unattended.

Hand Signals

Figure 5–24 shows the common hand signals:

HOIST: With forearm vertical, forefinger pointing up, move hand in small horizontal circles.

LOWER: With arm extended downward, forefinger pointing down, move hand in small horizontal circles.

BRIDGE TRAVEL: Arm extended forward, hand open and slightly raised, make pushing motions in direction of travel.

TROLLEY TRAVEL: Palm up, fingers closed, thumb pointing in direction of motion, jerk hand horizontally.

STOP: Arm extended, palm down, hold position rigidly.

Lower: Arm down, forefinger pointing down, move hand in small circular motion.

Raise (Hoist): Forearm vertical, forefinger up, move hand in small horizontal circles.

Emergency Stop: Arms extended, palms down, move hands back and forth horizontally.

Stop: Arm extended, palm down, move arm back and forth horizontally.

Move Slowly: One hand is used to give a motion signal, the other hand is placed motionless in front of other hand.

Trolley Travel: Palm is up, finger closed, thumb points in desired direction of motion, hand is jerked horizontally.

Bridge Travel: Arm is extended forward, hand open and raised, make pushing motion in desired direction of travel.

FIGURE 5–24
Common hand signals.

EMERGENCY STOP: Arm extended, palm down, move hand rapidly right and left. An emergency stop signal must be accepted from any person. It is important to immediately react because the person giving the signal may have knowledge that the crane operator does not, including: better view, knowledge of hazards, etc.

MOVE SLOWLY: Use one hand to give any motion signal and place other hand motionless in front of hand engaging in motion signal.

Obey all STOP signals regardless of who gives them.

Other than EMERGENCY STOP, signals should be accepted from only one person at a time (normally the spotter for the operator).

Use standard crane signals and do not move unless directions are clear. Know standard hand signals.

Questions

1. What are the three main types of slings.
2. List the items that should be inspected for before using a rope or wire sling.
3. What type of chain can be used for lifting and how can it be identified?
4. Name an advantage of synthetic web slings.
5. What is a hitch?
6. What does it mean to balance a load and why is it so important?
7. A rectangular shaped object is to be lifted. The length is 3 feet, the height is 6 inches, and the width is 1 inch. It is aluminum. How much does it weigh?
8. A bar of steel round stock is to be lifted. The bar is 6 inches in diameter and 4 feet long. How much does it weigh?
9. A piece of steel pipe is to be lifted. It is 8 inches in diameter. The wall of the pipe is 1 inch thick. How much does the pipe weigh?
10. Explain how you can determine how much can be safely lifted by slings that are used at an angle.

CHAPTER 6

Fasteners

This chapter will introduce the reader to fastener fundamentals, terminology, and technology. Thread terminology, as well as tapping, threading, and removing broken fasteners will be examined.

Objectives

Upon completion of this chapter, the reader will be able to:

✔ Explain terms such as thread pitch, crest, root, minor and major diameter, and so on.
✔ Explain the common thread systems.
✔ Describe various types of fasteners.
✔ Describe the process of making threads.
✔ Describe methods of fastener removal.

INTRODUCTION

We take threads for granted. For thousands of years screw threads have played a large role in history and in the advancement of technology. One of the first practical uses of the screw was by Archimedes in approximately 250 B.C. He used a turning shaft with screw threads to lift water from a river to provide irrigation. This method is still being used in many third world countries.

As far back as the ancient Romans, it was possible to make and match male and female threads to work together. There was no standardization, however; one threaded bolt would fit only the one nut that was made to fit it. Threads were made by filing a round piece of metal with a file. The file was also used to make a crude tap that would be used to cut the inside thread for the nut. Threads were not used much because of the difficulty involved in making them. Handmade nails and wooden pegs were most commonly used because they were easy to fabricate.

In the 1400s, machines were invented to make fasteners for the clock industry. An early lathe was invented by Jacques Besson in about 1550 that made screw production practical. Screw threads continued to be made by hand, however, and there was no standardization.

In 1841, Sir Joseph Whitworth wrote a paper titled "The Uniform System of Screw-Threads." In this paper he suggested what later became the first national standard for screw threads and 55 degrees for the thread angle. His system became known as the Whitworth thread. Meanwhile, in America, William Sellers was developing his own thread standard based on a 60 degree thread angle. Mr. Seller's system was adopted by the United States as a standard in 1864. It was called the American National Standard thread. This meant that there were two different standards between the US and Great Britain. The lack of a common standard in trade between the countries, especially during WWI and WWII, caused a lot of problems.

In 1948 representatives from the US, Canada, and Great Britain agreed on a common standard, which was based on the best of both countries' previous standards and called the Unified Screw Thread System.

The US has not adopted the metric system; Great Britain adopted the metric system in 1965. The International Standards Organization (ISO) adopted an international screw thread standard in 1969. The ISO metric standard is the most widely used in the world.

SCREW THREADS AND FASTENERS

Most precision machined parts are ineffectual until they are assembled into mechanical components. These assemblies require many different types of fasteners. This chapter will introduce different types of fasteners and their proper uses. Threaded fasteners take on many different shapes and forms, but they all have one thing in common: a thread. Although threads may be used for adjustment, measurement, and the transmission of power, the main application of a thread is as a fastening device.

THREAD TERMINOLOGY

Some of the more commonly used thread terms are:

Angle of Thread—The angle of the thread is the included angle between the sides of the thread (Figure 6–1). For example, the thread angle for Unified Screw Thread forms is 60 degrees.

Figure 6–1
Major parts of a thread.

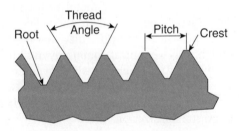

Major Diameter—Commonly known as the outside diameter (Figure 6–1). On a straight screw thread, the major diameter is the largest diameter of the thread on the screw or nut.

Minor Diameter—Commonly known as the root diameter (Figure 6–1). On a straight screw thread, the minor diameter is the smallest diameter of the thread on the screw or nut.

Number of Threads—Refers to the number of threads per inch of length.

Pitch—The distance from a given point on one thread to a corresponding point on the next thread (Figure 6–1).

Lead is the distance that a screw thread advances in one revolution. The lead and the pitch of a single lead thread are the same amount. On double lead threads, the lead is twice the pitch. A double lead thread has two thread start points.

THREAD FORMS

There are many thread forms. The Unified Screw Thread System is an attempt to standardize thread forms in the United States, Canada, and Great Britain. Unified threads are divided into the following series: UNC (Unified National Coarse), UNF (Unified National Fine), and UNS (Unified National Special).

Unified National Coarse and Unified National Fine refer to the number of threads per inch on fasteners (see Figure 6–2). A specific diameter of bolt or nut will have a specific number of threads per inch of length.

EXAMPLE

A 1/4 inch diameter Unified National Coarse bolt with 20 threads per inch is identified as 1/4-20 UNC, 1/4 being the major diameter and 20 the number of threads per inch.

A 1/4 inch diameter bolt with a fine thread would be identified as 1/4-28 UNF, 1/4 being the major diameter and 28 the number of threads per inch.

The Unified National Special series is identified the same way. For example, a 1/4 inch diameter UNS bolt may have 24 or 27 threads per inch.

Why is there a need for UNC and UNF series threads? Following are some of the principle uses of coarse and fine threads.

Coarse Thread Series

The coarse series, UNC, is the one most commonly used in the mass production of bolts, screws, nuts, and other general fastening applications. It is also used for threading into lower tensile strength materials (bronze, brass, aluminum, and

Inch

Diameter (Inch)	Pitch Coarse	Fine
No. 0 (.060")		80
No. 1 (.073")	64	72
No. 2 (.086")	56	64
No. 3 (.099")	48	56
No. 4 (.112")	40	48
No. 5 (.125")	40	44
No. 6 (.138")	32	40
No. 8 (.164")	32	36
* No. 10 (.190")	24	32
No. 12 (.216")	24	28
1/4	20	28
5/16	18	24
3/8	16	24
7/16	14	20
1/2	13	20
9/16	12	18
5/8	11	18
3/4	10	16
7/8	9	14
1	8	14
1 1/8	7	12
1 1/4	7	12
1 1/2	6	12
1 3/4	5	
2 in.	4 1/2	
2 1/4	4 1/2	
2 1/2	4	
2 3/4	4	
3	4	

* Equivalent to 3/16

Metric

Most Common

Diameter (mm)	Pitch Coarse	Fine
1	0.25	
1.2	0.25	
1.6	0.35	
2	0.4	
2.5	0.45	
3	0.5	
4	0.7	
5	0.8	
6	1	
8	1.25	1
10	1.5	1 (1.25)
12	1.75	1.25 (1.5)
16	2	1.5
20	2.5	1.5
24	3	2
30	3.5	2
36	4	3
42	4.5	3
48	5	3
56	5.5	4
64	6	4
72		6
80		6
90		6
100		6

Not Popular

Diameter (mm)	Pitch Coarse	Fine
1.4	0.3	
1.8	0.35	
2.2	0.45	
3.5	0.6	
14	2	1.5
18	2.5	1.5
22	2.5	1.5
27	3	2
33	3.5	2
45	4.5	3
52	5	3
60	5.5	4
68	6	4
76		6
85		6
95		6

Special Applications

Diameter	Pitch Coarse	Fine
7	1	
11	1.5	1
15		1
25		1.5
26		1.5
28		2
39	4	3

- **Note:** To determine the tap drill size for metric fasteners, simply subtract thread pitch from the fastener diameter and drop all but the first decimal place.

- **Example:** M12 - 1.75 = 10.25 is a 10.2 tap drill size

FIGURE 6–2
Screw thread chart.
Courtesy of Atlantic Fasteners, Inc.

plastics) to obtain the best resistance to stripping of the internal thread. UNC threads are often used for quick assembly or disassembly.

Fine Thread Series

The fine series, UNF, has greater tensile stress area than coarse threads of the same size. The fine series will resist stripping better than coarse threads where the external and mating internal threads are subjected to loads equal to or greater than the capacity of the screw or bolt. Fine threads are also used were the length of engagement is limited or where thin wall thickness requires a fine pitch.

FIGURE 6–3

How to determine the thread length of hex cap screws.

Courtesy of Atlantic Fasteners Inc.

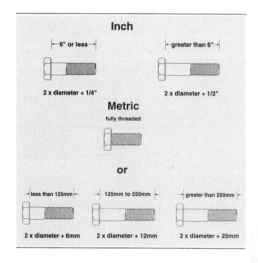

THREAD LENGTHS

Figure 6–3 shows formulas that can be used to determine the thread lengths of inch and metric cap screws. For example if we have a 1/4-20, 2 inch cap screw, the thread length would be 2 * diameter + 1/4 inch. 2 * 1/4 + 1/4 = 3/4 inch thread length.

CLASSES OF THREAD FITS

Some thread applications can utilize loose threads, while others require a tighter fit. An example of this would be the head of an engine. The head of your car or truck engine is held down by a threaded fastener called a stud. A stud is threaded on both ends. One end is threaded into the engine block. The other end uses a nut to tighten down the cylinder head. When the head is removed, you want the stud to remain in the engine block. This end requires a tighter fit than the end of the stud accepting the nut. If the fit on the nut is too tight, the stud will unscrew as the nut is removed.

Unified thread fits are classified as 1A, 2A, 3A . . . or 1B, 2B, 3B . . . (see Figure 6–4). The A indicates an external thread; the B indicates an internal thread. The numbers indicate the class of fit. The lower the number the looser the fit. Class 2 fits are used on the largest percentage of threaded fasteners. The tighter the fit the closer the tolerance of the sizes of the thread and hence the more expensive to purchase. A typical notation of a unified thread form with fit tolerance would be:

1/4-28 UNF 2A

In this particular case the class of fit would be a 2. The symbol A indicates that it is an external thread.

FIGURE 6–4
Table of fits for
the unified series
of threads.

UNIFIED THREAD FITS		
	EXTERNAL	INTERNAL
LOOSE	1A	1B
MEDIUM	2A	2B
TIGHT	3A	3B

FIGURE 6–5
Specification example
for a unified system
thread.

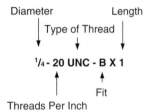

Figure 6–5 shows a typical unified screw thread designation. The nominal diameter of this thread is 1/4 inch. Next the number of threads per inch is specified. There are 20 threads per inch in this case. The UNC means that this is a Unified National Coarse thread. Next the fit (tolerance) is specified. In this example it is B, or a medium fit. Last the length of the fastener is specified: 1 inch in this example.

METRIC THREADS

With the importation and exportation of goods, especially in the automotive industry, metric threads have become the prevalent thread type on many kinds of equipment. The metric thread form is similar to the unified thread form in that they are based on a 60-degree thread angle. Metric thread series take the following form:

M10 × 1.5-6g

where M10 is the major diameter in millimeters and 1.5 (millimeters) is the pitch (distance from one thread to the next thread), 6 is the class of fit with the g symbolizing external thread. This external thread would have a major diameter of 10 millimeters, a pitch of 1.5 millimeters, and a "medium" thread fit.

Figure 6–6 shows an example of the total metric thread specification. The M specifies that this is a metric thread. The next descriptor is the diameter of the thread in millimeters. The thread diameter is 10 mm, and the pitch of the thread is specified as 1.25 mm. Next, the length of the fastener is specified in millimeters. The last two designations are for the fit of the thread.

The fit designations under the ISO system are more descriptive. There can be two number/letter combinations to describe the fit (see Figure 6–7). The first number/letter combination describes the pitch diameter tolerance. The second letter/number

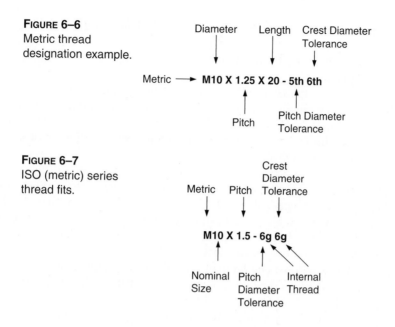

FIGURE 6–6
Metric thread designation example.

FIGURE 6–7
ISO (metric) series thread fits.

ISO ALLOWANCE	EXTERNAL	INTERNAL
LARGE ALLOWANCE	e	E
SMALL ALLOWANCE	g	G
NO ALLOWANCE	h	H

FIGURE 6–8
ISO allowance designations.

combination describes the crest diameter tolerance. The number in each specification denotes the grade tolerance. The lower-case "g" means it is an internal thread. A capital "H" is used to represent an external thread. The tolerance grade is designated by the numbers 3 through 9 for external threads, and 4 through 8 for internal threads. The larger the number is, the larger the tolerance. The larger the number is, the looser the fit. A smaller number means it is a tighter tolerance and a tighter fit.

The equivalent of unified system nominal fit 2A/2B would be the ISO 6g/6H. The 6 is a medium tolerance. See Figure 6–8.

TAPER PIPE THREADS

Standard taper pipe threads (NPT) are unlike standard threads in the way they are cut and in the way they are used. Taper pipe threads are one of the more challenging thread cutting operations. Taper pipe thread cutting requires greater accuracy than standard threads. Pipe threads are designed and cut to mechanically seal a threaded joint for pressure and to prevent leakage. To create the mechanical seal, 100 percent of the thread height must be cut by the tap to maintain the standard thread profile.

The NPT standard limits for both external and internal are identical, so that hand-tight engagement can result in thread flank contact only, or crest and root contact only. Wrench tightening normally produces an assembly with crest and root clearances. Excessive tightening with a wrench without the use of some kind of tape or chemical sealer is not advisable.

Taper threads are also unlike standard threads in their designations. A machinist looking at a 1/4-18 NPT would immediately notice that the outside diameter of the tap is much bigger than 1/4 of an inch. This is because the tap size designation refers to the inside diameter of the standard iron pipe it is designed to fit. Another major difference between taper pipe tapping and standard thread tapping is the way in which the machinist must control the thread diameter. Since the tap is tapered, the machinist can control the thread diameter by adjusting the depth that the tap threads into the hole. To get the proper thread depth, a machinist drives the tap into the work 12 full turns. More turns may result in the tapped hole being too big. Fewer turns may result in the tapped hole being too small. There is no one particular class of fit for NPT threads, such as 3A or 3B. Proper taper thread gauging is the only acceptable method for determining proper depth for taper pipe tapping. One method of monitoring tap depth is to wrap a piece of wire or tape around the tap as a line to serve as a depth indicator. This practice works well to get you "close." This trick works all right for hand tapping where cutting fluids are applied by hand. When machine taper tapping, this practice may prevent cutting fluids from reaching the cut zone. Because a lack of cutting fluid reaching the cut zone may result in early tap failure or torn threads during machine tapping, the practice of wrapping anything around the base of the tap should be avoided.

Taper pipe tapping is more difficult than tapping regular straight threads because the taper tap, instead of cutting 60 to 70 percent threads, is cutting in excess of 100 percent threads. The hole size for a normal straight thread was designed to create 60-70 percent threads, which means that less material is removed when cutting straight threads. When tapping a pipe thread, the tap is removing 100 percent or more of the material to make the threads. The tap will be extremely tight in the hole when the tap has been driven to the required depth. The tap becomes wedged and considerable force may have to be used to reverse the tap and break it loose. Since a taper tap will become entirely loose after just 1/8 reverse turn, great care must be used to avoid cross threading when starting the tap back into the hole. Starting the tap also becomes more critical. When starting a taper tap, pressure needs to be applied at the start to ensure that the first threads are not torn. When hand tapping, a spring center in a drill press or other tapping device should be used to ensure tap alignment. A taper tap that is not started straight will almost always result in a broken tap or poor mating thread engagement.

One practice that is almost always associated with taper pipe tapping is taper reaming. The drilled hole is reamed with a taper ream consisting of a taper of 3/4 inch per foot. Taper pipe reaming prior to taper tapping, though widely viewed as an accepted and/or necessary operation, can cause a tap to cut unacceptable threads if not done properly. Starting a taper tap in a tapered hole can also be very difficult because all of the teeth on the tap are trying to cut at once. When a taper tap cuts threads in a straight hole, the cutting forces start at the chamfered end of the tap and progress one thread at a time. In most cases, especially in smaller tap sizes, taper tapping a nontapered hole will produce enough 100 percent threads to seal properly. Unless there are absolute reasons for taper reaming prior to taper pipe tapping, it is best to straight drill the holes. Then you can hand tap the holes to depth according to the NPT gauge for this tap size, using a mechanical means of keeping the tap straight.

FASTENER IDENTIFICATION

Bolts

A general definition of a bolt is "an externally threaded fastener that is inserted through holes in an assembly." A bolt is tightened with a nut. A screw is an externally threaded fastener that is inserted into a threaded hole and is tightened by turning the head. From these general definitions a bolt can become a screw or a screw can be used as a bolt. This depends on how they are used. Bolts and screws are the most common of threaded fasteners; machine bolts are made with hexagonal or square heads.

The strength of an assembly in large part depends on the diameter of the bolt or the thread engagement of the screw. Thread engagement is the distance a screw extends into the threaded hole. For maximum strength, the minimum thread engagement should be equal to 1-1/2 times the screw diameter. This would assure that the screw or stud would break before the threads pulled out of the mating part. This is the desirable condition because it is easier to remove a broken stud than it is to drill and tap for a larger screw thread.

The body diameter—the diameter of the unthreaded portion of the bolt below the head—is typically slightly larger than the nominal or standard size of the bolt. A hole that is to accept a bolt must be drilled slightly larger than the body diameter.

Carriage Bolts

Carriage bolts are used to fasten wood and metal parts together. Carriage bolts have round heads with a square body under the head. The square part of the bolt, when pulled into the wood, keeps it from turning while the nut is tightened (see Figure 6–9).

Socket Head Cap Screws

Socket heads are the strongest of the head styles. The height of the head is equal to the shank diameter. These screws should not be used with a regular hex nut as the hex nut is not as strong as the screw.

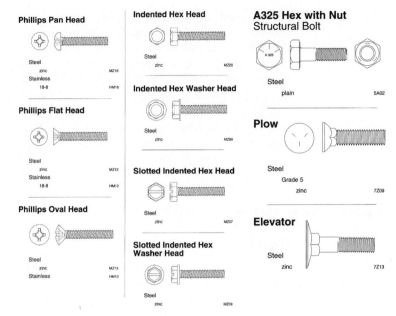

Figure 6–9
Common bolts.

Courtesy of Atlantic
Fasteners Inc.

Low Head Cap Screws

These screws are designed for applications in which head height is limited. They are not as strong as the regular socket head, as the head height is approximately half the shank diameter.

Flat Head Cap Screws

These screws are designed for flush mount applications. Note that the inch and metric screws have different countersink angles. The correct countersink must be applied or the screw may fail.

Button Head Cap Screws

The button head cap screw is particularly suited to holding thin materials because of its large head. It is good for tamper-proofing because the hex drive style makes it less susceptible to tampering with a screwdriver. It is also good for applications where a regular screw (slotted or Phillips) is prone to stripping.

Socket Shoulder Screws

Socket shoulder screws are typically used as pivot points or axles for other parts to rotate around. The shoulders are ground to tight tolerances.

Machine Screws

The machine screw is used for general assembly work. It is manufactured in both fine and coarse thread series and fitted with either a slotted or recessed head.

Machine screw sizes vary from No. 0 (0.060) to 1/2 inch (0.500) in diameter, and come in many different lengths. Figure 6–10 shows examples of common machine screws.

Socket head screws are made with a variety of different head shapes and are used where precision bolts or screws are needed. Socket head screws are manufactured to close tolerances and have a finished appearance. Cap screws can have round heads, low heads, flat heads, button heads, or can be in a shoulder screw configuration (see Figure 6–11).

Set screws

Set screws are used to lock pulleys, collars, or shafts in place. Figure 6–12 shows typical set screw types. Set screws are available with different points. The flat head set screws are used where minimal indentation to the part is needed and are used where frequent adjustments are made. They are also used to provide a jam screw effect: a second set screw is added to prevent the first set screw from vibrating loose. A dog point set screw is used to hold a collar to a shaft. Alignment is always maintained with a dog point set screw because the shaft is drilled with a hole of the

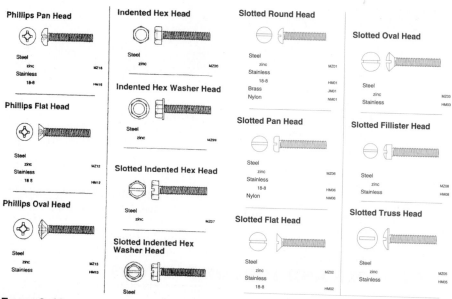

FIGURE 6–10
Common machine screws.
Courtesy of Atlantic Fasteners Inc.

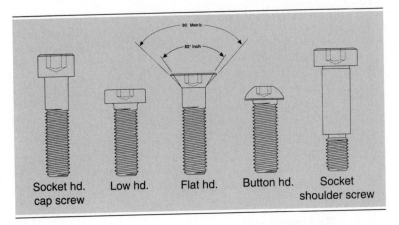

FIGURE 6–11
Various socket head screws.
Courtesy of Atlantic Fasteners Inc.

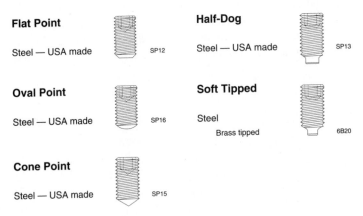

FIGURE 6–12
Typical set screw types.
Courtesy of Atlantic Fasteners Inc.

same diameter as the dog point. A cup pointed set screw will give a very good slip resistant connection. Figure 6–13 shows suggested uses for various types of set screws.

BOLT GRADES AND TORQUE FACTORS

In some instances bolts need to be fastened with the correct amount of pressure (torque). In these instances the manufacturer of the product will recommend a certain torque be applied to a particular fastener. Insufficient torque will usually result

Application Strength	Knurled Cup	Plain Cup	Flat	Oval	Cone	Half Dog	Soft Tip
Vibration Resistant	*						
Can be Readjusted Without Damaging a Surface			*	*			*
Use Against Thin Wall			*				
Soft or Hard Surfaces	*	*			*		
Act as Adjusting Screw			*		*		
Fits Well in Locating Holes						*	
Fits Well in Locating Grooves				*		*	
Creates Most Compressive Force					*		
Permanently Holds Part	*	*			*	*	

FIGURE 6–13
Suggested uses for various set screw types.

in parts working loose and causing a malfunction due to misalignment. Over-tightening, on the other hand, can cause stress or warpage which also might disturb the alignment of assemblies. The "armstrong" (overtightening) method of tightening fasteners can also cause broken castings, broken bolts, or stretching of the fastener.

Steel has excellent elasticity. Elasticity is the ability of an object to stretch slightly, like a spring, and then return to its original shape. Any fastener must reach its limits of stretching in order to exert clamping force. But also like a spring, an overstretched fastener takes on a set, loses its elasticity, and cannot snap back to its original shape. Proper torquing will prevent this condition.

A popping or snapping sound is sometimes heard during the final tightening of a fastener. This popping sound indicates that the fastener is undergoing set. When a new fastener is being used and the popping occurs, the remedy is to back it off and retighten to the proper torquing specifications. When an old fastener is being used and you hear popping, take the fastener out and clean the bolt and internal threads completely. The safer and more economical thing to do is replace the old fastener with a new one.

Rolled Versus Conventionally Cut Threads

There are two main thread processes to make threaded fasteners: cutting and rolling. Figure 6–14 shows the process of forming and rolling a bolt. The process begins with a piece of steel that has been cut to the correct length. Then the forming process begins. The cut off steel is upset to begin to form the thread. The next process, heading, forms the head and marking. The bolt is then extruded to create the correct diameter on which to form the thread. Next the bolt head is trimmed to form the hex on the head, and then the thread end is chamfered. The last step in the process is rolling the thread.

Figure 6–15 shows the grain structure in the steel of the bolt after each process and the actual rolling process. Rolled threads are much stronger because the

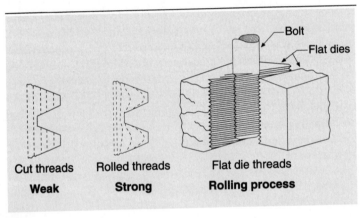

FIGURE 6–14
Thread rolling process.
Courtesy of Atlantic Fasteners Inc.

FIGURE 6–15
Comparison between cut threads and rolled threads and the rolling process.
Courtesy of Atlantic Fasteners Inc.

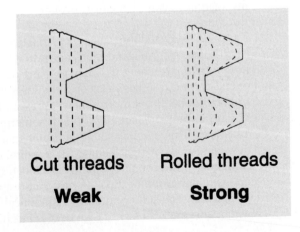

threads are formed and the grain follows the thread shape. Rolled threads are also smoother than cut threads.

SCREW GRADES

Just as critical as proper torquing is the selection of the right grade of fastener for the job. The grades of bolts, or cap screws, in Figure 6–16 are identified by the markings on the heads. The grade indicates the strength of the fastener. Use a manufacturer's chart as a guide for the proper torque of fasteners.

Metric Property Classes

The correct terminology for the grade of a metric fastener is property class. Figure 6–17 shows a comparison between a unified series bolt and a metric bolt. Figure 6–18 shows a comparison of ISO and SAE grades.

Figure 6–19 shows a comparison of a few inch and metric screws from the weakest to the strongest grade.

Figure 6–20 shows a comparison of the strength of various grades of screws. Note that when the screw reaches a certain tensile strength, it starts to elongate. Note also that the stronger screws tend to experience less stretch. It is important to torque screws properly.

Figure 6–21 shows a few examples of common stainless steel hex head markings. Note that the two lines on the top example represent an 18-8 industry marking that can be 302, 303, or 304 stainless. The numbers reflect 18% chromium and 8% nickel content. The 316 example is a stainless alloy that contains molybdenum, which makes it more corrosion resistant. The 18-8 and the 316 heads have the same tensile strength.

In the metric example, A2 means 18-8 stainless and A4 means that it is 316 stainless. The classes of 50, 70, and 80 designate increasingly higher tensile strength. A manufacturer's mark is required on metric M5 or larger diameters.

NUTS

Nuts have either a hexagonal or square head, and are used with bolts with the same shaped head. They are available in different degrees of finish. A regular finish nut is not machined (except for the threads). Regular semifinished nuts are machined on the bearing face to provide a straight, flat surface for the washer. Heavy semifinished nuts have the same finish as the regular semifinished; however, the body is thicker for greater strength.

There are many types of nuts. Figure 6–22 shows some of the more common types of available nuts. Castellated or slotted nuts have milled slots across the flats so they can be locked in place using a cotter pin or wire. Acorn nuts are used when appearance is important or where projecting threads must be protected. Wing nuts are used when frequent adjustments or removal is necessary.

Grade Marking	Specification	Material
	SAE - Grade 0	Steel
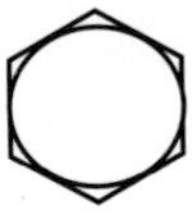	SAE Grade 1 ASTM - A 307	Low Carbon Steel
	SAE Grade 2	Low Carbon Steel
	SAE Grade 3	Medium Carbon Steel - Cold Worked
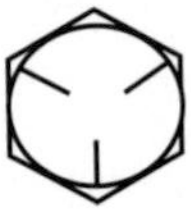	SAE Grade 5 ASTM - A 449	Medium Carbon Steel - Quenched and Tempered
 A 325	ASTM - A 325	Low Alloy Steel - Quenched and Tempered
 BB	ASTM - A 354 Grade BB	Low Alloy Steel - Quenched and Tempered
 BC	ASTM - A 354 Grade BC	Medium Carbon Alloy Steel - Quenched and Tempered - Roll Threaded after Heat Treatment
	SAE Grade 7	Medium Carbon Alloy Steel - Quenched and Tempered
	SAE Grade 8	Alloy Steel - Quenched and Tempered
	ASTM - A 354 Grade BD	
 A 490	ASTM - A 490	Alloy Steel - Quenched and Tempered

FIGURE 6–16
A few examples of fastener grades.

FIGURE 6–17
Comparison of the marking of a unified series grade 5 bolt and a metric 8.8 grade bolt.

Unified

Metric

FIGURE 6–18
Comparison of ISO and SAE grades.

Comparison of ISO and SAE Grades				
METRIC	1	2	5	8
SAE	4.6, 4.8	5.8	8.8	10.9

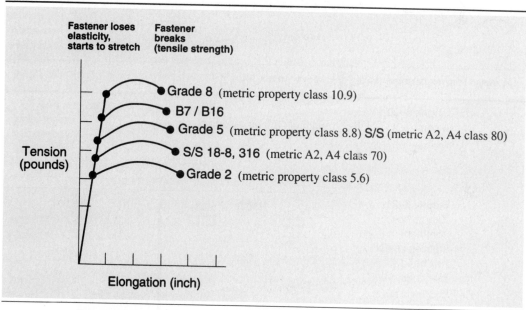

FIGURE 6–19
Chart showing strength of screws.
Courtesy of Atlantic Fasteners Inc.

WASHERS

Washers are used to distribute the clamping pressure over a larger area, and to prevent marring. They can also be used to provide a larger bearing surface for bolt heads and nuts. Figure 6–23 shows some of the common types of washers. Not shown is the lock washer, which is used to prevent a bolt or nut from loosening under vibration.

Hex Cap Screws

	Inch system			Metric system			
Grade	Head marking	For inch diameters	Tensile strength (PSI) — amount of force required to pull apart fastener	Property class	Head marking	For metric diameters	Tensile strength (PSI)[1] — amount of force required to pull apart fastener
2	No marking	1/4 - 3/4	74,000 PSI	5.6	For M5 and above	M12 - M24	72,500 PSI
		7/8 - 1 1/2	60,000 PSI				
5		1/4 - 1	120,000 PSI	8.8	For M5 and above	M17 - M36	120,350 PSI
		over 1 - 1 1/2	105,000 PSI				
8		1/4 - 1 1/2	150,000 PSI	10.9	For M5 and above	M6 - M36	150,800 PSI

[1] Converted from megapascals (MPa) $\frac{PSI}{145}$ = MPa or MPa x 145 = PSI

FIGURE 6–20
A comparison of the strength of various grades of screws.
Courtesy of Atlantic Fasteners Inc.

FIGURE 6–21
Examples of common stainless steel hex head markings.
Courtesy of Atlantic Fasteners Inc.

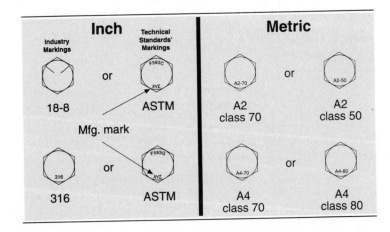

NON-THREADED FASTENING DEVICES

Non-threaded fasteners make up a large group of fastening devices. Figures 6–24 and 6–25 show examples of nonthreaded fasteners.

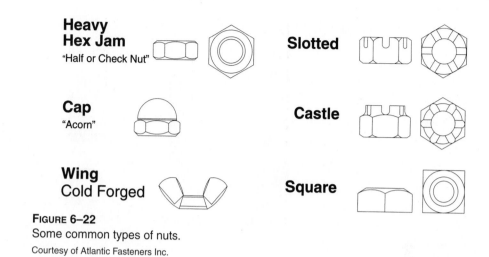

FIGURE 6–22
Some common types of nuts.
Courtesy of Atlantic Fasteners Inc.

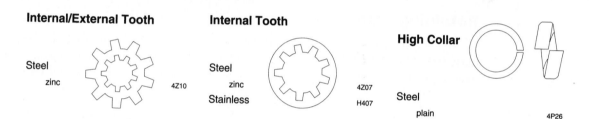

FIGURE 6–23
Common washers.
Courtesy of Atlantic Fasteners Inc.

Dowel Pins

Dowel pins are made of treated alloy steel and are used in assemblies where parts must be accurately positioned and held in absolute relation to one another (see Figure 6–24). They assure perfect alignment and facilitate quicker disassembly and assembly of parts in exact relationships.

Cotter Pins

Cotter pins are fitted into holes that are drilled crosswise in shafts to prevent parts from slipping or loosening.

FIGURE 6–24
Dowel pins.

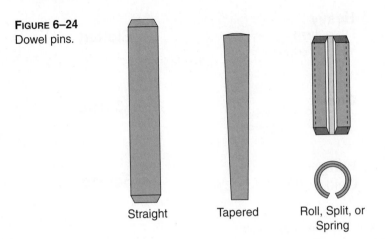

Straight   Tapered   Roll, Split, or
Spring

Retaining Rings

Retaining rings are stamped rings, both internal and external, that are used to keep parts from slipping or sliding apart (see Figure 6–25). While most retaining rings need a groove to seal them in position, some types are self-locking and do not require a recess.

Keys

A key is a small piece of metal imbedded partially in the shaft and partially in the hub to prevent rotation of a gear or pulley on a shaft. Following are some different types of keys (see Figure 6–26).

FIGURE 6–25
Various types of
retaining (snap) rings.

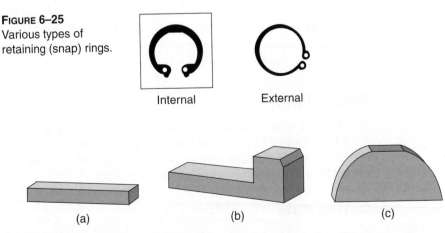

Internal   External

(a)   (b)   (c)

FIGURE 6–26
Keys: (a) Square, (b) Gib head, (c) Woodruff.

FIGURE 6–27
A keyway is broached into the hub of the part. A keyseat is machined in the shaft.

Square Keys—The width of a key is usually one fourth the shaft diameter. One half of the key is fitted into the shaft and one half is fitted into the hub.

Gib Head Key—Except for the gib head, this key is identical to the square key. The gib head provides easy removal.

Woodruff Key—A Woodruff key is semicircular in shape and fits into a keyseat of the same shape. The top of the key fits into a keyway in the mating part (see also Figure 6–27).

CUTTING THREADS

The term tapping is used to describe cutting an internal thread such as that found in a nut. Threading is the term used to describe cutting an external thread.

Tapping

Tapping is the use of a tap to cut an internal thread. The most important part of tapping a hole is drilling the correct size hole. The correct size drill is found in a drill chart that shows correct drill sizes for various sizes of tapped holes. Figure 6–28 shows a tap drill size chart; for example, a 1/4-20 hole would use a number 7 drill. The drills in the drill charts were sized to provide 70 to 75% threads.

There are several styles of taps available to fill various application needs. Taps are made from tool steel that is hardened, which makes them brittle and easily broken. Flutes are cut in the sides of the tap to allow cutting fluid to get into the cut and to allow chips to exit the hole as the threads are being cut. The ends of taps are square to allow a tap wrench to grip the tap.

There are three basic types of taps: bottom, plug, and taper. Figure 6–29 shows a taper tap, a plug tap, and a bottom tap. The difference between the three types is the number of beginning threads that are tapered. A bottom tap has between one and two threads tapered. A plug tap has between three and five threads tapered. A taper tap has the most threads tapered and the smallest end. This makes the taper tap the easiest to start and turn.

A bottom tap is used when you must have threads almost all the way to the bottom of a blind hole. You should start the threads with a plug or taper tap and

Tap Drill Size Chart

Tap Size	Drill Size	Probable Thread %	Tap Size	Drill Size	Probable Thread %
10-24	25	69-75	1/2-13	27/64	73-78
10-32	21	68-76	1/2-20	29/64	65-72
M5 X .8	4.2 mm	69-77	M14 X 2	15/32	76-81
12-24	16	66-72	9/16-12	31/64	68-72
12-28	15	70-78	9/16-18	33/64	58-65
M6 X 1	10	76-84	5/8-11	17/32	75-79
1/4-20	7	70-75	5/8-18	37/64	58-65
1/4-28	3	72-80	M16 X 2	35/64	76-81
5/16-18	F	72-77	3/4-10	21/32	68-72
5/16-24	I	67-75	3/4-16	11/16	71-77
M8 X 1.25	6.7 mm	74-80	M20 X 2.5	11/16	74-78
3/8-16	5/16	72-77	7/8-9	49/64	72-76
3/8-24	Q	71-79	7/8-14	13/16	62-67
M10 X 1.5	8.4 mm	76-82	M24 X 3	53/64	72-76
7/16-14	U	70-75	1-8	7/8	73-77
7/16-20	25/64	65-72	1-12	59/64	67-72
M12 X 1.75	10.2 mm	69-74	1-14	15/16	61-67

FIGURE 6–28
A drill chart shows the correct size of drill for various tap sizes.

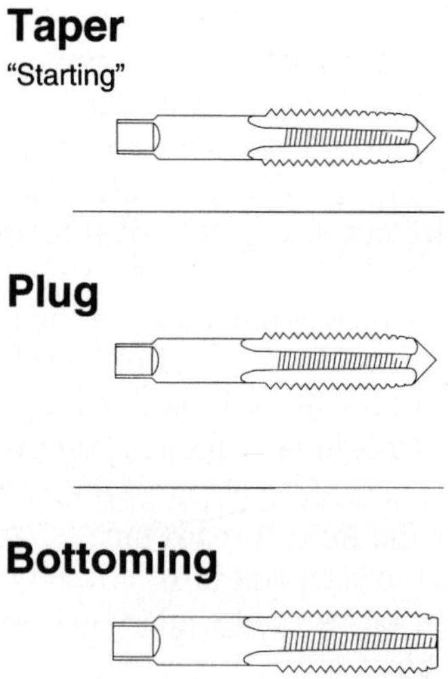

FIGURE 6–29
A taper tap, a plug tap and a bottom tap.

Courtesy of Atlantic Fasteners Inc.

Taper
"Starting"

Plug

Bottoming

finish a few threads with a bottom tap. The bottom tap requires more pressure and is easier to break. In most cases, a plug tap will tap enough threads in a blind hole.

Some taper taps for through holes are designed to send the chips ahead of the tap and out the bottom of the hole. Some are designed to force the chips up and out the hole. When tapping a through hole, choose a tap that sends the chips ahead of the tap and through the hole. In a blind hole, choose one that brings the chips up and out of the hole.

When tapping a hole by hand, be careful to guide the tap carefully into the hole to be sure it is perpendicular to the top surface. It is wise to make a few turns and then back the tap out one turn to break the chip and free the tap again. Be sure to use the correct tapping fluid. Different materials require different tapping fluids. Make sure proper attention is devoted to the tap. Keep both eyes on the tap and keep even pressure on both handles of the tap wrench. A broken tap can cause a great deal of anguish and a ruined part.

Threading an External Thread with a Die

Dies are used to cut threads on the outside of a bolt or stud. The die is held in a die stock and is turned onto the bolt or studs. As the die stock is rotated, it cuts the threads. Dies come in two varieties: solid and adjustable (see Figure 6–30). The solid die is often used to clean up threads that already exist on a stud or bolt. The adjustable is most often used to cut new threads. The adjustable die usually has a set screw to open or close the die so that the user can adjust how deeply the threads are cut. This allows loose or tight threads to be cut.

When cutting new threads, the end of the bolt or stud should be chamfered to permit the die to start more easily. Many dies have one side of the die where the treads are more tapered. This is the side that should be used for cutting, as it is easier to start. While turning the die, occasionally reverse the direction to break the chips. Cutting fluid should be used at all times when cutting threads.

FIGURE 6–30
Two dies and a die stock. The die on the right is an adjustable die and the die on the left is a solid die.

FASTENER REMOVAL AND REPAIR

Broken Bolt or Stud Removal

When a bolt or stud breaks, the reason most often is rust or corrosion between the bolt and mating thread. If time is not an immediate issue, the visible threads should be cleaned with a wire bush and good penetrating oil applied and allowed to soak into the threads. Then try a vise grip, or better yet, a stud remover to take it out. An impact wrench can increase the chances of success, as the "hammering" action may help break the threads free.

If this doesn't work, heat may be applied with a torch, first ensuring that heat will not damage the part. If the part has bearings or seals heat should be avoided. The correct method is to heat the outside of the part first. If the area immediately around the stud were heated first, the stud would get hot much more rapidly than its surrounding area and the stud would actually be tighter. This happens because the heat that is applied close to the stud goes into the stud and is also transferred away from the area around the stud to the outside of the part. Preheating the outside of the part first reduces this possibility.

If the bolt has broken off below the surface, a different method must be tried. The first method to try is to use a small chisel or centerpunch with a ball peen hammer. Position the chisel point on the outer edge of the thread and at an angle so that when the hammer strikes the punch it will apply a counterclockwise (for a right-hand thread) shock to the thread and hopefully free the thread.

A second method that can be tried is a Drill-Out tool. Figure 6–31 shows how a Drill-Out tool is used. The teeth on the tool drill a hole. The center on the tool keeps the tool centered and prevents it from walking off center. The tool is run in the opposite direction of the thread. Once the hole is started, the tapered, threaded

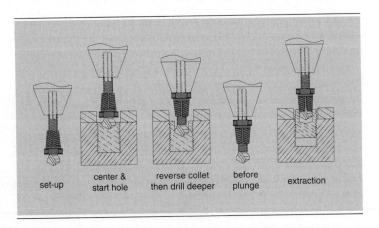

set-up   center &   reverse collet   before   extraction
         start hole   then drill deeper   plunge

FIGURE 6–31
A Drill-Out tool.

Courtesy of Atlantic Fasteners Inc.

FIGURE 6–32
Typical screw
extractors.

nut around the drill point is started into the hole. As the nut is tightened it removes the stud.

If that does not work, use a file or chisel to flatten the top of the bolt or stud. Centerpunch the fastener as close as possible to the center. Next select the correct size screw extractor for the fastener. Figure 6–32 shows typical screw extractors. Choose the correct drill for the extractor. The drill should be slightly larger than the small end of the extractor. There may be manufacturer data with your extractor set explaining which size extractor to use. Use a left-hand drill, if available. A left-hand drill rotates counterclockwise and will not further tighten the fastener, allowing the cutting forces to free and remove the stud. If possible, drill completely though the fastener to help reduce pressure on the threads and make them easier to remove.

After the hole is drilled, insert the extractor and rotate it counterclockwise while applying downward pressure. A tap wrench may be helpful. This should remove the fastener. Extractors are very hard and brittle. Be careful when you use them. A broken extractor is very difficult to remove.

If this does not work, you can drill the hole slightly smaller than the minor thread diameter, thoroughly clean the hole, and chase (recut to clean) the threads carefully with a tap.

Alternatively, the hole can be drilled larger and a Heli-Coil can be installed. The Heli-Coil, which looks like a spring (see Figure 6–33), is used to form a new thread that is the same size as the thread that was ruined. First choose the correct size Heli-Coil to match the ruined thread size and drill the hole to that size. (The Heli-Coil will specify a drill size to be used.) Next, use the correct Heli-Coil tap to tap the hole and, last, insert the coil and create new threads to match the old size.

FIGURE 6–33
Heli-Coil tool and Heli-Coils.
Courtesy of Atlantic Fasteners Inc.

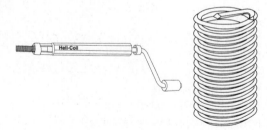

Removal of a Frozen Nut

To remove a frozen nut, visible threads should be cleaned with a wire bush, and good penetrating oil applied and allowed to soak into the threads. Then try to remove the nut with the correct wrench.

There are specialized tools to remove nuts. One of these is called the E-Z OFF salvage tool (see Figure 6–34). This figure shows how the tool is used. The tool is placed over the nut and hammered on. A standard 3/8 ratchet is used to loosen the nut. This tool comes in a set to fit various sizes of nuts and can be used on inch and metric nuts.

If this does not loosen the nut, heat may be applied to the outside edges of the nut. The heat will force the nut to grow and hopefully break free from the bolt or stud. If this does not work, a nut splitter can be tried. A nut splitter is a clamp-like device with a chisel point that splits the nut as it is tightened. Hydraulic nut splitters are also available.

If a nut splitter is not available, a chisel can be used with care not to damage the threads of the bolt or stud. A hacksaw can also be used to cut through the nut.

Removal of Broken Taps

Broken taps are troublesome and always break at the worst possible time. The part is almost done and ready for use when the tap is broken. This usually causes a great deal of frustration. The first thing to do is relax. Next the hole and tap should be thoroughly cleaned out. Solvent and an airhose will help. Make sure all of the chips and broken pieces of the tap are out of the hole. The next thing to try is a tap

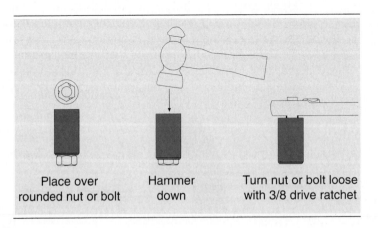

Place over   Hammer   Turn nut or bolt loose
rounded nut or bolt   down   with 3/8 drive ratchet

FIGURE 6–34
E-Z OFF salvage tool.

Courtesy of Atlantic Fasteners Inc.

FIGURE 6–35
Typical tap extractor.

extractor. Figure 6–35 shows a typical tap extractor. These are available for all tap sizes and for the correct number of flutes. For example, if a two-flute 1/4-20 tap is broken, a 1/4-20-2 tap extractor is available for that size thread. The fingers that move down into the flutes of the tap are also sized correctly. Place the extractor above the broken tap, slide the fingers in as far as possible, and slide the finger retainer (ferrule) down on top of the broken tap to provide as much lateral support as possible for the fingers. Using a tap wrench, carefully rotate the tap extractor in a counterclockwise (for right-hand threads) direction.

If this does not work, try the method previously described using the chisel and ball peen hammer to loosen and rotate it out.

The last but surest way to remove a broken tap or fastener is to utilize an electrical discharge machine (EDM) or an electrical tap extractor. These machines remove metal with electrical energy. An electrode is chosen, or made, that is slightly smaller than the minor thread diameter. The electrode is hollow to allow fluid to flow through the electrode. The fluid is used to remove the particles and also as an insulator to prevent direct shorts between the electrode and part. The electrode is slowly fed into the part and burns away the fastener or tap. The hole must then be cleaned and the threads carefully chased (recut) with a tap.

Questions

1. What is the term pitch as it applies to a thread?
2. What is the term lead as it applies to a thread?
3. If a thread has 40 threads per inch what is its pitch?
4. If a thread has 20 threads per inch what is its lead?
5. Describe the uniform thread series.
6. Thoroughly describe a 1/4-20 3B thread.
7. Thoroughly describe a 1/2-13 2A thread.
8. Thoroughly describe a M10 thread.
9. Describe the proper method to tap a hole.
10. What size hole should be drilled to tap a 1/4-20 tapped hole?
11. What size hole should be drilled to tap a 1/2-13 tapped hole?
12. What size hole should be drilled to tap a 5/16-24 tapped hole?
13. What size hole should be drilled to tap a 3/8-16 tapped hole?
14. What size hole should be drilled to tap a 3/8-24 tapped hole?
15. Describe at least 2 methods to remove a broken stud from a hole.

CHAPTER 7

Shafting and Shaft Bushing Components

This chapter covers steel shafting and bushing devices used with mechanical power transmission components, as well as types and sizes of keyways. The selection and sizing of steel shafting, the fitting of mechanical components onto a shaft, and the terms used to describe the general types of fit classifications are addressed. Concerns regarding various loads acting on the shaft during operation will also be reviewed.

Objectives

Upon completion of this chapter, the student will be able to:
✔ Understand the basic purpose of shafts within machinery.
✔ Understand the importance of proper shaft selection.
✔ Describe the various loads that shafting is subjected to.
✔ Accurately measure the diameter of standard shafting.
✔ Explain the various basic types of shaft fits.
✔ Calculate the amount of linear expansion for a given length of steel shafting.
✔ Identify the various types of tapered and keyless compression shaft locking devices.

INTRODUCTION

Shafts are an integral part of any mechanical system. Bearings are supported by the shaft and machine housing, and rotate within the mechanism. Shafts deliver the power, torque, and rotary movement. In the case of power transmission components—such as gears, pulleys, sheaves, and sprockets—the shaft is delivering the required power, torque, and rotary movement by an engaging or connected device to another shaft and component assembly. In many applications a power transmission component is mounted to the shaft by a tapered bushing mechanism. The wheel, in the form of a bearing, gear, roller, sheave, or sprocket—and the shaft it is mounted on—are a basic part of every machine.

SHAFT DESIGN, MATERIALS, AND SELECTION

The steel shafting commonly used on machines is selected to withstand the loads, speeds, and application variables and yield a satisfactory service life. The maintenance technician relies on the expertise of the design engineer to properly select the proper shaft. Certain situations and applications require the maintenance mechanic to modify an existing piece of shafting or assemble a shaft assembly. It is a general industry standard to use AISI grade 1045 steel shafting for most normally loaded power transmission components. Standard shafting is reasonably smooth and straight. Straight continuous shafts are common and their weights can be determined by referring to Figure 7–1.

Certain applications require that a short length at the ends of the shaft be turned to a different diameter for mounting bearings or other components. These necked down ends are called "turn downs" (see Figure 7–2). When it is necessary to use turn down shaft ends, a large shaft fillet should be used to keep the stresses minimal. A shaft fillet is the small curved portion at the step where diameters

| Shaft Size | Weight of Shafting for Various Lengths in feet | | | | | | | | | | | | | | | | | Weight Per Inc |
	1	2	3	4	5	6	7	8	9	10	12	14	16	18	20	22	24	
3/4	1.5	3.0	4.5	6.0	7.5	9.0	10.5	12.0	13.5	15	18	21	24	27	30	33	36	.125
7/8	2.0	4.0	6.1	8.1	10.2	12.2	14.3	16.3	18.4	20	25	29	33	37	41	45	49	.170
*15/16	2.3	4.7	7.0	9.4	11.7	14.1	16.5	18.8	21.2	23	28	33	38	42	47	52	56	.195
1	2.7	5.3	8.0	10.6	13.3	16.0	18.6	21.3	24.0	27	32	37	43	48	53	59	64	.223
1-1/8	3.4	6.8	10.0	13.4	16.7	20.1	23.4	26.7	30.1	34	41	47	54	61	68	74	81	.281
*1-3/16	3.8	7.6	11.3	15.1	18.9	22.6	26.4	30.1	34.0	38	45	53	60	68	75	83	90	.314
1-1/4	4.2	8.3	12.5	16.7	20.8	25.0	29.2	33.3	37.5	42	50	58	67	75	83	92	100	.348
1-3/8	5.0	10.1	15.3	20.2	25.3	30.3	35.4	40.4	45.4	50	60	71	81	91	101	111	121	.420
*1-7/16	5.5	11	17	22	28	33	39	44	50	55	66	77	88	99	110	121	133	.460
1-1/2	6.0	12	18	24	30	36	42	48	54	60	72	84	96	108	120	132	144	.500
*1-11/16	7.6	15	23	30	38	46	53	61	68	76	91	107	122	137	152	167	183	.634
*1-15/16	10.0	20	30	40	50	60	70	80	90	100	120	140	161	181	201	221	241	.835
2	10.7	21	32	43	53	64	75	85	96	107	128	150	171	192	214	235	256	.890
*2-3/16	12.8	26	38	51	64	77	90	102	115	128	153	179	205	230	256	281	307	1.06
*2-7/16	15.9	32	48	63	79	95	111	127	143	159	190	222	254	286	317	349	381	1.32
2-1/2	16.7	34	50	67	83	100	117	134	150	167	200	234	267	301	334	367	401	1.39
*2-11/16	19.3	39	58	77	97	116	135	154	174	193	232	270	309	348	386	425	463	1.61
*2-15/16	23.0	46	69	92	115	138	161	184	208	231	277	323	369	415	461	507	553	1.92
*3-7/16	31.6	63	95	126	158	189	221	253	284	316	379	442	505	568	631	695	758	2.63
*3-15/16	41.4	83	124	166	207	248	290	331	373	414	497	580	662	745	828	911	994	3.45
*4-7/16	52.6	105	158	210	263	315	368	421	473	526	631	736	841	946	1052	1157	1262	4.38
*4-15/6	65.1	130	195	260	326	391	456	521	586	651	781	911	1041	1172	1302	1432	1562	5.42
*5-7/16	79.0	158	237	316	395	474	553	632	711	790	947	1105	1263	1421	1579	1737	1894	6.58
*6	96	192	288	384	481	577	673	769	865	961	1154	1346	1538	1730	1923	2115	2307	8.01

*   **Recommended Diameters** These shaft diameters are recommended for use whenever possible as various transmission items such as couplings, collars, clutches, pulleys, etc., are carried in stock in these sizes, at least up to 3-15/16″, in the principal cities throughout the United States.

FIGURE 7–1
Steel shaft weights based on length and diameter.

Courtesy of Dodge Rockwell Automation.

FIGURE 7–2
Shafts shown with turned-down ends.

Courtesy of Precision Pulley & Idler.

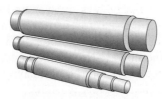

FIGURE 7–3
Shaft end showing the
fillet, turndown, and
chamfers.

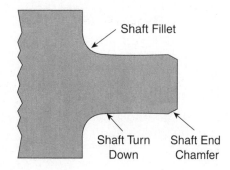

Shaft Fillet

Shaft Turn
Down

Shaft End
Chamfer

change (see Figure 7–3). The radius of the fillet should be not less than half the difference in the two diameters joined by the fillet. The fillet should be smooth and free of nicks and scratches that could produce a crack and result in failure. Generally the greater the fillet radius, the more favorable the stress distribution. Large diameter shafts that are heavily loaded require a large radius. If a power transmission component or bearing must abut directly against the shaft shoulder for support, the fillet must allow for flush mounting. The radius of the fillet must clear the corner radius of the mounted component. On certain bearing seats the corner may even be undercut. Shaft ends, bearings, and other power transmission components should have corners rounded off. This beveled edge, known as a "chamfer," allows for safe and easy assembly of parts (see Figure 7–3). Chamfering the corners of bearings allows flush and tight mounting against the shoulder.

An important factor that engineers must consider when selecting the proper shaft size is that it will be subjected to various forces. These forces on a shaft include bending moments, axial compression, axial tension, shear, and torque (see Figure 7–4). All rotating shafts are subjected to cyclic loading because the shaft is revolving. The bending moments may be continuously reversing.

Selection of shaft diameters to be used in a machine must take into account the numerous forces that it will be subjected to when running. Critical calculations involving section modulus and the moment of inertia need to be solved to determine cross section strengths. Other concerns are the mechanical properties, physical properties, and the service environment. Questions such as the following need to be addressed: How much static strength is necessary; is there impact loading; is wear resistance required; how much can the material bend, stretch, or compress; is machinability important; are standard sizes and shapes specified; what type of environment will it be operating in? All of these concerns, along with numerous others, should be reviewed prior to selection. Selection of shafting is best left to design and mechanical engineers for service life and safety reasons.

Figure 7–5 can be used in understanding how load, operational service factors, and service conditions affect the selection of the shaft diameter for various conditions. Engineers are usually generous when selecting diameters to increase the life of the system. This table uses a safe shear stress of 6,000 PSI for standard keyseated shafting. It is recommended that design engineers be responsible for the selection of steel shafting for service and safety reasons.

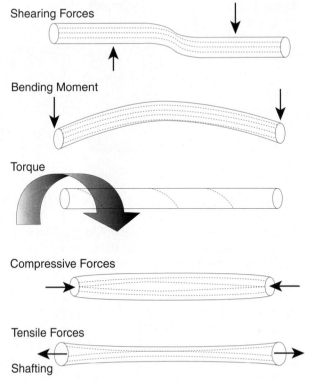

Figure 7–4
Various forces and moments (arrows) that can be imposed on shafting.

Shaft Keys/Keyseats/Keyways

Most shafts have a machined key slot or keyway cut into the end of the shaft. This slot, or groove, matches a similar slot that has been machined or cut into the component that is fitted onto the shaft. The rectangular or square-shaped void created by key slots is filled with a piece of metal key stock, also called a key. The key stock should be cut and shaped to fit snugly. This, in effect, joins the shaft and assists in preventing movement of the mounted component, providing a positive means to transmit torque. The key also assists in positioning and holding the mounted component onto the shaft. There are several different shapes of keys and keyways, including tapered and "woodruff." A "woodruff" key is a round bottom key that fits into a circular pocket. The most common key and keyway shape is square. Rectangular, round end, and other shapes are available (see Figure 7–6). Industry has established standard inch-size keyway dimensions for square and rectangular shaped keyways (see Figure 7–7). Key material should be softer than the shaft material. This will allow the key to shear in the event of overloads, rather than catastrophic failure of the shaft. Most manufacturers of key stock supply a "file-to-fit" material to allow for the easy dimensional alteration of the material.

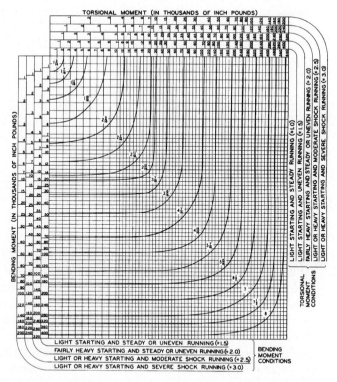

FIGURE 7–5
Shaft selection table.

Courtesy of Rockwell Automation.

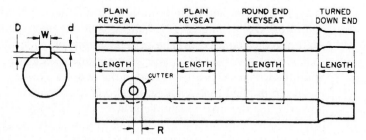

FIGURE 7–6
Standard key seat shapes.

Courtesy of Rockwell Automation.

Shaft Fits

When two machined objects are mated, there will be a difference between their actual size and the dimensional tolerance limit. This difference is the "fit" or amount of tightness. Whenever a component such as a hub, sprocket, gear, and so forth

FIGURE 7–7
Standard shaft
key sizes table.

Courtesy of
Rockwell
Automation.

Shaft Size	W width	D, d Depth		R R Cutter Run-out
		Regular	Shallow	
5/16 to 7/16	3/32	3/64		1/2
1/2 to 3/16	1/8	1/16		9/16
5/8 to 7/8	3/16	3/32		11/16
13/16 to 1-1/4	1/4	1/8		13/16
1-5/16 to 1-3/8	5/16	3/32		15/16
1-7/16 to 1-3/4	3/8	3/16	1/8	1-1/16
1-13/16 to 2-1/4	1/2	1/4	3/16	1-3/16
2-5/16 to 2-3/4	5/8	5/16	3/16	1-5/16
2-15/16 to 3-l/4	3/4	3/8		1-9/16
3-5/16 to 3-3/4	7/8	7/16	1/4	1-11/16
3-13/16 to 4-1/2	1	1/2	1/4	1-3/4
4-9/16 to 5-1/2	1-1/4	5/8	1/4	1-15/16
5-9/16 to 6-1/2	1-1/2	3/4	1/4	2-1/8
6-9/16 to 7-1/2	1-3/4	3/4	1/4	2-1/8
7-9/16 to 9	2	3/4	3/8	2-1/8
9-1/16 to 11	2-1/2	7/8	3/8	2-5/16
11-1/16 to 13	3	1	3/8	2-7/16

is mounted onto a shaft, it will have a certain "fit." The amount of fit is the difference between the shaft diameter and the diameter of the bore or hole in the object. When a bearing or power transmission component is mounted onto a shaft, it must be fitted or seated properly to prevent it from coming loose. All manufactured and machined items are produced to some tolerance limit. A tolerance limit is the variation, positive or negative, by which a size is permitted to depart from the intended design size. Shafting also has a tolerance limit. As a general rule, shafting is supplied approximately .0005″ to .002″ under nominal size. The tolerance is dependent on the class of shafting as well as its diameter. As an example, a 1″ nominal diameter standard power transmission piece of shafting will be 1.000″ + .0000″ to −0005″.

There are numerous terms used to describe a variety of different types or levels of fit. Many of these terms are used interchangeably by maintenance mechanics even though they have exact definitions. Machinist handbooks will cover in-depth the various terms and their definitions. Some fit categories overlap and can result in confusion. Design engineers should consult engineer's handbooks when choosing a shaft and the required fit. The fit selection is dependent on variables such as load, speed, rotational direction, type of component, heat, and numerous other factors. Shaft fits for bearings are covered in the bearing chapter of this book.

Keep in mind that many applications require an exact specific dimensional fit; however, for the sake of simplicity we can group the abundant types of fits into a few categories. From a practical standpoint, shaft fits can be grouped into three types: interference, transition, and clearance.

Interference fits are those fits where the diameter of the hole in the component to be mounted is smaller than the shaft diameter. An interference fit is often required in circumstances of heavy loads, fast acceleration, reversing directions, and large diameters. Interference fits are often referred to as press, tight, shrink, forced, and drive. The term "shrink" fit implies that heat is required to expand the hole to allow for mounting. It is advised that a machinist's handbook be consulted for definitive explanations of these terms.

Transition or line-to-line fits are those where the shaft and component bore are nearly the same. In reality there will be some dimensional difference between the two, but it is negligible. Line-to-line or transition fits are sometimes specified for outer rings of bearings pressed into the housing, which locates and holds them in place.

Clearance fits will have limits where the mating parts will always have some clearance between them. No force or heat is required to assemble the parts. Clearance fits are sometimes referred to as slip, slide, and loose. Clearance fits are often recommended for installation of small light-duty power transmission components. A clearance fit allows for easy installation. Generally power transmission components with shaft sizes under 4″ will be shipped standard with a clearance fit. Figure 7–8 is a hub-to-shaft fit tolerance table.

The size and the condition of the shaft and component bore must be correct to ensure proper location and fit. When assembling any machined parts onto a shaft, it is recommended that measurements be taken at multiple locations of both the shaft and the bore of the mating component. A minimum of twelve and a maximum of sixteen measuring points are advised. These measuring points should be located on the shaft where the respective component is to be mounted (see Figure 7–9). The shaft and bore should be checked with inside and outside micrometers for roundness

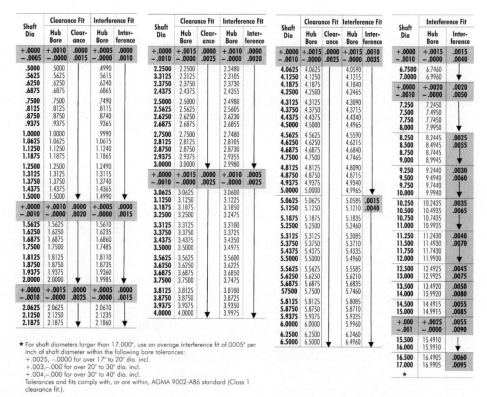

Block 1

Shaft Dia	Clearance Fit Hub Bore	Clearance	Interference Fit Hub Bore	Interference
+.0000 −.0005	+.0010 −.0000	.0000 .0015	+.0005 −.0000	.0000 .0010
.5000	.5000		.4990	
.5625	.5625		.5615	
.6250	.6250		.6240	
.6875	.6875		.6865	
.7500	.7500		.7490	
.8125	.8125		.8115	
.8750	.8750		.8740	
.9375	.9375		.9365	
1.0000	1.0000		.9990	
1.0625	1.0625		1.0615	
1.1250	1.1250		1.1240	
1.1875	1.1875		1.1865	
1.2500	1.2500		1.2490	
1.3125	1.3125		1.3115	
1.3750	1.3750		1.3740	
1.4375	1.4375		1.4365	
1.5000	1.5000		1.4990	
+.0000 −.0010	+.0010 −.0000	.0000 .0020	+.0005 −.0000	.0000 .0015
1.5625	1.5625		1.5610	
1.6250	1.6250		1.6235	
1.6875	1.6875		1.6860	
1.7500	1.7500		1.7485	
1.8125	1.8125		1.8110	
1.8750	1.8750		1.8735	
1.9375	1.9375		1.9360	
2.0000	2.0000		1.9985	
+.0000 −.0010	+.0015 −.0000	.0000 .0025	+.0005 −.0000	.0000 .0015
2.0625	2.0625		2.0610	
2.1250	2.1250		2.1235	
2.1875	2.1875		2.1860	

Block 2

Shaft Dia	Clearance Fit Hub Bore	Clearance	Interference Fit Hub Bore	Interference
+.0000 −.0010	+.0015 −.0000	.0000 .0025	+.0010 −.0000	.0000 .0020
2.2500	2.2500		2.2480	
3.3125	2.3125		2.3105	
2.3750	2.3750		2.3730	
2.4375	2.4375		2.4355	
2.5000	2.5000		2.4980	
2.5625	2.5625		2.5605	
2.6250	2.6250		2.6230	
2.6875	2.6875		2.6855	
2.7500	2.7500		2.7480	
2.8125	2.8125		2.8105	
2.8750	2.8750		2.8730	
2.9375	2.9375		2.9355	
3.0000	3.0000		2.9980	
+.0000 −.0010	+.0015 −.0000	.0000 .0025	+.0010 −.0000	.0005 .0025
3.0625	3.0625		3.0600	
3.1250	3.1250		3.1225	
3.1875	3.1875		3.1850	
3.2500	3.2500		3.2475	
3.3125	3.3125		3.3100	
3.3750	3.3750		3.3725	
3.4375	3.4375		3.4350	
3.5000	3.5000		3.4975	
3.5625	3.5625		3.5600	
3.6250	3.6250		3.6225	
3.6875	3.6875		3.6850	
3.7500	3.7500		3.7475	
3.8125	3.8125		3.8100	
3.8750	3.8750		3.8725	
3.9375	3.9375		3.9350	
4.0000	4.0000		3.9975	

Block 3

Shaft Dia	Clearance Fit Hub Bore	Clearance	Interference Fit Hub Bore	Interference
+.0000 −.0010	+.0015 −.0000	.0000 .0025	+.0015 −.0000	.0010 .0035
4.0625	4.0625		4.0590	
4.1250	4.1250		4.1215	
4.1875	4.1875		4.1840	
4.2500	4.2500		4.2465	
4.3125	4.3125		4.3090	
4.3750	4.3750		4.3715	
4.4375	4.4375		4.4340	
4.5000	4.5000		4.4965	
4.5625	4.5625		4.5590	
4.6250	4.6250		4.6215	
4.6875	4.6875		4.6840	
4.7500	4.7500		4.7465	
4.8125	4.8125		4.8090	
4.8750	4.8750		4.8715	
4.9375	4.9375		4.9340	
5.0000	5.0000		4.9965	
5.0625	5.0625		5.0585	.0015
5.1250	5.1250		5.1210	.0040
5.1875	5.1875		5.1835	
5.2500	5.2500		5.2460	
5.3125	5.3125		5.3085	
5.3750	5.3750		5.3710	
5.4375	5.4375		5.4335	
5.5000	5.5000		5.4960	
5.5625	5.5625		5.5585	
5.6250	5.6250		5.6210	
5.6875	5.6875		5.6835	
57500	5.7500		5.7460	
5.8125	5.8125		5.8085	
5.8750	5.8750		5.8710	
5.9375	5.9375		5.9335	
6.0000	6.0000		5.9960	
6.2500	6.2500		6.2460	
6.5000	6.5000		6.4960	

Block 4

Shaft Dia	Interference Fit Hub Bore	Interference
+.0000 −.0010	+.0015 −.0000	.0015 .0040
6.7500	6.7460	
7.0000	6.9960	
+.0000 −.0010	+.0020 −.0000	.0020 .0050
7.250	7.2450	
7.500	7.4950	
7.750	7.7450	
8.000	7.9950	
8.250	8.2445	.0025
8.500	8.4945	.0055
8.750	8.7445	
9.000	8.9945	
9.250	9.2440	.0030
9.500	9.4940	.0060
9.750	9.7440	
10.000	9.9940	
10.250	10.2435	.0035
10.500	10.4935	.0065
10.750	10.7435	
11.000	10.9935	
11.250	11.2430	.0040
11.500	11.4930	.0070
11.750	11.7430	
12.000	11.9930	
12.500	12.4925	.0045
13.000	12.9925	.0075
13.500	13.4920	.0050
14.000	13.9920	.0080
14.500	14.4915	.0055
15.000	14.9915	.0085
+.000 −.001	+.0025 −.0000	.0055 .0090
15.500	15.4910	
16.000	15.9910	
16.500	16.4905	.0060
17.000	16.9905	.0095
★		

★ For shaft diameters larger than 17.000″, use an average interference fit of .0005″ per inch of shaft diameter within the following bore tolerances:
+.0025, −.0000 for over 17″ to 20″ dia. incl.
+.003,−.000 for over 20″ to 30″ dia. incl.
+.004,−.000 for over 30″ to 40″ dia. incl.
Tolerances and fits comply with, or are within, AGMA 9002-A86 standard (Class 1 clearance fit.).

FIGURE 7–8
Shaft-to-hub fit tolerance table.

Courtesy of Falk Corporation.

THE TWELVE-POINT MEASUREMENT PROCEDURE:

1) Use two-point gauges that are accurate to .0001". It is recommended that gauges with accuracy to ¹⁄₁₀ of the units that are being inspected to be used (resolution to .00001"). We recognize readily available gauges read to .0001", however.

2) Measure four position at 0°, 45°, 90°, and 135° in three different planes of the mating surface (that is in direct contact with the bearing). The three planes should be evenly spaced across the contact area. The outboard measurements should be ⅛" to ½" in from each end.

3) Record the measurements on a chart like the one listed below. Keep all three sets of measurements oriented with respect to each other. Take an average of each plane.

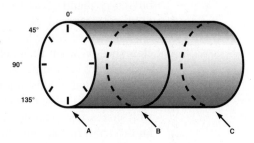

	0°	45°	90°	135°	AVERAGES
PLANE A	_____	_____	_____	_____	A = _____
PLANE B	_____	_____	_____	_____	B = _____
PLANE C	_____	_____	_____	_____	C = _____

FIGURE 7–9
Shaft measuring locations.

Courtesy of Timken/Torington.

and any taper. All dimensional and surface irregularities should be observed and noted. High points and small rises may be dressed off with a fine tooth file. If the shaft is extremely out of tolerance in any area or has been damaged from galling, fretting, or wear it should be discarded. Proper fits and dimensional accuracy will assist in preventing vibration, wear, unbalance, looseness, and catastrophic failure.

Shaft Expansion

Steel shafting expands or contracts with temperature change. Of primary concern is the linear amount of growth that will occur from start-up speed to the full running speed at normal operating temperatures. As the shaft temperature rises, the shaft will grow in length. This can cause misalignment and displacement of mating components. Provision should be made when thermal growth of a shaft will affect the supporting bearings. Normally one bearing will be fixed and the others will allow for shaft growth due to heat. Connected shafts that use flexible couplings need to be spaced properly to prevent shaft ends butting against one another and thrust-loading the bearings. Several shafts joined together, as in line shafts, act as a continuous piece of shafting and need to be treated as such. The amount of linear growth for standard steel shafts can be calculated using the following formula:

Expansion in inches = 0.0000063 × 12 × length in feet × temperature increase in degrees Fahrenheit (see Figure 7–10).

Different materials such as brass or aluminum will have their own coefficients of expansion. Machinist's handbooks can be consulted for the correct coefficient.

Length (Feet)	Temperature Increase—Degrees F.					Length (Feet)	Temperature Increase—Degrees F.				
	20°	40°	60°	80°	100°		20°	40°	60°	80°	100°
1	.0015"	.0030"	.0045"	.0060"	.0075"	40	.060"	.121"	.181"	.242"	.302"
2	.0030	.0060	.0091	.0121	.0151	45	.068	.136	.204	.272	.340
3	.0045	.0091	.0136	.0181	.0227	50	.076	.151	.227	.302	.378
4	.0060	.0121	.0181	.0242	.0302	55	.083	.166	.249	.333	.416
5	.0076	.0151	.0227	.0302	.0378	60	.091	.181	.272	.363	.454
6	.0091	.0181	.0272	.0363	.0454	65	.098	.197	.295	.393	.491
7	.0106	.0212	.0318	.0423	.0529	70	.106	.212	.317	.423	.529
8	.0121	.0242	.0363	.0484	.0605	75	.113	.227	.340	.454	.567
9	.0136	.0272	.0408	.0544	.0680	80	.121	.242	.363	.484	.605
10	.0151	.0302	.0454	.0605	.0756	85	.129	.257	.386	.514	.643
12	.0181	.0363	.0544	.0726	.0907	90	.136	.272	.408	.544	.680
14	.0212	.0423	.0635	.0847	.1058	95	.144	.287	.431	.575	.718
16	.024	.048	.073	.097	.121	100	.151	.302	.454	.605	.756
18	.027	.054	.082	.109	.136	110	.166	.333	.499	.665	.832
20	.030	.060	.091	.121	.151	120	.181	.363	.544	.726	.907
25	.038	.076	.113	.151	.189	130	.197	.393	.590	.786	.983
30	.045	.091	.136	.181	.227	140	.212	.423	.635	.847	1.058
35	.053	.106	.159	.212	.265	150	.227	.454	.680	.907	1.134

FIGURE 7–10
Linear expansion of steel shafting table.
Courtesy of Rockwell Automation.

SHAFT LOCKING DEVICES/BUSHINGS

Power transmission components such as V-belt sheaves, sprockets, and pulleys are often mounted to the shaft by the use of a tapered or compression bushing device of some type. The taper bushing, or taper-lock, utilizes the principle of the wedge to hold the bushing and component in place on the shaft. The combination of friction, clamping forces, and keys transfers the torque from shaft to bushing/hub. There are several advantages to using bushings with power transmission components. Tapered bushings are relatively easy to install and remove, they have gripping forces equal to or greater than a conventional fit to shaft, and they can be used on shafts that are slightly over or under size.

When mounting a tapered bushing, the preparation of all tapered mating surfaces is important. The process should include cleaning and checking for any nicks and burrs on the shaft and bushing that might interfere with the fit. They should be carefully removed. All surfaces should be free of grease and contaminants.

Tapered bushings use cap screws or bolts to mount the bushing to the hub. Torque wrenches should be used when assembling tapered bushings to prevent uneven mounting and overtightening. If the bushing is drawn up unevenly to the sheave hub, excessive axial run-out can be created.

Taper-Lock Bushings

The original "taper-lock" bushing has an 8° taper with no flange and is flush mounted to the sheave hub (see Figure 7–11). It is available in a variety of sizes. Inserting the screws into the bushing holes that are farthest apart and tightening to the proper wrench torque assembles the unit. It is vital to match holes and not

threads. Tightening the installation fasteners causes the bushing to squeeze the shaft as the bolts are drawn in. Removal is accomplished by removing the mounting screws and inserting one into the threaded bushing hole and tightening. This acts as a jackscrew, forcing separation to occur between the bushing and hub of the sheave (see Figure 7–12).

Quick Detachable (QD) Tapered Bushings

QD-style taper bushings work on the same principles as the "taper-lock" type only with a 4° taper angle. They have a flange as part of the bushing that is used for mounting and dismounting (see Figure 7–13). Hex head bolts are used rather than setscrews to draw the bushing and hub together. The bushing is drilled with through and tapped holes and can be mounted on the shaft either by standard or reverse mounting procedures (see Figure 7–14).

QD Bushing Installation Instructions The tapered, QD-type interchangeable bushing offers flexible and easy installation while providing exceptional holding power. To ensure that the bushing performs as specified, it must be installed properly. Before beginning, make sure the correct size and quantity of parts are available for the installation. The bushing has been manufactured to accept a setscrew over the key and its use is optional. It is packaged with the hardware on sizes SH to M and loosely installed in the bushing on sizes N to S.

Installation

1. Inspect the tapered bore of the sheave and the tapered surface of the bushing. Any paint, dirt, oil, or grease must be removed. Important: Do not use lubricants in this installation.

2. Select the type of mounting (see Figure 7–14) that best suits the application.

3. Standard mounting: Install shaft key. (Note: If key was furnished with bushing, you must use that key.) Install the bushing on clean shaft, flange end first. If bushing will not freely slide onto the shaft, insert a screwdriver or similar object into the flange saw cut to act as a wedge to open the bushing's bore.

Caution: Excessive wedging will split the bushing. If using the setscrew, tighten it just enough to prevent the bushing from sliding on the shaft. Do not over-tighten setscrew!

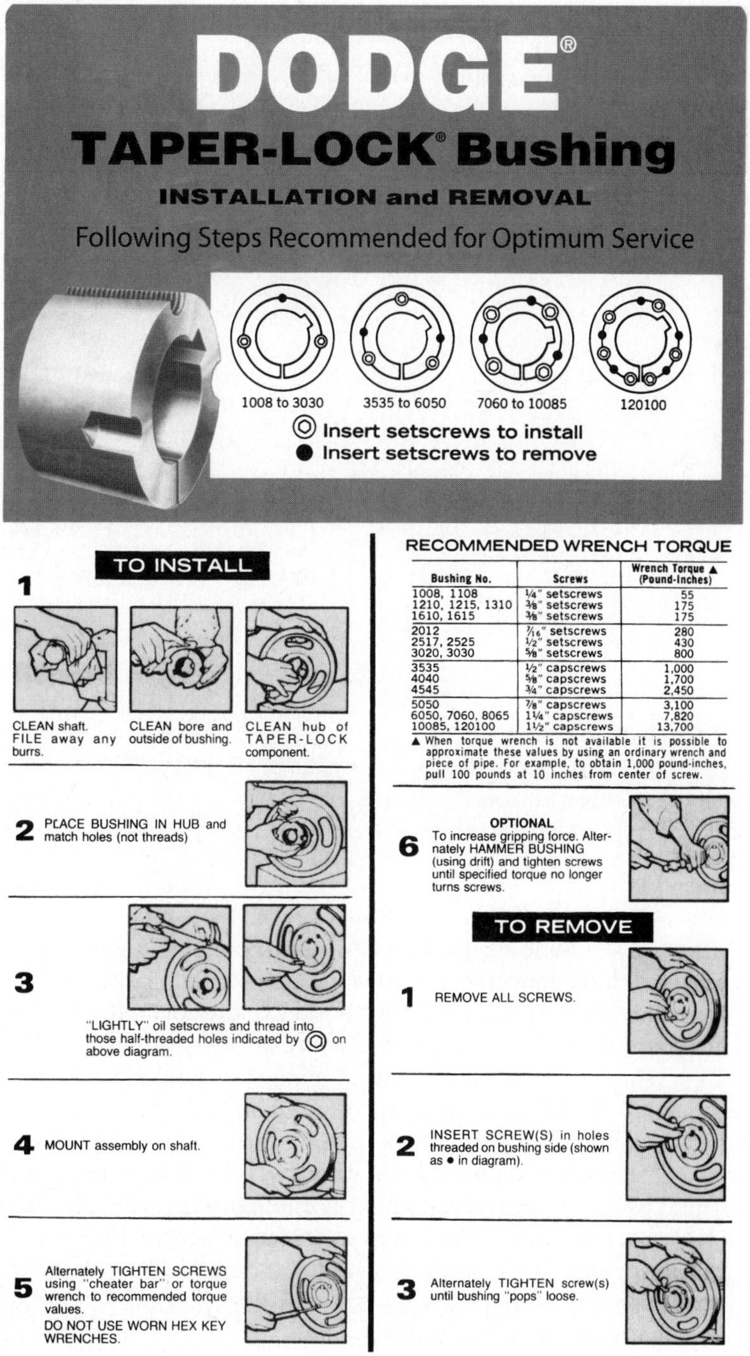

1

CLEAN shaft. FILE away any burrs.

CLEAN bore and outside of bushing.

CLEAN hub of TAPER-LOCK component.

TO INSTALL

RECOMMENDED WRENCH TORQUE

Bushing No.	Screws	Wrench Torque ▲ (Pound-Inches)
1008, 1108	¼″ setscrews	55
1210, 1215, 1310	⅜″ setscrews	175
1610, 1615	⅜″ setscrews	175
2012	⁷⁄₁₆″ setscrews	280
2517, 2525	½″ setscrews	430
3020, 3030	⅝″ setscrews	800
3535	½″ capscrews	1,000
4040	⅝″ capscrews	1,700
4545	¾″ capscrews	2,450
5050	⅞″ capscrews	3,100
6050, 7060, 8065	1¼″ capscrews	7,820
10085, 120100	1½″ capscrews	13,700

▲ When torque wrench is not available it is possible to approximate these values by using an ordinary wrench and piece of pipe. For example, to obtain 1,000 pound-inches, pull 100 pounds at 10 inches from center of screw.

2 PLACE BUSHING IN HUB and match holes (not threads)

3

"LIGHTLY" oil setscrews and thread into those half-threaded holes indicated by ◎ on above diagram.

4 MOUNT assembly on shaft.

5 Alternately TIGHTEN SCREWS using "cheater bar" or torque wrench to recommended torque values.
DO NOT USE WORN HEX KEY WRENCHES.

OPTIONAL

6 To increase gripping force. Alternately HAMMER BUSHING (using drift) and tighten screws until specified torque no longer turns screws.

TO REMOVE

1 REMOVE ALL SCREWS.

2 INSERT SCREW(S) in holes threaded on bushing side (shown as ● in diagram).

3 Alternately TIGHTEN screw(s) until bushing "pops" loose.

FIGURE 7–12

Taper-lock bushing mounting/dismounting instructions.

Courtesy of Rockwell Automation.

FIGURE 7–13
QD bushing.

Courtesy of T. B. Woods
Incorporated.

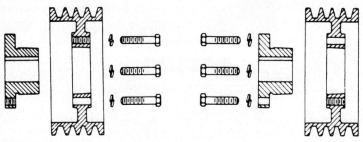

FIGURE 7–14
QD bushing standard and reverse mounting types.

Courtesy of T. B. Woods Incorporated.

Slide sheave into position on bushing aligning the drilled holes in the sheave with the tapped holes in the bushing flange. (Note: Install M thru S bushings so that the two tapped holes in the sheave are located as far away as possible from the bushing's saw-cut.) Loosely thread the cap screws with lock washers into the assembly. Do not use lubricant on the cap screws!

4. Reverse mounting: With large end of the taper out, slide sheave onto shaft as far as possible. Install shaft key (see shaft key note in #3 above). Install bushing onto shaft so tapered end will mate with sheave. (See wedging note in #3 above.) If using the setscrew, tighten it enough to prevent the bushing from sliding on the shaft. Caution: Do not overtighten setscrew! Pull the sheave up on the bushing, aligning the drilled holes in the bushing flange with the tapped holes in the sheave. Loosely thread the cap screws with lockwashers into the assembly. Do not use lubricant on the cap screws!

5. Using a torque wrench, tighten all cap screws evenly and progressively in rotation to the torque value in the following table. There must be a gap between the bushing flange and sheave hub when installation is complete. Do not overtorque! Do not attempt to close gap between bushing flange and sheave hub!

QD BUSHING SCREW TIGHTENING INFORMATION

Tapered Bushing	Size & Thread of Cap Screw	Ft.-Lbs. To Apply With a Torque Wrench
QT	1/4 × 1	9
JA	No. 10 − 24	5
SH-SDS-SD	1/4 − 20	9
SK	5/16 − 18	15
SF	3/8 − 16	30
E	1/2 − 13	60
F	9/16 − 12	110
J-JS	5/8 − 11	135
M-MS	3/4 − 10	225
N-NS	7/8 − 9	300
P-PS	1 − 8	450
W-WS	1-1/8 − 7	600
S-SS	1-1/4 − 7	750

Caution: The tightening force on the screws is multiplied many times by the wedging action of the tapered surface. If extreme tightening force is applied, or if a lubricant is used, bursting pressures will be created in the hub of the mating part.

To Remove:

a. Relieve drive tension by shortening the center distance between driver and driven sheaves.

b. Lift off belts.

c. Loosen and remove cap screws. If the bushings have keyway setscrews, loosen them.

d. As shown above, insert cap screws (three in JA through J bushings, two in QT and M thru W bushings, and four in S bushing) in tapped removal holes and progressively tighten each one until mating part is loose on bushing. (Exception: If mating part is installed with cap screw heads next to motor, with insufficient room to insert screws in tapped holes, loosen cap screws and use wedge between bushing flange and mating part.)

e. Remove mating part from bushing and, if necessary, bushing from shaft.

Split-Taper Bushing

The split-taper bushing system is similar to the QD type (see Figure 7–15). It has different mounting bolt patterns and flange dimensions. What makes this bushing system unique is that it is keyed to the shaft and keyed to the sheave hub which gives it a high torsional load capacity.

FIGURE 7–15
Split-taper bushing.
Courtesy of Emerson Power Transmission.

(a)

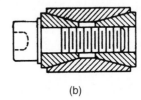

(b)

FIGURE 7–16
(a) Complete keyless locking mechanism assembly; (b) cross-section view.
Courtesy of Rockwell Automation.

Keyless/Compression Bushings

There are a variety of "keyless" or "compression" locking bushings that rely on compression and expansion of mating parts to hold the unit in place (see Figure 7–16a & 7–16b). Cap screws are used to draw the mating parts together, forcing inner and outer rings to contract and expand. Keyless bushings have high torque-holding capacities due to their design. Keyless locking devices allow more torque to be transmitted for a given shaft size without a keyway than if it had a keyway. This is because the holding forces are uniformly distributed around the outside diameter of the shaft and the inside diameter of the bushing ring. With shafts that use keyways, oftentimes the shearing stresses and torque are transmitted at the key. Under high or sudden shock loads, a failure can occur at the key. Compression or keyless bushings are frequently used on large shafts that see high torque. They are also used on indexing and reversing applications with frequent starts and stops. Proper torque settings and tightening procedures of the cap screws must be done when assembling a "keyless" bushing. This is to ensure that the forces are evenly distributed around the ring and that it is installed in a concentric manner.

Student Exercise

Determine the correct shaft sizes and hub bores based on standard tolerances. The nominal shaft diameters are given. Determine mounted component hub bores for both clearance and interference fits. Refer to recommended bore and shaft

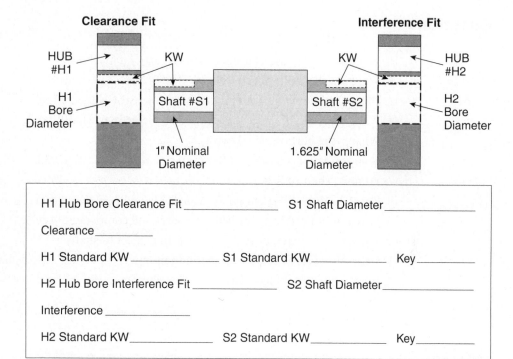

FIGURE 7–17
Student exercise worksheet.

tolerance tables. Determine the size of the hub bore of the mating components and the shaft diameters to within .0005″. One side of the shaft requires a clearance fit and the other side an interference fit. Determine the resultant clearance and interference. Specify the correct key and keyway sizes. Fill out the exercise worksheet in Figure 7–17.

Questions

1. What is the purpose of a steel shaft as part of a machine?
2. What is a shaft fillet?
3. Name five moments or loads that a shaft may be subjected to.
4. What are some of the considerations for selecting a shaft?
5. What is the purpose of a keyway?
6. Name and describe three different types of keys and keyways.
7. What is "shaft fit"?
8. Describe three different types of shaft fits and how they differ.
9. Why is it important to measure the shaft and component bore?
10. Why is linear expansion important with respect to shafts?
11. Name three types of bushings used to mount components onto a shaft.
12. What is the advantage of using keyless compression-type bushings?

CHAPTER 8

Seals and Sealing Devices

This chapter examines the various types of seals and sealing mechanisms used on machinery. Seals and sealing devices serve the dual purposes of retention of fluids and exclusion of contaminants. The different types of sealing devices will be covered as well as how they function. Issues pertaining to the proper installation and maintenance of seals and sealing devices will be addressed.

Objectives

Upon completion of this chapter, the student will be able to:
✔ Describe the differences and define the advantages/disadvantages between a contact and noncontact seal.
✔ Explain the purpose of seals and sealing devices.
✔ Explain the important issues that need to be considered when selecting a seal or sealing device.
✔ List the various types of seals and sealing mechanisms.
✔ Understand and list the various maintenance and installation procedures that pertain to seals and sealing devices.
✔ Describe which type of seal is appropriate for a given application.

INTRODUCTION

The basic function of all seals and sealing devices is to both retain and to exclude fluids or substances. In a machine the typical fluid might be grease or oil. Some sealing devices are used to retain fluids being moved from one location to another. Excluding foreign materials, such as water or dirt, and keeping them from contaminating the system is one of the primary purposes of seals and sealing devices. They also can be used to maintain pressure or vacuum. A variety of different types of sealing devices such as oil/grease seals, gaskets, stuffing boxes, o-rings, mechanical seals, packing, and others is used on machines.

Seals and sealing devices can be grouped into two basic types: static and dynamic. Static seals are used where there is little or no movement between surfaces. O-rings and gaskets are examples of this type. Dynamic seals and sealing devices are used when there is movement between mating machine parts. Radial lip seals, stuffing boxes, and mechanical seals are examples of dynamic types.

In order to function properly, seals, gaskets, packing, o-rings, and any form of sealing device must be designed and constructed to accommodate the rigors of the application. They should be capable of satisfying a number of important criteria: They must handle a wide range of temperatures, be compatible with the material being sealed, function under the equipment operating pressures, and accommodate the inherent motions of the machine. In short, they must not break down, wear out, or leak. That is why the selection of the type of sealing device, along with its design and the material it is made of, is determined by the conditions and expectations of the application.

RADIAL LIP SEALS

The purpose of dynamic radial lip seals is to create a barrier between surfaces in relative motion. Usually the "radial lip" seal is located in a stationary housing and sealing against a rotating shaft (see Figure 8–1). The radial lip seal is often referred to as a standard oil or grease seal. Some types are known as "shaft riding" seals and rotate with the shaft and seal against the housing (see Figure 8–2). A type of shaft

FIGURE 8–1
Radial lip seal retaining lubricant and excluding contaminants.
Courtesy of Chicago Rawhide.

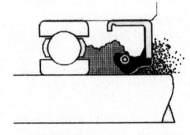

Dynamic seal on rotating shaft with bearing.

FIGURE 8–2
Shaft riding or V-ring seal excluding contaminants.
Courtesy of Chicago Rawhide.

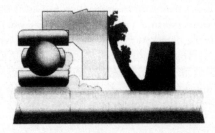

riding seal is a " V-ring" seal. Both the radial lip and shaft riding seals serve the purpose of retaining lubricating fluids and sealing out contaminants.

Radial Lip Seal Construction

The radial lip seal comes in a variety of shapes and designs. The dominant designs have a sealing element or lip, bonded to a shell or case. The sealing lip is made of a flexible elastomeric compound. One common version has a radial sealing lip held in place, against the shaft, by a garter spring (see Figure 8–3). The spring maintains pressure between the sealing element and sealing surface. Another form has metal fingers that act as a spring to apply pressure to the lip of the seal against the shaft (see Figure 8–4).

The most commonly used material for sealing lips is nitrile. Nitrile compounds have excellent compatibility with most mineral oils and greases. It has a recommended operating temperature limit of $-40°F$ to $250°F$ ($-40°C$ to $121°C$). A variety of compounds such as fluoroelastomers, silicones, Teflon, felt, leather, and many other materials are available for special applications. Oftentimes the lip material to be used is determined by its compatibility with the lubricant or fluid it is sealing out. If the seal is exposed to corrosive solutions, chemical and lip compound compatibility should be checked to prevent the seal material from deteriorating and leaking. The manufacturers of seals can be consulted for a specific recommendation of sealing lip materials that are best suited for applications that are outside of the norm.

Surface velocities of radial lip seals are an important issue. For example, a machine running at 1750 rpm with a 4.5″ diameter shaft will have a high shaft surface velocity measured in feet per minute ($.262 \times 4.5″ \times 1750$ rpm $= 2063$ fpm). Shafts with

FIGURE 8–3
Standard garter spring-type radial lip oil/grease seal.
Courtesy of Garlock Sealing Technologies.

FIGURE 8–4
Finger spring radial oil/grease seal.
Courtesy of Garlock Sealing Technologies.

surface speeds above 2000 feet per minute might require a special compound lip material to resist the heat generated from friction. The construction of the radial lip seal is relatively simple. The lips of the seal are bonded to a case or shell that is the backbone of the seal, and it is usually made of metal. The cup-shaped rigid shell or case serves the function of providing support, shielding, and a platform for the lip. Cases can be constructed from a single metal piece or from two different metal rings pressed together, affording more strength. The metal case usually will have a coating of an acrylic polymer to fill small imperfections in the housing bore, providing a better seal on the outside diameter. Some manufacturers make the seal entirely of a nonmetallic material. In some cases these nonmetallic, radial lip seals are formed in a mold.

Internal and external pressures are an important operating concern for radial lip seals. Most standard oil and grease radial lip seals are designed to have a maximum service life at 0 psi. Standard bonded type lip seals will satisfactorily function at pressures up to 10 psi. If internal air or fluid pressures become too high from heat, the seal will leak. Excessive pressure created by overzealous lubrication methods or high-pressure systems can also cause the seal to "blow out."

Radial lip seals come in many different types of construction, sizes, materials, and designs to fit the particulars of the application. Proper selection and installation will help prevent fluid leakage and the intrusion of contaminants.

Radial Lip Seal Installation

Properly installed radial lip oil/grease seals will reduce lubricant leakage and help prevent the intrusion of contaminants into the system. The following checklist should assist in installation:

- Lock out, tag-out, and verify that the equipment is electrically and mechanically neutralized.
- Confirm that the replacement seal size, design, and material are correct and properly selected for the application. Never reuse a worn or damaged seal. Measure the shaft diameter and housing bore, which are the boundary dimensions for the seal. The width of the seal should be no greater than the bore cavity depth. Check the seal lip compound for compatibility with the fluid being retained.
- Work with proper tools in a clean environment. Preferably move the equipment or machine that contains the seal to a well-equipped workshop.
- Inspect and measure the shaft. The shaft should be made from a high carbon steel or stainless steel with a hardness of Rockwell C 30 or harder. Check the run-out on the shaft to ensure it does not exceed 0.010″ TIR. For best sealing performance, the diameter should be within the following RMA and ISO recommended tolerances:

  Up to and including 4.000″ = plus or minus .003″

  4.001″ to 6.000″   = plus or minus .004″

  6.001″ to 10.000″   = plus or minus .005″

  10.001″ and larger   = plus or minus .006″

- The shaft should be free of any nicks, burrs, defects, or grooves that may have been cut into the shaft from the lip of a previous seal. If the shaft surface is in poor condition, a thin shaft sleeve of approximately 0.010 to 0.015″ thickness, commonly referred to as a "speedi-sleeve," may be placed over the damaged area. This will create a new surface for the seal lip to ride on (see Figure 8–5). It is helpful to have a shaft with a chamfered leading edge to assist in installation of the seal without cutting the lip. The sharp edges of the keyway also must be covered with a sleeve or tape to protect the seal lip from being cut during installation.

- Determine the direction the seal will face for most effective retention or exclusion. For retention of lubricating fluids the lip or spring should face the lubricant. If the primary purpose is for exclusion of contaminants, the sealing lip or spring should face out toward the source of foreign materials (see Figure 8–6).

- Inspect and measure the seal-housing bore. Make sure the housing bore dimensions are round within a general tolerance of plus or minus .001″ to .002″. A recommended press fit of no more than .005″ to .006″ is an approximate average for metal case seals. The housing bore must be free of all nicks, burrs, and defects to prevent damage occurring during installation.

- Prelubricate the sealing lip and the outside diameter with grease or oil. This will aid in the installation process and provide a more effective lubricant and contaminant barrier.

- Slide the seal over the shaft squarely and press it firmly and evenly into the housing bore with the proper installation tool (see Figure 8–7). The tool must not come into contact so as to damage the sealing lip, shaft, or the housing bore. Do not use a punch or screwdriver to force the seal into place. Never hammer directly on the face of the seal. Ensure that it is not cocked in the housing or that the lip is damaged during installation. If the seal uses a garter

FIGURE 8–5
Sleeved shaft with a standard radial lip seal and V-ring seal protecting a bearing.
Courtesy of Chicago Rawhide.

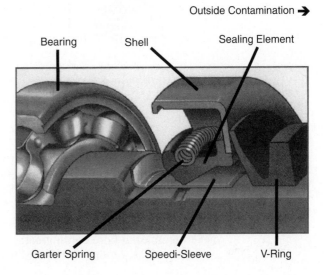

Outside Contamination →

Bearing   Shell   Sealing Element

Garter Spring   Speedi-Sleeve   V-Ring

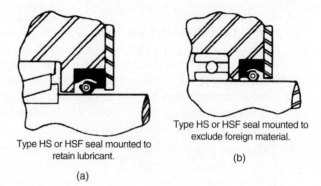

Type HS or HSF seal mounted to
retain lubricant.

(a)

Type HS or HSF seal mounted to
exclude foreign material.

(b)

FIGURE 8–6
(a) Radial seal position for retention; (b) radial seal position for exclusion.

Courtesy of Chicago Rawhide.

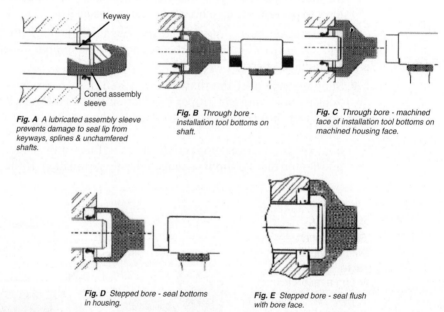

Fig. A *A lubricated assembly sleeve prevents damage to seal lip from keyways, splines & unchamfered shafts.*

Fig. B *Through bore - installation tool bottoms on shaft.*

Fig. C *Through bore - machined face of installation tool bottoms on machined housing face.*

Fig. D *Stepped bore - seal bottoms in housing.*

Fig. E *Stepped bore - seal flush with bore face.*

FIGURE 8–7
Seal installation procedures.

Courtesy of Chicago Rawhide.

spring to provide pressure on the lip to the shaft, take care not to dislodge the spring during installation. If a defective or damaged seal is installed, leakage will occur.

■ Seat the seal in the proper location. Manufacturers of machines will sometimes make the housing bore depth greater than the width of a standard lip seal. This allows for repositioning of the seal lip where it contacts the shaft surface if it should wear a groove over time. Pushing the seal deeper into the housing bore will provide a new shaft surface and prevent leaks.

- ■ Never paint over the seal lip or on the shaft surface where the seal rides.
- ■ Observe the seal during normal operating speeds and conditions for leaks.

MECHANICAL AND ROTARY SEALS

Mechanical and rotary seals are designed to eliminate leakage originating from inside the machine or pump, as well as excluding foreign contaminants. Mechanical seals are used in place of stuffing boxes and packing on some pumps. Rotary mechanical seals are also used in place of radial lip seals. The advantages that both types share over conventional stuffing box/packing arrangements and standard radial lip seals are as follows: relatively little leakage, high pressure capability, virtually no shaft or housing wear, no adjustment needed, minimal friction, and highly efficient. The major disadvantages are that they are expensive and they must be sized properly or they will leak and fail. Both types require dimensional fitting accuracy and proper installation to ensure sealing effectiveness.

Mechanical Seals

A mechanical seal is a sealing mechanism made up of several assembled parts (see Figure 8–8). It is designed to seal at three locations. First, it must act as a static seal between the housing and the stationary part of the seal. Second, it acts as a static seal between the shaft and the rotating mechanical seat. Finally, it functions as a dynamic seal between the rotating seal face and the stationary seal face. Oftentimes an o-ring is used to hold the stationary seal part in position within the housing. The stationary component is usually made from stainless steel or ceramic. The rotating portion of the seal is comprised of three parts: collar, spring, and wear face. The wear face is usually made from a carbon/graphite composite that is held to the shaft by another o-ring. A spring assembly provides contact pressure between the collar

FIGURE 8–8
Mechanical seal.
Courtesy of John Crane.

face portion of the seal assembly. Wear takes place between the faces of the stationary and rotating parts. The collar holds the spring in place and positions the assembly. Mechanical seals may be mounted internally or externally to the machine.

Rotary Seals

Rotary noncontact seals are essentially an assembly of static and dynamic components (see Figure 8–9). A typical configuration will include a "stator" which is the stationary component, and a "rotor" that is the rotating part. They are designed to exclude moisture and contaminants in severe operating environments. They are usually made from some form of high-impact polymer or graphite-filled PTFE that is heat resistant.

Rotary seals operate under the principle of a labyrinth seal. A labyrinth seal is a multichambered seal that relies on internal channels and pressures to inhibit the flow of fluids (see Figure 8–10). Primarily labyrinth seals are used on applications that use grease as a lubricant. The rotary seal is designed with channels and chambers that have minimal spacing between them. A typical rotary seal consists of two or three parts. Some manufacturers will have an inner stator and an outer stator; others have only one stator. The rotor ring will have multiple o-rings that are on the inside of the collar to hold it in place on the shaft. These o-rings and inner portion of the seal rotate with the shaft. The outer seal collar (stator) is positioned into the

FIGURE 8–9
Rotary seal.

Courtesy of Garlock Sealing
Technologies.

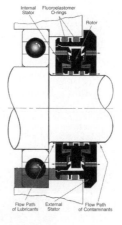

FIGURE 8–10
Labyrinth seal construction.

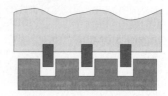

housing and held in place by o-rings. Inside of the assembly various passages and channels create a noncontacting seal that acts as a labyrinth to prevent leakage and exclude contaminants. The operating environment and type of fluids being retained determine the material composition of the seal. These seals are very effective but are expensive compared to standard radial lip seals.

Gaskets

A gasket is a pliable material that is shaped and placed between two mating machined surfaces to form a seal. A gasket usually serves a dual purpose of retaining fluids and sealing out contaminants. In flange gaskets used on piping, the object is to ensure the gasket is compressed enough to maintain sealing contact with the flange under the operating pressures in the system. The gasket material must be conformable to the facings of the closure so they can flow into the imperfections of the metal face. It usually will have cut-outs or holes to allow for various size fasteners and dowels to pass through (see Figure 8–11). The gasket material must not be porous or it will leak. Gasket materials are various and the application and the operating conditions determine their selection. Some gasketing materials are dispensed as a liquid and harden with time. Gaskets are often treated with other materials to achieve better surface properties, change their adhesion qualities, and resist chemical attack.

The selection of the gasket material is determined by considering three important variables. The first variable is the temperatures that the gasket will encounter when exposed to the fluid or medium it is sealing. The second variable is the internal pressures that the gasket must handle without bursting. The last variable is the compatibility of the gasket to the fluid/medium in order to remain impervious to breaking down and leaking. Manufacturers have selection tables that take these issues into account when selecting the right material for the job.

FIGURE 8–11
Typical gasket configuration
on a pipe flange.
Courtesy of J/M Clipper Corp.

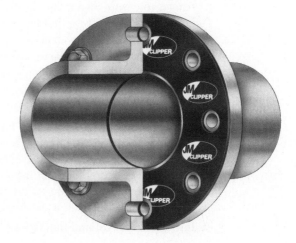

Gasket Replacement The following replacement procedures are designed primarily for nonmetallic gaskets used on standard pipe flanges.

■ Follow all lockout/tag-out procedures. Shut off the inlet and outlet, drain or purge the pipe.

■ On piping flanges where a full-face gasket is used, loosen and remove all the bolts. If the gasket is of the ring variety, loosen all the bolts but remove only enough to allow the removal of the old gasket. Examine the bolts for damaged threads or elongation and replace if necessary.

■ On piping applications, spring the flanges apart with flange spreaders or wedges. Take care not to let the wedges gouge the faces of the joint.

■ Carefully remove the old gasket and examine it for clues as to the nature of the failure. Look for signs of deterioration or hardening from chemical attack or excessive heat. Check for physical damage that might have occurred during installation from non-uniform bolting practices or for tool marks caused by improper assembly. A uniform, heavily defined joint face impression should be obvious. Never reuse old gaskets because they will leak.

■ Examine the joint facings for damage, corrosion, or warping that could have caused the leak. Clean the faces and the back side of the flange with a wire brush and remove all traces of the old gasket material.

■ Cut the new gasket to the proper size. Do not form the gasket by hammering or cutting it against the face. The bolt holes on a full-face gasket should be punched to the same diameter as the flange holes unless special recommendations are given. An arch punch tool should be used to cut holes. The gasket inside diameter should be slightly larger than the inside diameter of the bore of the joint face. Certain materials, such as rubber, will expand considerably under some conditions and require more clearance. The type of flange, welded neck, slip-on, lapped joint, or screw-on will dictate the size and shape of the gasket.

■ Locate and line up the gasket in position so it is not pinched or obstructing the bolt holes and the flange bore. Insert enough prelubricated bolts into one of the flanges to ensure alignment. Special gasket sprays or fluids can be applied to ease installation and assist in holding the gasket in place during assembly. These compounds will also act as an additional seal by filling voids and depressions on the face. They are not designed to replace a properly selected, cut, and installed gasket.

■ Bring the mating flanges together, making sure they are parallel and in line. Avoid using the bolts to draw the flanges into parallelism. Make sure that all tongues, pilots, or positioning seats are lined up and seated. Hand-tighten the bolts.

■ It is imperative that the proper bolt-tightening sequence and torque specifications be followed (see Figure 8–12). All mating surfaces must be drawn up evenly and uniformly. Initially tighten bolts in a cross-tightening pattern to 1/3 or 1/2 of the recommended torque values. Repeat the tightening sequence until they are drawn up to the recommended full torque specification. Check the bolts for looseness after a run-in period and retighten if needed.

■ Inspect the gasket and flange for leaks during the run-in time.

FIGURE 8–12
Typical bolt tightening sequence
for pipe flange.

PACKING AND STUFFING BOXES

Stuffing boxes and packing are used in a wide range of equipment in which centrifugal
and rotary pumps, reciprocating pumps, rods, plungers, and pistons are present. A
stuffing box is a sealing chamber filled with packing material to provide a sealing function. The stuffing box is used to control leakage at the point a shaft protrudes from a
machine, such as on a pump. It is contained within a confined space that is different in
pressure from the surrounding area. The stuffing box consists of three parts: the packing chamber (box), the packing, and the stuffing gland (see Figure 8–13). The stuffing
box is the stationary metal part of the machine that holds the packing. Stuffing boxes
on pumps where the suction pressure is below atmospheric pressure or where slurries
are being pumped will have a small inlet port for an introduced sealing and lubricating
fluid. The packing is usually made of multiple shaped rings that seal against the shaft
and box housing walls. The gland provides clamping pressure and compresses the
packing rings to seal against the walls of the box and the surface of the shaft.

FIGURE 8–13
Standard stuffing box with
braided packing. Note that the
darker lantern ring is positioned
between the sets of gasket rings
and aligned with the inlet port for
introduced fluid (seal water).

Courtesy of J/M Clipper Corp.

Packing and Stuffing Box Application

In a dynamic application such as pumps, packing must leak a small amount of fluid to perform properly. Its purpose is to control leakage, not prevent it. The packing within the stuffing box is subjected to some compressive forces, but not enough to prevent minor leakage. This minor leakage is needed during operation to assist in lubrication and carry off the heat generated from friction. Adjustment of the gland, done by tightening the bolts, is required as the packing wears and packing material is lost. The key is to adjust the gland to keep leakage to a minimum, but not to over-tighten it which will result in excessive heat and wear.

Most ordinary stuffing boxes use multiple rings of compression packing material. The purpose of using multiple rings is to reduce the pressure of the fluid being sealed to near zero at the end of the box. The first ring into the box, closest to the sealed fluid, does most of the sealing work. It is imperative that this ring be installed correctly. Practical stuffing box arrangements include five packing rings and a lantern ring. A lantern ring is a spacer device that allows passage of a fluid into the packing, onto the shaft, and eventually to the pump discharge. The lantern ring is positioned between the packing rings with the front edge in line with the inlet port. As the packing rings wear and the gland is tightened, the lantern ring will be moved forward. In some applications where slurry is being pumped an external source of clear liquid is pumped into the inlet at a higher pressure than the slurry. This is often referred to as the "seal water." The water serves as a lubricant for the packing and prevents air from entering the pump. The fluid also keeps abrasives and chemicals away from the packing, which would cause it to fail.

To meet the variety of application requirements, compression packing is manufactured in a variety of materials and constructions. Packing is constructed in many different types of weave. The type of service that the packing will be exposed to dictates the preferred weave style. Three common packing weave types are braid-over-braid, interlocked, and plaited (see Figure 8–14).

FIGURE 8–14
Packing weave types.
(top) Braid over braid;
(middle) interlocked;
(bottom) plaited.
Courtesy of J/M Clipper Corp.

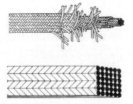

The most common materials for packing are carbon and graphite. Modern packing materials are self-lubricating, nonabrasive, nonhardening, resistant to chemicals, and do not contain lead or asbestos. Special packing material is made with impregnated compounds that resist various chemicals and withstand extreme temperatures. Packing materials should be selected by consulting the manufacturer with the particulars of the application.

Packing Installation

Packing must be installed correctly into the stuffing box in order for it to act as an effective seal and yield long service life for the equipment. If packing and stuffing boxes are not functioning properly, it is usually because of improper installation or damage that occurred during the procedure. While there are numerous different types of stuffing box arrangements, the following procedural guidelines can be applied for most of the standard packing installations.

1. Follow all basic safety procedures. Lockout, tag-out, and verify that the equipment is electrically and mechanically neutralized. All appropriate valves should be closed and all suction or head pressure must be fully discharged.

2. Use quality packing that has been selected based on the operating conditions of the application.

3. Remove the gland fasteners and the gland. Inspect them for damage.

4. Use a packing hook to remove the old packing rings. Do not damage the shaft or throat with the hook during the removal process. Inspect the packing for signs of problems. Charred and dry packing material indicates an inadequate lubricant supply. Wear on the outer surface of the packing rings possibly indicates that the rings are rotating with the shaft due to incorrect sizing or being stuck on a corroded shaft.

5. Clean and inspect the stuffing box throat. Remove all debris and material with a noncorrosive cleaning agent. The surfaces of the shaft and throat should be free of corrosion and scoring. A rough and pitted shaft/sleeve or throat will not seal effectively. Rework the surfaces if necessary.

6. Correctly size the required packing rings. Packing rings can sometimes be obtained in precut sections. Usually it comes on a roll and must be cut by the technician. To determine the correct size, measure the diameter of the shaft to get the inside diameter of the ring. Measure the inside diameter of the stuffing box to get the outside diameter of the ring. Subtract the inside measurement from the outside measurement and divide by two. The result is the required cross-section size packing.

7. Wrap the packing firmly around a mandrel of the same diameter as the pump shaft/sleeve without stretching it excessively. Cut the rings without crushing the packing material. Always cut separate rings. Never wind a continuous coil of packing into the throat. For most pumps and agitators, cut the packing rings with butt (square) ends. For certain valve stem and reciprocating applications

some manufacturers recommend a 45° skive cut end. Each ring can be cut in this manner or the first correctly sized ring can be used as a template for the next rings.

8. Install one ring at a time, carefully twisting it open and wrapping it around the shaft. Never bend open the packing ring because this will crack the rings. A small amount of clean, light oil will assist the process of installation. Firmly seat the first ring squarely in the bottom of the throat, using a bushing or sleeve to push against the ring. The proper seating of the first ring is critical to prevent leakage and damage to the packing and stuffing box.

9. Each additional ring should be firmly seated using the split bushing or tamping tool, with the split ends staggered at 90°. Staggering the joint ends does not create a direct path for leakage. Do not use the gland to seat the individual rings. If a lantern ring is required, place it in the proper position so the front end is in line with the inlet port. Rotate the shaft occasionally to make sure binding does not occur.

10. Install the gland follower after the seating of enough rings so the nose of the follower contacts the last ring and extends out approximately one-third of the total packing depth. Tighten the bolts evenly and with enough pressure to hold the assembly in place and allow for leakage.

11. After following correct and safe start-up procedures, start the system and flood the pump.

12. Allow the packing to leak freely and generously during the first hour of operation. After the run-in period, take up gradually on the gland follower until the leakage is controlled. Do not overtighten the gland or you will burn up the assembly.

13. Periodically check the gland for the amount of leakage and adjust as needed.

O-RINGS

O-rings are flexible synthetic rubber or elastomeric compound static seals. O-rings are used extensively in hydraulic and pneumatic mechanisms to seal fluids and air. They are available in a variety of shapes and sizes. The most prevalent shape is the circular cross section. Square cross-section rings are also made for high-pressure applications. Typical o-ring part numbers will designate the cross-section diameter, inner diameter, and the material it is made of. One of the most common materials used in o-ring construction is Buna-N, of a 70 to 90 durometer. Fluoroelastomer and silicone compounds are used in high temperature applications to 400°F.

The o-ring is designed to work under compression in a matching groove and seal between two mating surfaces. It is generally considered a static device but can be used to seal slow moving, low cycle rate rotating or reciprocating shafts.

The o-ring is fitted into a groove that is usually rectangular in shape. The groove is wider, but shallower, than the ring cross-section thickness. The cross-section

FIGURE 8–15
O-ring installation.

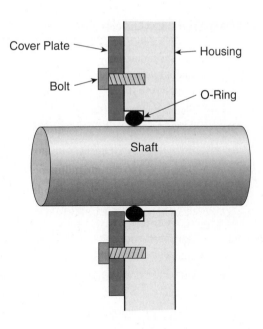

diameter deforms slightly from mechanical squeezing of the ring (see Figure 8–15). Pressures from the confined fluid also cause the ring to further deform and increase the sealing surface area. Under pressure the ring is forced to slide and roll away from the source and deform. A minimum amount of compression, approximately 10% of its cross-sectional dimension, is required. O-rings are manufactured with approximately 10% greater cross-section diameter than their nominal size. For example, a 1/8″ cross section is actually .139″. This builds in the required squeeze to ensure a tight seal. Any o-ring that exhibits signs of wear or damaged surface area should be replaced or it will fail to act as a seal.

Troubleshooting Quick Guide

Problem	Probable Cause	Corrective Action
Oil/Grease seal leaking at shaft	Damaged seal lip	Inspect seal and shaft, replace if needed. Do not damage during installation.
	Damaged shaft surface	Inspect shaft surface, sleeve or replace.
	Improper installation	Correctly install and seat seal.
	Excessive shaft run-out (bent shaft)	Replace shaft.

Troubleshooting Quick Guide *(continued)*

Problem	Probable Cause	Corrective Action
	Excessive pressure	Reduce pressure and/or replace with high-pressure type seal.
	Fluid and lip material noncompatible	Check fluid compatibility charts.
	Extreme temperatures (lip cracked)	Verify operating temperatures and change lip material to a compatible type.
	Worn lip from abrasives	Shield seal.
	Wrong size	Measure shaft diameter and housing bore. Properly size replacement seal.
Oil/grease seal leaking around seal shell	Damaged seal shell	Inspect seal, replace, and install carefully. Do not cock during installation.
	Damaged housing bore	Measure, clean and inspect, remove burrs. Check chamfers.
Packing leaking	Damaged shaft/journal	Inspect, clean, and repair if needed.
	Damaged stuffing box	Measure, clean and inspect, remove burrs.
	Deteriorated packing material	Replace packing.
	Incorrect size and installation of packing rings	Properly cut, position, and install correct quantity and type of packing rings.
	Fluid noncompatibility	Replace packing with compatible type.
	Loose packing rings	Tighten gland bolts.
	Packing glazed or burned	Excessive gland bolt pressure.
Gasket leaking	Wrong shape	Properly cut and size gasket material.
	Noncompatible with fluid	Check fluid compatibility charts.
	Excessive pressure	Reduce pressure.
	Improper bolting	Follow proper bolting torque tightening specifications and sequencing. Do not overtighten.

Student Exercise

The radial lip oil/grease seal installed in a gearbox to seal the input shaft is leaking. The 4.5″ diameter shaft rotates at 1750 RPM. The surface temperature of the shaft at the seal contact area is checked and it is 250°F. A dial indicator is used to check the radial run-out of the shaft and it is .006″. The retained fluid is synthetic gear oil with an EP (extreme pressure) additive. What might be the cause of the leaking seal and what are some recommendations that may lead to a solution? Fill out the worksheet in Figure 8–16.

Shaft Surface Velocity_____

Maximum Recommended PSI _____

Maximum Recommended Shaft Radial Run-out _____

CAUSES

_____

_____

_____

_____

RECOMMENDATIONS

_____

_____

FIGURE 8–16
Student exercise worksheet.

Questions

1. What is the difference between a contact and a noncontact seal?
2. What is a radial lip seal and how is it constructed?
3. Name three materials that radial lip seals can be made from.
4. What is a gasket?
5. Describe packing materials and where they are found.
6. What is a stuffing box and how does it work?
7. What is an o-ring and how does it seal?
8. What is a mechanical seal?
9. Name four advantages of a mechanical seal.
10. What is a rotary seal?
11. Describe how a labyrinth seal works.
12. List three safety procedures that must be followed when installing packing in a stuffing box.

CHAPTER 9

Bearings

This chapter addresses plain and antifriction bearings. Bearings are used to provide a reduced friction, supporting surface between mating machine parts. They serve the purpose of supporting rotating, oscillating, or laterally moving shafts. Many variations of bearings exist and an overview of the major types will be included in this section. Emphasis will be on understanding the purpose of bearings and being able to identify various types, along with how to maintain them. The topics of installation and maintenance will be covered in detail.

Objectives

Upon completion of this chapter, the student will be able to:
✔ Understand the purpose of bearings and how they react to various loads.
✔ Describe the differences between and list the various types of plain bearings and antifriction bearings.
✔ Describe the advantages and disadvantages of various types of bearings.
✔ Understand and list the various maintenance and installation procedures that pertain to bearings.
✔ Give the reasons why proper handling of bearings is important.
✔ Understand how bearings fitting practices and operational clearances affect the performance of bearings.
✔ Demonstrate the ability to properly remove and install various bearings.

INTRODUCTION

A bearing is a component that is designed to provide shaft support, positioning, guidance, free rotation, and reduce friction between moving parts. Bearings usually support a rotating, oscillating, or lateral moving shaft and are supported around the outside diameter by a housing of some form. Bearings of one form or another have been in use for thousands of years. Early bearings used wooden rollers to reduce the required force to move the supported object. As early as 700 BC, Assyrian bas-relief sculptures showed large stone statues being moved on rollers used as bearings. In ancient times simple wooden rollers lubricated with animal fats were used as bearings on carts. Leonardo da Vinci sketched ideas for a pivot bearing with rolling elements. In the late 1700s and early 1800s numerous patents around the world began to appear for rolling element bearings to be used in industrial applications. During the industrial revolution the need for bearings used in numerous types of machines created a demand for better bearings. The popularity of the bicycle and the usefulness of railroads for transportation helped drive the development of the bearing industry. In the late 1800s and early 1900s Swedish, Russian, and German engineers intensely studied how bearings reacted under load and speed while lubricated. This led to the science of tribology.

Tribology is the study of friction between interacting parts. Many theories and formulas representing how a bearing will react under certain conditions, and their resultant life, were formulated during this time and are still valid today. Innovative thinkers such as Friedrich Fischer, who invented the ball milling machine, Richard Stribeck, who developed the formula for the distribution of the externally applied loads on the individual rolling elements of a bearing, and Arvid Palmgren and Gustaff Lundberg who were primarily responsible for the fatigue life theory, led the way. In the United States, around the turn of the century, Henry Timken made the practical contribution by replacing plain bearings in horse-drawn carts with taper roller bearing units. He went on to found the Timken Roller Bearing Company. These trailbreakers turned their ideas into reality and began to turn the manufacturing of bearings into a science, and provided a supply for an ever-increasing market demand. With the development of numerous machines related to the production of bearings and their related components during the early to mid 1900s, the bearing industry matured.

Today in the twenty-first century their ideas and machines continue to be used as a foundation in the expanding market of bearings. Research and development of new materials and manufacturing methods continues at this time with the goal of producing frictionless bearings. As long as there continues to be moving parts within machines, there will be a need for bearings.

Bearings can be divided into two main groups: plain bearings (friction) and rolling element bearings (antifriction). Plain bearings rely on sliding action, and rolling element bearings primarily rely on rolling action. The primary purpose of both types of bearings, besides load support, is to reduce friction and thus require less energy to move a mass. Increased resistance due to friction requires more energy to move an object. By placing the object on sliding or rolling bearings, the resistance to movement is reduced. The coefficient of friction (μ) for sliding or rolling surfaces is in part related to the surface smoothness. The smoother the surface the lowered the coefficient. The need for bearings that allowed less friction and lowered the energy requirements to perform work has led to the development of the antifriction bearing. Antifriction bearings generally roll rather than slide, which results in less friction. The force required to overcome the friction resistance of two separate surfaces in sliding or rolling contact is reduced when separated by a lubricating film. Therefore surfaces in close proximity to each other, but separated by lubricated rolling or sliding bearings, will require less effort to be moved.

PLAIN BEARINGS

Plain bearings are referred to by numerous names. Sleeve, journal, bushed, or friction bearings are some of the common names used to describe plain bearings. The plain bearing is generally cylindrical in shape and designed to carry high loads at relatively low speeds. The original plain bearings were made from wood. Many different types of materials are used in current designs, some of which are discussed later in this chapter. Plain bearings are of a simple sleeve construction. Plain bearings can be mounted into a support block that is bolted to a base, or fit into a cylindrical bore of a machine. A portion of a shaft referred to as the "journal" rides on the bearing. Plain bearings provide a low cost, easily assembled, reduced friction bearing surface that supports rotating or oscillating shafts. Plain bearings are available in both metric and inch dimensional specifications.

Plain Bearing Application Concerns

Plain or sleeve bearings are selected and applied based on analysis of loads, speeds, operating conditions, size, and service requirements. Plain/sleeve bearings have numerous advantages. Generally sleeve bearings will handle relatively low to moderate loads that can be cyclic or shocking in nature. Most plain bearing applications are relatively slow speed. Certain types of adequately lubricated plain bearings will handle speeds in the thousand revolutions per minute range. The correct sleeve material can handle operating environments with extreme temperatures as high as 750°F or more. Abrasive and dusty applications are also a place where certain plain bearings are well-suited. Excessive moisture and exposure to corrosive liquids can be accommodated by sleeve bearings of the right material. Sleeve bearings are inherently quiet in operation because of the lack of rolling elements. Plain bearings are also suited for applications where a high level of service and maintenance might not be possible for reasons of accessibility or environment. Another advantage of plain bearings is they require minimal space and are relatively inexpensive. One of the major advantages of plain bearings is that they rarely fail suddenly and catastrophically. If all operating, load, speed, installation, lubrication, size, and environmental concerns have been addressed, the bearing will wear gradually. It is difficult to precisely predict the life of a plain bearing because it will see gradual wear until the amount exceeds allowable limits. The manufacturer of the machine that the bearing is installed in establishes these limits. Because of the above-mentioned facts, selection and performance criteria are based in part on history, risk factors, and the engineer's best judgment.

Precise alignment of the supported shaft/journal operating in most sleeve bearings is imperative. Shafting deflection and operating misalignment can be detrimental to the system and will limit the functional life of the bearing. If the bearing is required to operate with slight misalignment capability, allowances must be made in the design of the supporting bearing housing. This is typically accomplished by having a bearing cartridge or bearing outside diameter that is spherical in shape.

A major factor contributing to plain bearing operation is the quality of the surface finish of both the journal and the bearing. The rougher the finish the more lubricant film thickness will be required to separate the two surfaces. Roughly finished surfaces have numerous minute peaks and valleys of varying heights and distances between the peaks. These peaks and valleys are known as surface asperities. As the bearing operates it will wear-in over time as the surface aspirates are reduced in size and quantity by the smoothing action created by the rotation of the journal on the bearing. This wear-in or breaking in period, if short lived, can be a natural part of plain bearing function. If excessive due to poorly finished journals and bearing surfaces, elevated temperatures will result along with extreme wear resulting in a loss of bearing service life. It is imperative during this wear-in period that an adequate supply of lubricant is provided to the bearing.

Plain bearings will have a radial load capacity at various revolutions per minute for a particular shaft size. This capacity is listed in manufacturer's tables usually in pounds. Another one of the few valid performance and selection factors used to measure and select a plain bearing is the "PV" factor. "P" is the pounds per inch (psi) or pressure the bearing will be subjected to. P is equal to the load on the bearing in pounds divided by the projected area in square inches. For sleeve bearings, the projected area is length multiplied by the inside diameter. "V" is the velocity in feet per minute of the wear surface/shaft. Certain manufacturers of plain/sleeve bearings will publish maximum allowable PV factors for a specific material that can assist the selection of the bearing.

EXAMPLE

A 3/4″ diameter shaft turning at 341 revolutions per minute (rpm) has a load of 90 pounds on it. The bearing length is 1″. The velocity, in feet per minute (fpm), is equal to .262 × rpm × shaft diameter. Thus, .262 × 341 × .750″ = 67 fpm. P is equal to the total load divided by the projected area (area = .750″ × 1″ = .750″ sq). Thus, 90 lbs ÷ .75 = 120 psi. PV = 120 psi × 67 fpm = 8040 PV. A sleeve bearing material that is used in this application must have a rating of at least 8040 PV.

Plain Bearing Types

Plain bearings are available as solid (one-piece) and split (two-piece) types. There are numerous variations of both types and the terminology used to describe them is often used interchangeably.

Solid plain bearings are often referred to as bushed or sleeve bearings (see Figure 9–1). Bushed bearings are generally thin walled. This means that the inside diameter is close to the outside diameter, which results in a thin wall. Because the bushing is thin, the surrounding bearing housing must be strong to accommodate loads. Oftentimes the machine housing acts as the bearing support. Once excessive wear occurs within the inside diameter, the bushing's outside diameter

FIGURE 9–1
Solid plain
bearing/bushings
made from bronze.
Courtesy of Boston Gear.

FIGURE 9–2
Split plain bearing block
with a babbitt lining.
Courtesy of Rockwell
Automation.

typically can be easily pressed in and out of the supporting housing. Sleeve bearings are generally thicker walled than bushed types. Both sleeve and bushed bearings may have their own block housing, which will be mounted on or within a machine.

Split plain bearings are usually split in half (see Figure 9–2). Splitting the bearing allows for easy removal and installation in difficult locations. In certain applications, split bearings can be inspected, removed, and installed without removing the shaft or connected components. This is a cost and time saving feature. Generally matched and machined halves are assembled in pairs and should not be mixed with other halves. Some precision machined halves are available that are interchangeable because of the close manufacturing tolerances that are held during production. If the split sleeve bearing has match marks indicating correct order of assembly, they must be properly oriented during installation to ensure long service life.

Plain Bearing Materials and Properties

Plain/sleeve bearings are available in a wide range of metallic and nonmetallic materials. The selection of the materials is based on the same factors that are addressed in the preceding paragraphs. Adequate bearing material must have certain characteristics that will meet or exceed the requirements of the application.

A certain amount of trade-off is necessary and typically the material selected is a compromise of desirable characteristics. Important material characteristics are:

Score Resistance—Score resistance is that quality of a material which prevents damage to the journal/shaft during varied lubricant film conditions. The bearing must not seize or weld to the journal/shaft.

Compressive Strength—Compressive strength is the ability of the bearing material to carry the load without disintegration or excessive deformation.

Fatigue Strength—Fatigue strength is the ability of the material to give adequate service life when subjected to diverse stresses.

Deformability—Deformability is the characteristic of the material that allows it to yield slightly under normal operating loads without failing.

Conformability—Conformability is the ability of the material to wear away or conform to deflected loads without creating high temperatures.

Embeddability—Embeddability is a desirable quality in an abrasive environment. Foreign material must be allowed to embed into the bearing material, which is relatively soft as compared to the shaft. If the particle is jammed against the bearing and shaft it will eventually score both the bearing and shaft.

Corrosion Resistance—Corrosion resistance is the quality of the material to resist oxidation and deterioration of the material due to exposure to various fluids. These fluids can be moisture, chemicals, or acids and varnishes formed from lubricants.

Shear Strength—Shear strength is the ability of the bearing material to resist movement of the various layers relative to one another.

The following materials are a few of the common types used with plain bearings:

Babbitt Babbitt is an alloy of several types of metals such as tin, cadmium, copper, and lead, named for Isaac Babbitt, who patented it in 1839. Babbitt sleeve bearings are one of the most common materials employed in plain bearings. It is a relatively soft material that exhibits good embeddability characteristics. Scoring and scratching of shaft surfaces is minimized with the use of babbitt bearings. One of the main reasons babbitt bearings were used in early bearing applications was that they were easily repaired in the field. The repair was accomplished by removing the older, worn material and pouring a new liner into the shell/housing. The maximum operating temperature for babbitt-type bearings is approximately 180 to 200°F. The strength of most babbitt materials decreases as the operating temperature increases. Applications with light to moderate loads utilize babbitt bearings. Babbitt bearings are often bonded to an exterior shell of steel, cast, or bronze, which increases their fatigue strength capacity. Large electric motors will often use babbitt bearings to allow for free axial movement of the rotor and shaft.

Leaded, Aluminum, and Tin Bronze Bronzed, bushed sleeve bearings are suited for higher loads than babbitt types. In some cases they can accommodate twice the loads that a soft babbitt bearing will handle. Alloy bronzes have good shock resistant qualities and will operate in temperatures as high as 300 to 400°F. Speed limits are in the low to moderate range. Because bronze is harder than most babbitt materials, bronze alloys have a tendency to scratch the shaft if exposed to excessive abrasive contaminants. Lead–bronze sleeve bearings have fair embeddability characteristics and will operate properly in a moderately contaminated operation. Bronze alloy plain bearings have almost no conformability, which means shaft alignment is critical. Common tin bronzes include SAE 62, SAE 620, and SAE 63. Common leaded bronzes include SAE 660, SAE 67, and SAE 62.

Sintered Powdered Metal Sintered powdered metal bearings are available in a variety of materials including iron and bronze. Sintered bronze bushings have the characteristic of being porous. During the manufacturing process of sintered bearings, a powdered metal alloy is pressed under extreme pressure in a die to control the density. The bearing is then sintered at a specific temperature in a reducing atmosphere. This process creates a sponge-like structure capable of absorbing significant quantities of oil lubrication. Special oils are used that are nongumming. These bearings are referred to as oil-impregnated bushings. During operation, the heat created from friction and pressure between the shaft and bearing surface causes a capillary action to occur. This capillary action causes small amounts of oil to be released, creating a lubricating film. These bearings are used in locations where continuous lubrication cannot be provided. Additional lubrication or periodic flooding of the bearing with oil will assist in replenishing the oil supply. Avoid using soap-based grease, which may plug the pores of the bearing. Re-machining of these types of bearings should be avoided unless special controlled techniques are utilized, or the open pore structure will be smeared over. The maximum operating temperature for oil-impregnated sintered bushings is 200°F (93°C).

Carbon-Graphite Carbon-graphite or composite material bearings are becoming increasingly popular in adverse environment applications. They function in temperatures up to 700°F (371°C). Oftentimes carbon-graphite composite-type bearings can be found in ovens, dryers, and applications subjected to extreme temperatures but low loads because they are generally resistant to most chemicals and moisture. The carbon material essentially acts as a solid lubricant. Their chief disadvantage is they have almost no tolerance for operating in a dirty environment. Abrasive materials will quickly wear away the material. Another form of plain bronze bushing bearings will have plugs or grooves filled with a solid lubricant such as graphite (see Figure 9–3).

Plastics, Rubbers, and Synthetic Bearings A mixture of plastics such as polypropylene, polyethylene, nylon, and Teflon can be found in applications that are exposed to moisture and chemicals. Speed and loads must be kept minimal for most of these materials. Their chief advantage is they require no lubrication because of their very low coefficient of friction. In some cases water acts as the lubricating agent.

FIGURE 9–3
Graphite-plugged plain
bearing/bushing.

Plain Bearing Installation Considerations

Plain bearings should be inspected periodically for wear and damage to the load-bearing surfaces. It is also important to inspect the supporting housing or block that the bushing/sleeve is mounted in for cracks or damage. Any evidence of excessive internal clearances, wear, scoring, or damage to the bearing and shaft requires that replacement take place.

A major concern of plain bearing operation is wear. Due to the inevitable wearing away of material from lack of adequate lubricant film, the clearance between the shaft/journal and the bearing surface will increase. This clearance or air gap will have installation and operating recommended limits provided by the manufacturer or established by past experience. Some machine manufacturers will specify an initial installed clearance as well as a maximum allowable clearance due to progressive wear. This wear can be from loss of bearing material, shaft material, or a combination of both. Excessive clearance results in a lowering or positional change of the shaft axis relative to the geometric center of the bearing. This change can cause problems with alignment of connected equipment and motor rotor clearances, and yield vibration and out-of-tolerance production. With split-type plain bearings, the initial and ongoing clearance between the shaft/journal and bearing inside diameter can be checked by shutting down the equipment and employing the use of plastic gauge. Plastic gauge is a small diameter, crushable material that is placed the length of the shaft within the bearing. The bearing halves are reassembled and clamped tight. The bearing is opened again, the plastic gauge material is removed, and its thickness is measured. Feeler gauges can also be used on larger plain bearings to measure the gap between the shaft and bearing bore. A general rule of thumb is when the clearance is two to three times the original new installation, a bearing or shaft repair and replacement are required. It is important to note that the wear might not be even along the bearing surface due to off-center loads and misalignment. The degree and amount of allowable clearance will be decided by taking into account the manufacturer's recommendation, history, speeds, loads, and critical nature of the machine.

Bronze bushings and sleeves can be removed with a press or a bearing puller. Care must be taken not to damage the surrounding housing when removing the sleeve. Never hammer the bearing out of the housing. The press-out should be done in a clean, continuous movement. Care should be taken to observe any anti-rotation pins or dowels that hold the bushing into the housing.

The shaft/journal must be free of nicks, burrs, and scoring. Minor cleanup of the shaft may be done by using a light crocus cloth. The shaft/journal diameter should be measured to ensure a proper size and fit.

The replacement of the babbitt in a plain bearing is best left to the original manufacturer and qualified service shops specializing in the process. In past times, partly because of the quantity of plain bearings in use, it was a common practice to burn out the babbitt linings from used bearings and pour a new liner into the old housing. Once a new liner was poured and machined to the correct inside diameter, it was "scraped" to allow for proper shaft contact. Scraping involved the craftsman using special tools to scrape away material in specific areas on the bearing surface to yield the correct amount of shaft/journal contact. This contact was checked with dyes to reveal the degree of contact and contact pattern shape. The contact pattern is required to be fairly uniform and not carrot-shaped across the length of the bearing. This process of scraping is not an exact science and few people are qualified to perform the task. Improper scraping of a babbitt bearing can result in excessive removal of bearing material and premature failure of the liner.

For the most part, it is no longer cost effective for the plant maintenance technician to perform the functions of removal, pouring, measuring, machining, and fitting of the liner. Environmental, safety, and design considerations all must be taken into account when performing these tasks. Only through the use of computer-aided design and modern manufacturing technology, in conjunction with experience, can these processes be accomplished properly. Issues such as the preparation of the surfaces, the porosity of the cast backing, and the bonding of a rebabbitted liner all need to be addressed. The shape of certain babbitt bearings involved in moderate to high speed applications is not necessarily perfectly round. Allowances have been made to accommodate lubricant film and lift of the shaft at operating speeds. Manufacturers have the original dimensional specifications and tolerances. Only the original manufacturer of the babbitt bearing and certain qualified repair agents should attempt to repair babbitt bearings.

Installing sleeve bearings into the machine housing or supporting block should proceed after the housing bore has been measured. Measuring the bearing outside diameter and comparing it to the housing bore measurement will ensure a proper tight fit. Generally the sleeve bearing will require a press fit into the housing and a relatively loose fit to the shaft. The bushing will be manufactured to a size slightly larger than the nominal outside diameter by approximately .001″. This slight oversizing of the bearing outside diameter guarantees that the bearing will not spin or come loose in the housing. If the housing is undersized by an extreme amount, it will have the effect of crushing the bushing. A certain amount of "close-in" will occur during the press. Close-in is the reduction of the inside diameter during the fitting that is determined by factors such as wall thickness, press fit tolerances, and material characteristics. Excessive close-in may require that the inside diameter of the bearing be re-bored.

A mechanical arbor press or hydraulic press may be used for the pressing operation. If the bushing does not enter the housing bore squarely and aligned, it may

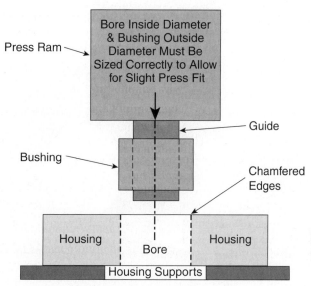

Press Ram

Bore Inside Diameter
& Bushing Outside
Diameter Must Be
Sized Correctly to Allow
for Slight Press Fit

Guide

Bushing

Chamfered
Edges

Housing

Bore

Housing

Housing Supports

FIGURE 9–4
Proper setup for pressing a bearing/bushing into the bore of a housing that is correctly
supported. Bushing must be centered with the housing.

be damaged or a portion of the material may be shaved off. By chamfering the edge
of the housing bore a smoother entry can take place. Mating surfaces must be free
of nicks, burrs, and defects. The depth of the housing bore and the length of the
bushing should match. Do not overpress the bushing into the bore or it may swage
the end of the bushing. Never use a metal hammer to directly strike the bushing.
Always use a guide and proper fitting tools when installing the bushing to avoid
damaging the bearing and causing a potential safety hazard (see Figure 9–4).

When assembling split-type plain bearings, observe and align all match marks.
Install any anti-rotation pins, keys, or dowels that are required to hold the liner in
place. Usually these split liners are precision machined to ensure a proper fit in the
housing. Certain split plain bearings are designed to have a slight amount of
"crush" and/or "spread." Crush is a term used to describe an extra amount of lining
material that makes each insert half slightly larger than a half circle. Spread is a
minor oversizing of the diameter of the insert across the open-end width (see
Figure 9–5). Both of these conditions are usually associated with thin-walled sleeve
bearings. This crush and spread ensures a proper fit.

Plain Bearing Lubrication

Proper lubrication is essential to plain bearings to yield adequate life. Lubrication
provides a film barrier between mating parts reducing friction, dissipating heat, and
preventing wear. The lubrication of plain bearings may be hydrodynamic and/or

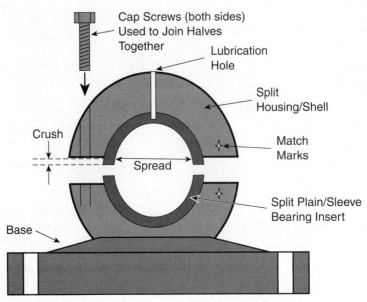

FIGURE 9–5
This drawing illustrates the concepts of "crush" and "spread" in a split plain bearing to ensure proper fit to the shaft.

hydrostatic (see Lubrication chapter). Hydrodynamic lubrication relies on the formation of a continuous fluid film under sufficient pressure and the rotating shaft to prevent any surface contact. With hydrostatic lubrication, the film is maintained by external pressure from an outside source such as a lube pump.

During the operating cycle of a plain bearing—from static to full operational speed—it will progress through several different lubricated conditions. When the shaft is at rest, the weight of the load and journal squeezes out most of the lubricant. At this stage metal-to-metal contact is pronounced. This can be avoided by using a lubricant lift pump to place the bearing in a hydrostatic state. Without the aid of external pressure only a boundary lubrication condition exists—only a minimal film of lubricant separates the shaft and bearing surface. As shaft revolutions increase, a mixed film condition will occur. Mixed film lubrication is a state between boundary film and full-film lubrication. Once the bearing reaches its full operational speed, a full-film mode takes over with a layer of lubrication of adequate thickness to prevent metal-to-metal contact. Full-film thickness is an ideal state that requires less energy to rotate, supports full loads, and generates minimal friction. The lubricant film thickness is determined by numerous factors, such as oil viscosity, temperature, speed, and load. Of course the film thickness and the preferred operational state of full-film lubrication can occur only if these issues are correctly addressed.

Plain/sleeve bearings can be classified by their method of lubrication. Class 1 bearings are those which require oil, grease, or other lubricating agents to operate.

The lubricant is delivered from an outside source, such as an oil-drip system or periodic shots of grease from a grease gun. Class 2 bearings are those that contain a lubricant within the walls of the sleeve. Oil-impregnated sintered metal bearings are an example of this type. Class 3 sleeve bearings are those that are in themselves a lubricant. Sleeve bearings such as carbon, graphite, and Teflon are examples of class 3 type.

It is a common practice with plain bearings to machine or cut internal grooves on the bearing surface (see Figure 9–6). Grooves are generally required for oil lubricated bearings with a length-to-diameter ratio of more than 1:1 and grease lubricated bearings with a ratio over 1.5:1. These grooves provide a path for the lubricant over the entire load bearing area as well as a reservoir during shutdown. The correct groove is the smallest size that directs adequate oil into the load zone. Some bushings will use multiple grooves and are available in diverse shapes. In some small, lightly loaded bushings a single oil hole in the low-pressure area provides an entry point for oil to be distributed around the bearing and shaft. This oil will only flow axially approximately 1/2″ from the hole in either direction. If the bearing sleeve is longer than 1″ a groove may be necessary. If the grooves are not properly located with relation to high-pressure and low-pressure areas, the load capacity of the bearing can be significantly reduced. Groove design must carry the grease over the entire width of the bearing. The groove should be placed so that the shaft makes contact with the lubricant. The edges of the groove should be rounded off so they do not scrape away the grease from the journal, and the depth of the groove should not exceed one-third the sleeve wall thickness.

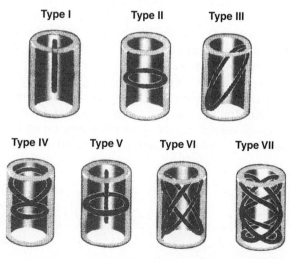

FIGURE 9–6

A variety of lubrication grooves in plain bearings/bushings machined to allow the free flow of a lubricant.

Courtesy of Boston Gear.

ANTIFRICTION/ROLLING ELEMENT BEARINGS

Antifriction bearings use some form of rolling element, such as balls or rollers, to carry the load and reduce friction between two surfaces. They are also referred to as rolling element or rolling bearings. The purpose of antifriction bearings is similar to that of plain bearings in that they provide an extremely low friction support for rotating shafts. Because of their low friction design, power requirements to move or rotate an object are less than that of plain bearings; the result is an increase in over-all efficiency within the machine.

The rolling elements can take numerous different shapes such as balls, taper rollers, spherical rollers, cylindrical rollers, and other forms (see Figure 9–7). Rolling element bearings are available in a number of variations, which meet the needs of specific application requirements. The characteristics and properties of a given type of bearing depend on its design. For example, a spherical roller bearing has the capacity to handle significant radial loads as well as being capable of handling modest amounts of misalignment. Whereas a straight roller bearing is also capable of handling heavy radial loads, but it will not operate properly in a misaligned condition.

Many factors must be considered when selecting an antifriction bearing for a particular application. Load, speed, dimensional size, life requirements, lubrication, and numerous other issues must be considered during the selection process. For example, is the selected bearing required to permit axial displacement or permit angular movement? The actual selection of a bearing to be used in a particular machine is best left to qualified bearing engineers that rely on calculations and experience to correctly choose a bearing.

The basic parts of an antifriction bearing are the inner ring, outer ring, rolling element, and a retainer (see Figure 9–8). The inner ring will typically be mounted onto a shaft and have a race groove or raceway on its inside diameter. This raceway provides a smooth finished path for the rolling elements. The outer ring will be mounted into a machined housing or machine bore. It will also have a corresponding internal raceway for the rolling elements. The rolling elements will be separated by a cage or retainer to guide, position, and separate the rolling elements. The cage can be manufactured in materials such as steel, brass, and polyamides.

Antifriction bearings can be mounted directly into the machine's housing, such as those supporting the shaft of an electric motor. They also can be placed into a block or housing that will be bolted to a surface. This type of bearing is often

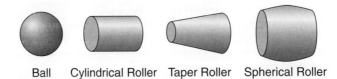

Ball   Cylindrical Roller   Taper Roller   Spherical Roller

FIGURE 9–7
Common rolling element shapes used in antifriction bearings.

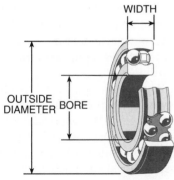

WIDTH

OUTSIDE
DIAMETER BORE

**Self-Aligning
Ball Bearing**

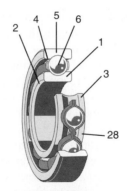

2 4 5 6

1

3

28

**Single Row Deep
Groove Ball Bearing**

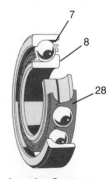

7

8

28

**Angular Contact
Ball Bearing**

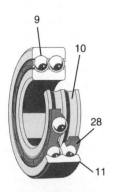

9

10

28

11

**Double Row Deep
Groove Ball Bearing**

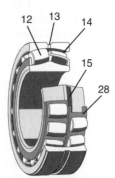

12 13 14

15

28

**Spherical
Roller Bearing**

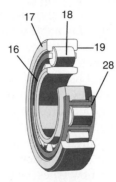

17 18

16

19

28

**Cylindrical
Roller Bearing**

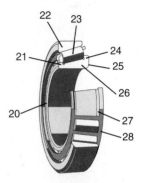

22 23

21

24

25

26

20

27

28

**Tapered
Roller Bearing**

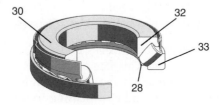

30

32

33

28

**Spherical Roller
Thrust Bearing**

1. Inner Ring
2. Inner Ring Corner
3. Inner Ring Land
4. Outer Ring Land
5. Outer Ring
6. Ball
7. Counter Bore
8. Thrust Face

9. Outer Ring Raceway
10. Inner Ring Raceway
11. Outer Ring Corner
12. Spherical Roller
13. Lubrication Feature
    (Holes and Groove) (W33)
14. Spherical Outer Ring Raceway
15. Floating Guide Ring
16. Inner Ring Side Face

17. Outer Ring Side Face
18. Cylindrical Roller
19. Outer Ring Flange
20. Cone Front Face
21. Cone Front Face Flange
22. Cup (Outer Ring)
23. Tapered Roller
24. Cone Back Face Flange

25. Cone Back Face
26. Cone (Inner Ring)
27. Undercut
28. Cage
30. Face
32. Shaft Washer (Inner Ring)
33. Housing Washer (Outer Ring)

FIGURE 9–8
Bearing terminology.

Courtesy of SKF.

referred to as a housed or mounted bearing. Mounted or housed bearings are available in pillow blocks, cartridges, flanged blocks, take-up units, and various configurations that allow it to be bolted to a structure outside of a machine. For the sake of discussion, the description of bearings later in this chapter is separated into non-housed and housed antifriction bearings.

Bearing rings are typically made out of high carbon, chrome alloy steel that has excellent fatigue strength and wear resistance. SAE 52100 steel is the dominant bearing steel. A variety of newer steel alloys of varying levels of hardness have been developed to meet the demands of specific industries, such as papermaking. The single overriding issue, regardless of hardness or composition, is that the steel must be clean. Impurities known as inclusions within the steel can result in premature failure and inadequate service life. Inclusions represent weak points in the steel structure and lead to fatigue after a certain number of stress cycles. Steelmaking and refining technology has made great strides in recent years in providing extraordinary pure steel for the production of bearings. The steels used in bearings posses high strength, good hardenability, favorable wear properties, and dimensional stability.

Ceramics of various types have been recently introduced as a material for bearing component parts. Bearings with ceramic rolling elements and steel rings are referred to as hybrid bearings. Ceramics offer numerous advantages such as reduced weight; nonconductive, inert properties; and reduction in rolling friction, to name just a few. Their chief disadvantage is cost and availability as compared to traditional bearing materials. As the demand for lightweight, efficient bearings increases—such as those required for the needs of the aerospace industry—and manufacturing technology improves, production costs will decrease. Bearing manufacturers continue to develop new types, materials, and designs to meet the increasing demand for frictionless movement within a machine.

Antifriction Bearing Loads, Speed, and Life

Calculating the varying loads/forces that a bearing may be subjected to is a complex task. Although specific loading calculations are outside the realm of this book, it is important for the maintenance technician to have a basic and practical understanding of applied loads and how bearings react to those loads. Bearings react to loads based on the direction, magnitude, and the point of application of the force, along with the geometry of the bearing. The geometry or shape of the rolling elements and races will dictate how the bearing handles the load. Controlling the geometry of the races and rolling elements redistributes the stress concentrations and provides for uniform distribution of applied loads. Rolling element bearings, the ball, and the rollers carry the load in a similar basic manner. The load passes through the shaft to the inner ring, through the rolling elements to the outer ring, and into the housing. In some applications this condition can be reversed, such as those with revolving housings. In a pure radial application less than 50% of the rolling elements will be sustaining the load. There is a distinct difference in the manner in which a ball reacts and handles a load versus the way a roller handles a load. A ball bearing has what is known as "point contact." The point of contact where the ball touches the race under load is actually a small elliptical area. This small area of contact allows for less rolling friction than a roller. It also gives the bearing the capability to handle minimal misalignment within the system. A cylindrical roller bearing has what is known as "line contact," which is the contact area between the race and roller. The result is a higher load-carrying capacity than a ball bearing, size-for-size, but an inability to handle misalignment. Bearings are selected in part as to how they will react to the applied loads specific to the application and also based on their overall dimensional size. The load-carrying capacity is directly related to its size. Generally speaking, the bigger the bearing and more rolling elements it has, the more load it will carry.

Bearings are required to handle three major types of loads: radial loads, which are loads that are at a right angle to the shaft axis; axial or thrust loads, which are exerted parallel to the shaft axis; and combined or angular loads (see Figure 9–9). Most loads in real life applications are combined loads. Seldom are loads purely radial or axial. For example a bearing supporting a gear shaft within an enclosed gear drive will be subjected to the radial loads created by the weight of the shaft and gear, as well as the axial loads caused by the mesh of the helical gears and thermal expansion of the steel.

Exact calculation of bearing loads requires data on all applied forces, vectors, and their magnitude, along with the weight of the supported components. It is usually a sum of weights and tangential forces. Sometimes it is desirable to calculate the approximate basic bearing loads created by connected power transmission components such as a V-belt drive. Figure 9–10 illustrates an application with the force or load applied between the bearings, commonly referred to as "straddle mounting." An example of this type of application would be a sheave supported between two bearings. The radial load is a combination of sheave weight and belt tension. The bearings must react to the radial load in the opposite direction of the applied load. Figure 9–10 also illustrates pure radial loads in an overhung mounting application where the load is applied outside of the two bearings. Because the load is applied outside of the supporting bearings in an overhung mounting application, the bearing nearest to the load will be supporting a load greater than the original radial load and will react to it in the opposite direction. On overhung applications the bearing farthest from the load is smaller and must react in the same direction as the original applied load. These drawings and examples illustrate the importance of the amount of load, as well as the relationship between the

FIGURE 9–9
Basic bearing load directions.

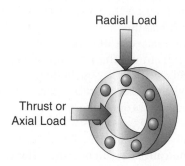

Radial Load

Thrust or
Axial Load

Straddle and Overhung Mountings

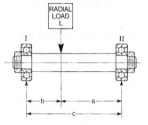 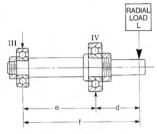

Straddle Mounting	Overhung Mounting

Radial load on Brg. I: $\dfrac{L \times a}{c}$ Radial load on Brg. III: $\dfrac{L \times d}{e}$

Radial load on Brg. II: $\dfrac{L \times b}{c}$ Radial load on Brg. IV: $\dfrac{L \times f}{e}$

Example

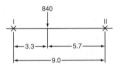

Radial load on I: $\dfrac{840 \times 5.7}{9.0} = 530 \text{ lbs.}$ Radial load on III: $\dfrac{840 \times 3.9}{6.4} = 510 \text{ lbs.}$

Radial load on II: $\dfrac{840 \times 3.3}{9.0} = 310 \text{ lbs.}$ Radial load on IV: $\dfrac{840 \times 10.3}{6.4} = 1350 \text{ lbs.}$

(530 + 310 = 840 check)   (1350 − 510 = 840 check)

FIGURE 9–10
The above drawing illustrates the two common types of loads applied to a two bearing/shaft assembly. An example is given for calculating "straddled" and "overhung" loads.
Courtesy of MRC Bearing Service.

location in which the load is applied relative to the position of the supporting shaft bearings. It is imperative that the maintenance technician comprehend that the placement, size, and weight of the component that is being installed will have an effect on the load on the supporting bearings. These loads, if excessive, will result in significantly less life.

The life and operating reliability of the bearing is related to the bearings load rating and the forces that are imposed on it. Numerical values are used in the calculation of life that expresses the load-carrying capacity of the bearing. Load ratings of bearings are grouped into two categories: dynamic and static. The manufacturer of a bearing will provide these ratings in Newtons or pounds in tables found in their catalog. Many of the charts will list a rating at various speeds resulting in a certain number of hours of life. The basic dynamic rating is used in calculations for rotating and dynamically stressed bearings. The rating is a value that will give a basic rating life of one million revolutions. The basic static load rating is used in calculations for stationary, slowly rotating, or oscillating movements. The static rating must be taken into account when heavy shock loads act on the bearing.

Rolling bearings all have speed limits based on such factors as design, internal clearances, material, loads, lubrication, and cooling capacity. The speed limits for most bearings are generally related to the permissible operational temperature limits of the lubricant being used and bearing component material limits. Higher speed usually means an increase in temperature within the bearing. The ambient operating environment temperature and the frictional heat generated within the bearing combine to produce temperature rise. At high speeds, such as those above typical electrical motor speeds, it is preferable to have some means to carry away the generated heat. This can be accomplished through the use of circulating oil systems. Generally, bearing speed limits are higher for the same bearing lubricated with oil versus grease. Bearing speed limits should be checked when designing and applying bearings.

The life of a bearing is a function of two groups of variables. One group represents the conditions of the application such as load, speed, temperature, installation methods, maintenance, and lubrication. The other group includes the bearing design, material, and manufacturing methods. In the real world of man and machine, it is difficult to predict the life of a particular bearing with accuracy due to the numerous variables. However it is possible to predict or calculate the life of the majority of a given group of identical bearings operating under the same conditions from statistical sampling. The life of a rolling bearing is defined as the number of revolutions or a number of operating hours at a given speed that the bearing is capable of enduring before the first sign of fatigue (flaking, spalling) occurs on the race or rolling elements. The basic rating life of the bearing is the number of operating hours of life that 90% of a large group of identical bearings operating under identical conditions, properly installed, are expected to attain or exceed. This life, which can be calculated, is referred to as the L_{10} life of the bearing and is due to the fatigue of the steel (see Figure 9–11). Bearing steel fatigues, in part, because of the high cyclic loading stresses imposed on it in a typical application. Therefore the theoretical life of the bearing is related to loads, speeds, steel fatigue strength, application, and design.

The service life of the bearing is the actual life of a specific bearing up to the point of failure. The service life is determined by variables such as installation,

FIGURE 9–11
A simplified L_{10} bearing life calculation formula.

$$L_{10} \text{ Life hours} = \frac{\left(\dfrac{C}{P}\right)^{B} \times 10^{6}}{\text{RPM} \times 60}$$

C = Bearing basic dynamic load rating

P = Equivalent dynamic bearing load

B = Exponent of life equation
    B = 3 for ball bearings
    B = 10/3 for roller bearings

lubrication, operating environment, contamination, sealing arrangements, alignment, etc. If these application variables are closely controlled and the bearing is properly selected, it is possible to see nearly infinite life with the modern rolling element bearing.

Antifriction Bearing Tolerances and Precision

Antifriction bearings are generally made to very exacting dimensional tolerances and are considered an extremely precise machine component. In fact, rolling element bearings and their parts are one of the most exact machined components made by man outside of the world of nanotechnology. The steel ball used as a rolling element can be made in batches to a uniform diameter within 0.00001″ of each other for size and within 0.000005″ for roundness. This is a nearly perfect sphere. Dimensional tolerance is a prescribed allowable deviation from a target dimension. Precision is the quality of being exact. Holding close tolerances and making precise bearings yields increased accuracy of the mating parts, reduced friction, quiet operation, better fits, and smooth running performance.

It is imperative that the maintenance technician and engineer understand that the use of precision bearings within a machine makes sense only if the surrounding housing and shaft are correspondingly exact. Factors such as shaft and housing roundness, surface quality, shaft position, fitting tolerances, and installation methods all play a role in yielding overall accuracy of the bearing arrangement.

Organizations such as the ABMA, ABEC, RBEC, DIN, and ISO have determined acceptable tolerances for inner ring diameters, outer ring diameters, and ring widths, along with radial and axial run-out limits. Inner ring and outer ring diameter deviations are kept within minute tolerances. Bearings are manufactured and grouped into various classes of precision by their exactness (see Figure 9–12). Higher precision bearings are generally defined by tighter dimensional tolerances

Bearing Precision Classes

Precision Level	ABMA	ISO	NSK (ISO)	NSK (ABEC)	NSK Miniature	ABMA Roller	ABMA Tapered
Lowest	ABEC 1	Normal	P0	PA 1	–	RBEC 1	Class 4
	ABEC 3	Class 6	P6	PA 3	3	RBEC 3	Class 2
	ABEC 5	Class 5	P5	PA 5	5 P	RBEC 5	Class 3
	ABEC 7	Class 4	P4	PA 7	7 P		Class 0
Highest	ABEC 9	Class 2	P2	PA 9	9 P		Class 00

FIGURE 9–12
Chart showing bearing classes of precision and their alphanumeric designations for various organizations.
Courtesy of NSK.

and increased running accuracy. Classes of precision have been designated with a numbering system such as the ABEC classes 1, 3, 5, 7, and 9. ABEC class 1 bearings are considered an acceptable tolerance level for normal usage. Depending on the application requirements various degrees of accuracy may be required. Higher precision bearings such as ABEC 3, RBEC 3 (ISO P6), and higher are used for applications where extreme precision is required, such as machine tool spindles and measuring instruments. Bearings are made precisely to assist in achieving running accuracy of the machine. As an example of how precisely made a standard ABEC class 1 bearing is, the inner ring tolerance for a 25 mm (.9843″) bore ball bearing will have a tolerance of plus 0.000 mm (0.000″) to minus 0.010 mm (0.0004″).

High precision bearings will often have the high point of eccentricity marked on the face of the ring with a small dot. This dot allows the installer to offset the marks of the paired bearings to yield the least amount of radial run-out. On applications that call for extra precision, the shaft run-out must also be taken into account.

Bearing dimensional tolerances are difficult to measure without digital, calibrated, certified, precision measuring instruments in a clean and controlled environment. Because manufacturers of precision brand bearings have intense inspection procedures and carefully control the manufacturing process, there is no practical need for the maintenance technician to precisely measure the bearing prior to installation.

Nonhoused Antifriction Bearing Types

For the sake of easy classification, rolling element bearings can be considered nonhoused if they do not come as a unit with a supporting housing. Although all bearings will be placed into a machine housing prior to operation, many are available as a "naked" bearing without a supporting block/housing. Some of the most common configurations available are the deep groove ball, single row ball angular contact, thrust ball, thrust roller, spherical roller, cylindrical roller, taper roller, and needle roller. The various terms and bearing component descriptions are illustrated in Figure 9–8. The basic configuration for all nonhoused antifriction bearings is similar. They will consist of an inner and outer ring of hardened steel that will have a machined raceway for the rolling elements to travel in or on. In addition, the rolling elements of a particular shape will separate the rings and be held in place by a retainer or cage. The cage material will usually be steel, brass, or plastic (polyamide).

Deep Groove Ball Bearings The single-row, deep groove ball bearing is available in two basic configurations, the Conrad type and the maximum type. The Conrad type is designed to handle light to moderate loads in the radial direction at relatively high speeds (see Figure 9–13). It is capable of handling light axial or thrust loads. It is not considered a self-aligning bearing although it will handle minute amounts of misalignment. The maximum type single-row, deep groove ball bearing is similar in construction and dimensions to the Conrad type except the

FIGURE 9–13
A single row deep groove radial
ball bearing.
Courtesy of SKF.

rings have a filling slot machined into their sides that allows the installation of an
additional ball. This gives the maximum type ball bearing additional radial load carrying capacity over the Conrad type.

Radial ball bearings are also available in a commercial grade or unground
version. The commercial grade bearings have less precise tolerances and race surfaces that have not been ground and finished. These unground radial ball bearings
are inexpensive and used where speeds and loads are low and precision is not
required.

Radial ball bearings are available in sealed or shielded versions. The seals and
shields provide protection against foreign matter such as dirt and water from entering the bearing. Contamination of the bearing will result in premature failure. Seals
and shields also assist in retaining lubrication, which is placed within the bearing at
the time of its manufacture. An assortment of different types of seals and shields
made from different materials is available (see Figure 9–14).

Seals have a significant difference from shields. Seals generally make contact
with a portion of the inner ring, whereas shielded bearings are noncontact. Sealed
bearings will have a synthetic material clad over a metal disc that is pressed into a
small groove in the outer ring. A contacting lip rides on the outer edge of the inner
ring. Seals are better suited for lubricant retention and contaminant exclusion. The
downside to sealed bearings is their potential to create higher temperatures from

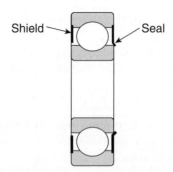

Shield — Seal

FIGURE 9–14
Cross section of a ball bearing showing seal and shield. Notice the shield has a small gap
between the inner ring of the bearing and the shield edge. The sealed side of the bearing
has no gap because the surface of the rubber seal rides against the inner ring.

friction due to the rubbing action against the ring surface. They also cannot be relubricated unless a pinhole for lubricant entry is drilled in the outer ring that corresponds with a lubricant groove and fitting in the housing. The bearings that cannot be relubricated are referred to as "sealed for life" radial ball bearings.

Shielded antifriction bearings utilize a metal shield, with no rubbing or contacting parts, that is fitted into the outer ring and has minimal clearance above the inner ring. It can take the form of a single plate or a labyrinth with multiple channels and plates. Shields exclude large contaminants, allow oil to flow into the bearing, and can operate at high speeds without friction-generated heat.

Wide Inner Ring Ball Bearings A version of the deep groove, single-row ball bearing is the wide inner ring type (see Figure 9–15). It is designed to accommodate the same radial, axial, and combined loads as the single-row ball bearing. The outer ring shape can be spherical or cylindrical. The spherical-shaped outer ring is mounted into a corresponding spherical seat to allow for misalignment. The cylindrical extended ring bearing can be pressed into a straight bored housing or tube. The extended inner ring ball bearings are often used in housed units such as ball bearing pillow blocks or flanges. Their inner rings accommodate setscrews or locking collars to mount onto straight shafts and can be positioned without the aid of shoulders, adapters, or lock nuts. The most common type will use two setscrews placed approximately 120° apart through the inner ring for effective holding force. These setscrews hold the ring onto the shaft, preventing unseating of the bearing. Another version uses concentric squeeze collars or eccentric locking collars. The concentric type will often rely on the concentric collar to squeeze the inner ring against the shaft to produce a 360° clamping force. An eccentric self-locking collar works as a locking cam against the shaft (see Figure 9–16).

FIGURE 9–15
A cutaway drawing of a wide inner ring ball bearing with a setscrew for locking the bearing to the shaft .
Courtesy of SKF.

FIGURE 9–16
A cutaway drawing of a wide inner ring ball bearing with a setscrew and an eccentric locking collar for locking the bearing to the shaft.
Courtesy of SKF.

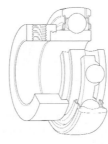

FIGURE 9–17
Angular contact ball bearing. Notice
the shape of the rings.
Courtesy of SKF.

Single Row Angular Contact Ball Bearing The single-row angular contact ball bearing is designed to accommodate combined radial and axial (thrust) loads (see Figure 9–17). The raceways are arranged so that the applied forces are transmitted from one race to the other at a predetermined contact angle. Because of the shape of the raceways, the bearing can tolerate significant axial loads in one direction. This is accomplished by combining high ring shoulders on the thrust side of the bearing with high operating angles. These bearings are marked on the rings with a "Thrust here" note indicating the proper position relative to the thrust load. Many angular contact bearings are used in pairs. The ring faces will be machined extremely flat; bearings that have undergone this process are referred to as "flush-ground." Flush-ground bearings are intended for mounting in tandem. If they are mounted in a tandem opposing position (back-to-back, face-to-face), they will be capable of tolerating thrust loading in either direction. Flush-ground angular contact bearings are referred to as duplex pairs. Various mounting arrangements include back-to-back, face-to-face, or tandem (see Figure 9–18). When mounted alone or in pairs, care must be taken to properly orient the bearings to handle the thrust loads inherent in the application. When installed as a single piece or as a duplex pair, adjustment is required to clamp the bearing axially. Certain high precision flush-ground bearings will have rings that have been intentionally machined slightly offset to one another. When the rings are clamped with the faces parallel to one another, they become a rigid assembly (preload). Because the bearing is relatively loose axially prior to mounting, some means must be provided to move the rings axially into the correct position relative to one another to prevent excessive internal looseness or excessive preloading. Common methods of doing this are through the use of springs or shims, or specific torque adjustment of an assembly nut. Applications such as pumps and machine tool spindles are common for angular contact ball bearings.

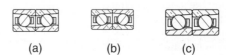

(a)   (b)   (c)

FIGURE 9–18
(a) Back-to-back, (b) Face-to-face, and (c) duplex angular contact bearing arrangements.
Courtesy of Torrington.

Figure 9–19
Self-aligning double row ball
bearing.
Courtesy of SKF.

Self-Aligning Ball Bearing The self-aligning ball bearing was one of the first types of antifriction bearings used in industry. It consists of two rows of balls retained within a cage, separating an inner and outer ring (see Figure 9–19). By machining the outer ring race surface in a spherical shape the bearing is capable of handling small amounts of tilting or misalignment. It is placed in applications that require high speeds and moderate radial loads that are subjected to misalignment.

Double-Row Deep Groove Ball Bearings Double-row ball bearings that are not self-aligning are manufactured to accommodate higher radial loads than single-row ball bearings (see Figure 9–20). As the name implies, they have a double row of balls. They are capable of handling low to moderate amounts of axial load, compared to single-row ball bearings. Because of their race design and construction they are essentially angular contact bearings.

Pure Thrust Rolling Element Bearings Pure thrust bearings are designed to accommodate only axial/thrust loading. Rollers or balls can be used in thrust bearings. Thrust bearings are used in applications for shaft guidance and where extreme axial forces are present. There are two basic types: the single acting and the double acting. The single-acting bearing consists of two grooved washers (rings) and a set of rolling elements within a cage (see Figure 9–21). The double-acting type, which is

Figure 9–20
Double-row angular contact ball
bearing.
Courtesy of SKF.

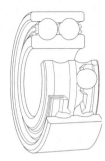

Figure 9–21
Single-acting thrust ball bearing.
Courtesy of SKF.

FIGURE 9–22
Cylindrical roller bearing.
Courtesy of SKF.

used for reversing axial forces, consists of two housing washers (rings), two rolling element sets with cages, and an intermediate shaft washer. Both flat or grooved washer races are available.

Cylindrical Roller Bearings Cylindrical or straight roller bearings use a symmetrically shaped rolling element in the shape of a cylinder (see Figure 9–22). This type of rolling element bearing uses line contact along the roller and race surface to carry the load. This design provides a significantly higher radial load capacity, size-for-size, than a ball bearing. They are available in separable inner or outer ring versions. Many types use a flange or rib on one of the rings to help guide and axially position the rollers. Any axial loads, in the wrong direction, will push the rollers into the guiding flange/rib and create metal-to-metal contact that will generate friction and heat. Because of this, it is recommended that cylindrical roller bearings be used in conjunction with a different type of bearing that is capable of handling axial forces. The chief advantage of this type of bearing is that it is capable of high radial loads. Versions with one ring being nonflanged can accommodate small amounts of lateral travel of that ring without a significant decrease in load-carrying capacity.

Spherical Roller Bearings Spherical roller bearings have barrel-shaped rollers used as the rolling elements (see Figure 9–23). The shape of the rollers and races is designed to handle high radial loads and moderate amounts of axial loads. It is capable of operating in a misaligned condition up to 1 or 2 degrees without reduction in load carrying capacity. Two-row and single-row versions are made. Cages and ring guide flanges will guide the rollers between the rings. The capability to handle misalignment is advantageous when shaft deflection or housing misalignment is inevitable. Many configurations, such as tapered bores and special cages, are available to suit the needs of the application. This bearing is used extensively in applications such as papermaking machines, conveyor pulley shaft bearings, and large gearboxes.

FIGURE 9–23
Spherical roller bearing.
Courtesy of SKF.

FIGURE 9–24
Taper roller bearing.
Courtesy of SKF.

Taper Roller Bearing Taper roller bearings are one of the workhorses of the bearing industry (see Figure 9–24). Usually taper roller bearings are available as two separate parts, the outer ring/race known as the "cup" and the inner ring/race with the rollers and cage, known as the "cone." The roller shape is a straight-sided, tapered roller. The rollers and races are tapered and if a line is extended beyond the contact line, they will converge at a common point on the axis of rotation (see Figure 9–25). This "on-apex" design means that any point along the cup, cone, or roller is subjected to the same circumferential speed. The benefit of this design drastically reduces any sliding or skewing of the rollers during operation. Skewing

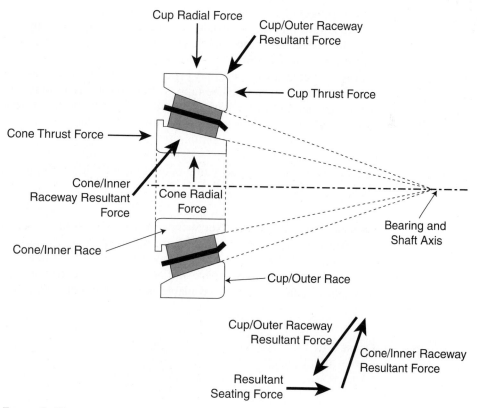

FIGURE 9–25
The drawing illustrates the taper roller "on-apex" design and how it works under load/force.

FIGURE 9–26
Two-row taper roller assembly.
Courtesy of SKF.

and sliding detracts from the satisfactory performance and load capacity of the bearing. The tapered configuration produces a "seating force" that pushes the roller against a rib or flange on the inner race. The seating force is a function of the different angles of the outer and inner races and also prevents roller skew.

Taper rollers are capable of handling significant amounts of radial loads combined with axial loads. In fact, they require some axial loads to function properly. The line contact pattern and roller require some axial force to prevent excessive internal clearance and edge loading of the rollers. Edge loading is when a small portion (edge) of the rolling element is carrying the load, which will result in premature failure. Significant misalignment cannot be tolerated by these bearings.

The internal clearances of these bearings can be adjusted by the placement of variable amounts of shims, or tightening a clamping nut to a specific torque. This operational clearance (end-play) is established by the manufacturer and by the nature of the application. The taper roller bearing functions optimally at a near zero clearance or with a slight preload. Care must be taken to consult machine manuals when setting the proper operational clearance/preload limits or bearing failure will occur. Two-row versions are available as sets with double cups and single cones, or double cones and single cups (see Figure 9–26). These assemblies can be made with a center spacer ring made to a specific width and tolerance that will yield a predetermined end-play.

Needle Roller Bearings Needle bearings are a form of cylindrical roller bearings (see Figure 9–27). Roller diameters are relatively small in relation to their length, typically 1/4″ or smaller. The chief advantage of needle roller bearings is that they can be installed in a small space and are capable of carrying high radial loads for their size. Needle bearings are available as radial types designed to handle pure radial

FIGURE 9–27
Needle roller bearing.
Courtesy of SKF.

Figure 9–28
Thrust needle roller bearing.
Courtesy of SKF.

loads, and as thrust types designed strictly for pure axial loads (see Figure 9–28). The needle rollers are contained within a drawn cup or outer ring/housing. This housing contains the rollers and acts as both a housing and outer ring, allowing the bearing to be installed in a compact space where it is not practical to harden and grind the machine's housing bore. Typically they are installed in machine housings that have low rigidity such as cast iron, aluminum, or plastics. Needle bearings are made in full complement or caged forms. Full complement types have no cage to separate the rollers. Cageless types have high radial load capacity compared to the caged form. The increased friction caused by contact with the adjacent roller limits their speed. Caged needle roller bearings have fewer rollers than a cageless type, which permits higher speeds but lower radial loads. Hardened inner rings are available for the bearing rollers to operate on, but most use the hardened shaft as the inner ring. It is recommended that the shaft be hardened to a recommended value of Rockwell 60C-58C.

Cam Follower Bearings Cam follower bearings are a type of needle roller bearing with a thicker cross-section outer ring. This outer ring is typically flat but may have a V shape to the outer surface. Cam followers are designed to run on or in a guide or channel. This heavy-duty needle roller bearing is available with an extending stud for cantilever mounting (see Figure 9–29). It is also available with a hollow bore used for yoke mounting. Slight crowning of the outer race compensates for modest amounts of misalignment and prevents edge loading of the rollers and races. Cam followers are designed to roll in a track and can act as a linear guide.

Rod-End Bearings Rod-end bearings consist of a bored, spherical-shaped ball held within a metal housing. This housing may be a round cylindrical sleeve or a threaded stud (see Figure 9–30). The stud can be a male threaded rod or a female threaded collar. Rod-end bearings use the spherical shape of the ball to accommodate large amounts of misalignment. Typically they are used in linkages and oscillating shafts that are not rotating. Some versions that are essentially plain bearings can accommodate very slowly rotating shafts. An antifriction type utilizing rolling elements is also available.

Figure 9–29
Cam follower bearing.
Courtesy of SKF.

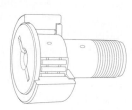

Antifriction Bearing Numbering System and Configuration

Most ball and roller bearings are made to industry accepted standards and have a common base numbering system. The number given to the bearing generally corresponds to the size and type of bearing. The boundary dimensions, which include the inside diameter (bore), outside diameter, and width, determine the size of the bearing. These external dimensions are generally standardized in the bearing industry and were first developed at the turn of the twentieth century by the major bearing manufacturers. Standard external specifications allow for interchanging amongst different brands and the installation of different types of bearings within the same mounting space. Internal design, such as the number of rolling elements, race geometry, and cage composition, are not standardized but may be similar. The majority of precision nonhoused bearings are made to metric specifications, with the exception of taper roller bearings. Taper rolling bearings are commonly available in both inch and metric sizes.

Radial ball bearings are available in extra light, light, medium, and heavy series. The series' names refer to their relative proportions and load-carrying capacity. Manufacturers have designated these as a 100, 200, 300, and 400 series. Manufacturers of cylindrical and spherical rollers use a similar system. A bearing may have the same bore size with four different widths and outside diameters. For example a 108, 208, 308, and 408 bearing will have a 40 mm inside diameter but correspondingly larger outside diameters and widths. The boundary dimensions of precision ball, spherical, and cylindrical roller bearings are standardized as metric throughout the industry. These metric dimensions are often converted into inch system measurements (25 mm = .9843″), but the boundary dimensions for the vast majority of unmounted precision antifriction bearings are in metric units.

Housed Type Antifriction Bearings

Antifriction rolling element bearings are available as housed units in a premounted version. Both ball and roller bearings are used in the housed units, with their use and selection based on load, speed, space, and life requirements. The housing is made from materials such as steel, cast, or even plastics. Special coatings and platings are available, such as nickel, to accommodate operation in hostile environments. The housing allows the bearing to be bolted outside of the machine or to an external supporting structure. The housing provides support and alignment of the internally mounted bearing. Housed or mounted bearings are made as pillow blocks, flanges, cylindrical cartridges, and take-up units. Many housed bearings used in the United States have inch dimensional specifications. Important housing dimensions like the shaft size, center height, bolt spacing, etc., are in increments of 1/16″. Increasingly the marketplace is seeing the use of blocks with metric dimensional specifications. Most housed units require additional lubrication at some point. Lubrication is provided to the bearing through the housing by some form of grease fitting.

Pillow Blocks Pillow blocks are one of the most common types of housed bearings (see Figure 9–31). The bearing is contained within a housing that has mounting pads or feet. The mounting surface for the pads is parallel to the supported shaft's axis. Pillow blocks are made in ball, roller, or plain bearing versions. Both metric and inch dimension specifications are available.

Ball bearing pillow blocks use an extended inner ring bearing with some form of shaft locking device such as a setscrew to hold and lock them in place. Applications that operate at moderate to high speeds and with light to moderate loads are ideal for ball bearing pillow blocks. These applications include fans, rollers, packaging machinery, and light duty conveyors.

Roller bearing pillow blocks are made with spherical, tapered, and other shaped rolling elements. Roller bearing blocks are used on heavier loaded applications such as bulk material handling conveyors and large blowers. Roller bearing pillow blocks are available in two basic versions: the unit block and the component block. The unit block comes as a self-contained preassembled unit with all bearings, adapters, seals, sleeves, etc., contained within the assembly (see Figure 9–32). Internal bearing clearances are preset at the factory with only a simple mounting

FIGURE 9–31
Ball bearing pillow block.
Courtesy of Timken/Torrington.

FIGURE 9–32
Unit roller bearing
pillow block.
Courtesy of Rockwell
Automation.

FIGURE 9–33
Dissembled component roller bearing pillow block. The complete assembly is made up, from left to right: seal ring, adapter sleeve, bearing, split housing, stabilizing ring, lock-washer, lock-nut, and a second seal ring.
Courtesy of Timken/Torrington.

procedure needed to mount the block. The component roller bearing pillow block comes as a group of parts that requires adjustment and assembly for mounting (see Figure 9–33). Both types have their advantages and disadvantages. The major advantage of unit blocks is that minimal effort is required to install the block and, because it is sealed at the factory, it will not become contaminated during assembly in the field. The chief advantage of component blocks is that worn parts, such as the bearing, can be easily replaced in the field. Most of the component pillow blocks have a split housing that allows for ease of assembly.

Flange Blocks Flanged blocks are used on mounting surfaces that are perpendicular to the shaft axis. They are available in two-, three-, and four-bolt versions (see Figure 9–34). A piloted version of the four-bolt base types provides better positioning accuracy. Both ball and roller bearing versions are available as well as metric and inch specifications.

Take-Up Blocks Take-up blocks are ball or roller bearing units that are used to adjust the center distance of shafts, or on a terminal pulley to adjust belt tension of a flat belt (see Figure 9–35). They can be found on the tail pulley shafts of light to medium sized material conveyors.

Cylindrical Cartridges Cylindrical cartridges are bearing units that provide shaft support like flange blocks where the shaft axis is perpendicular to and passing through a machined housing (see Figure 9–36). The machine housing provides the

FIGURE 9–34
A four-bolt flange block bearing.
Courtesy of Timken/Torrington.

FIGURE 9–35
Center pull type take-up
block bearing.
Courtesy of Timken/Torrington.

FIGURE 9–36
Cartridge block bearing.
Courtesy of Timken/Torrington.

rigidity to the cartridge mount. The cartridge is pressed into the through bore housing of the machine.

Antifriction Bearing Arrangements

Bearing arrangements usually require at least two bearings to support and locate a shaft both radially and axially (see Figure 9–37). Typically one of these bearings will be referred to as the located bearing and the other as the nonlocated bearing. Certain applications require that both bearings be responsible for locating the shaft axially and are referred to as cross-located bearings. In most applications the shaft will move axially due to such contributing factors as thermal growth, shaft and housing tolerances, and thrust loads, and during initial start-up when the shaft positions itself in its normal operating center. In many instances this movement is unavoidable and needs to be controlled. Locating one of the bearings helps to prevent damage to connected components or stressing of the internal machine parts in an unwanted manner. Excessive thrusting against bearings or the improper running position of components, such as gears, will result in premature failure or even catastrophic failure.

Determining what type of bearing arrangement best suits the demands of the application will assist in controlling the axial movement, which will in turn prevent damage to bearings and surrounding equipment. For example, if a pan conveyor drive shaft operating in an oven is directly coupled to a gearbox output shaft under

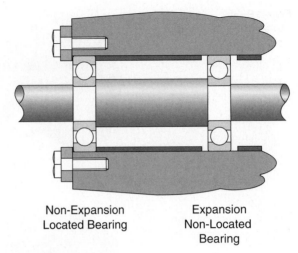

Non-Expansion   Expansion
Located Bearing   Non-Located
                        Bearing

FIGURE 9–37
Two bearings mounted on a shaft and retained in a housing. The left side bearing is
located or held in place by the housing and cover plate. The right side bearing is a
nonlocated bearing that is free to move axially.

the extreme ambient temperatures present in the system, significant linear growth
of the drive shaft can occur. The coupling gap could become closed and the bearing's thrust loaded. This axial movement should be directed away from the driving
mechanism. For this reason, on many applications the located bearing is on the
drive side. The rule of placing the held bearing on the drive side is a generalization.
Each application must be looked at carefully and the location of the fixed bearing
determined by considering numerous variables specific to each machine.

The located bearing is called the "held," "fixed," or "nonexpansion" bearing. It
provides radial support and axial guidance in both directions. It is locked in place
laterally on both the shaft and within the housing. In many applications this is
accomplished by placing the bearing squarely against a shaft shoulder on the inner
ring and holding it in place with a press fit (interference) or by use of a lock nut
holding it tightly against the shoulder. Interference fits of the inner ring to the shaft
or outer ring to the housing are not sufficient alone to secure a bearing position
axially when subjected to moderate or greater axial forces. The outer ring should be
clamped within a housing of a depth equal to the width of the outer ring or by use
of snap rings, stabilizing rings, or shouldered machined end covers.

The nonlocated bearing is referred to as the "free," "floating," or "expansion"
bearing. It provides radial support and is free to move axially. The floating bearing
has to compensate for minor variations in length distances between shaft and
housing shoulders due to machining tolerances. It also must be able to compensate
lateral growth along the shaft from increased operating temperatures. This is accomplished in one of two ways. In the case of an application that uses a cylindrical
roller bearing with removable inner or outer rings, one of the races is free to float
axially during the rotation of the bearing. This internally floating design will still

have the outer surface of the outer ring shouldered and clamped and the inner ring inner diameter fitted to the shaft, but the rollers and their associated containing ring can move axially within the bearing.

Applications that use spherical rollers or ball bearings may have their inner rings fitted to a shaft by a press fit or held in place by a sleeve clamped tight with a locking nut. The outer ring outside diameter will be free to float axially within a housing that has a slightly wider seat width than the outer ring of the bearing. For example, split-type spherical roller bearing pillow blocks will use the same housing outer ring seat width dimension for both expansion and nonexpansion bearings. This housing bearing ring seat is made slightly wider than the bearing outer ring to accommodate a nonlocated bearing. Placing what is commonly referred to as a "stab" ring—which is a "C" shaped spacer ring—that has the same outside diameter as the bearing into the housing seat will secure the bearing axially. This way the same standard housing can be used for either fixed or floating units. The simple installation or removal of the ring makes the bearing either free or fixed. Certain types of roller bearing blocks design permissible limited axial movement into the internal construction of the housing. These unit blocks are preset at the bearing factory. With these types, the block must be designated expansion or nonexpansion by the manufacturer.

A cross-located bearing arrangement is used when axial loads in one direction, and oftentimes both directions, are prevalent. This arrangement is common when using taper roller bearings that are clamped in place with cup followers or bearing cover plates that are held in place with cap screws. Behind the end plates, next to the housing, will be a pack of shaped shims. This method is used with press fitted cones and loose fitting cups. By shimming the bearing cover end plates with more or fewer shims, the amount of lateral movement or end-play can be adjusted easily. This arrangement is called an adjustable mounting. Applications such as gearboxes using helical gearing will often use this arrangement. It allows the installed clearances to be set based on predictable changes in shaft and housing expansion and ensures proper gear mesh. This method also allows preloading of the bearings.

Determining which bearing should be located, nonlocated, or cross-located must take into account numerous variables. Thermal growth of the shaft and housing, thrust loads created by gearing, forces created or caused by the load being moved, housing design and tolerances, bearing type, and the nature of the application are just a few of the variables. The design engineer, taking into account operating experience, should determine the proper bearing arrangement.

Bearing Shaft and Housing Fits

The rolling bearing must be properly fitted onto a shaft and into the housing. Fit is a general term that is used to signify the range of tightness or looseness that may result from the assembly of a combination of allowances and tolerances of joining parts. Without a proper fit of the inner ring to the shaft and the outer ring into the housing, the bearing will not see a satisfactory service life. In fact one might say that the bearing is only as good as the shaft it is mounted on and housing it is placed into. The dimensions and tolerance of the bearing, shaft, and housing determine the actual fit of the bearing. Because the manufacturer predetermines the bearing dimensions and tolerances, controlling the shaft and housing dimensions is the key to determining the proper fit.

Proper fit of the bearing is required for the obvious reason that it must be radially supported and not come loose. In certain applications it is desirable to have the shaft and inner ring of the bearing act as a unit. Loosening of the inner ring to shaft or circumferential movement within the bearing housing is often referred to as "creep." Creeping of the rings on the shaft or within the housing can result in damage to the bearing rings as well as the shaft and housing surfaces. In extreme cases, where the rings spin on the shaft or within the housing, fretting and galling will occur and an operating temperature increase will result. Removed metal from the shaft or housing from the galling process will end up inside the bearing and result in premature failure. In some rare special applications, the outer ring is allowed to creep minimally within the housing for reasons of load. The rings of a typical rolling element bearing are relatively thin for the loads they must carry. Therefore the rings can take on the shape of the shaft and housing. If the shaft or housing is not truly round, neither will be the bearing. The concentricity of the bearing must be maintained to allow free rotation of the rolling elements, correct internal clearances, and optimum load handling characteristics. Proper fit also provides that the bearing is aligned and not tilted on the shaft or within the housing. Last of all, proper fit produces running accuracy. If the bearing rings are moving around while a machine is turning, its product will not be accurate.

The system of shaft and housing fits and their limits that has been accepted by industry for rolling bearings (except tapers) is ISO Standard 286. It is an alphanumeric system that contains numerous choices for shaft and housing fit tolerances. Bearing handbooks will list tables with recommended fits (see Figure 9–38). The letter (lower case for shafts, upper case for housing bores) locates the tolerance zone in relation to the nominal dimensions, and the number gives the magnitude of the tolerance. Many of these fit zones overlap so that the high end of one fit designation is the low end of another. Bearing manufacturers or machinist handbooks can be consulted for specific designations and explanation of fit terms. The numerous terms used to describe varying degrees and conditions of fit can be confusing. Although exact definitions exist for most of these terms, such as sweat, press, slip, etc., many of these names are used interchangeably among maintenance technicians.

For the sake of clarity, fits can be grouped into three categories. It is probably an oversimplification to do this from an engineering standpoint, but from the bearing maintenance view it provides easy comprehension of the fit concept. The three general fit categories are tight, transitional, and loose (see Shafting chapter). Tight fits are known by numerous terms such as press, interference, sweat, and shrink, to name a few. This type of fit requires force, pressure, or heat to mate the parts. For example the shaft will have a slightly larger diameter than the inner ring bore of the bearing, requiring either controlled expansion of the inner ring from heat or the use of mechanical pressure from a device such as an arbor press. Tight fits are generally associated with the rotating ring of the bearing and moderate to high loads. As a general rule, the heavier the load and larger the diameter, the greater the amount of interference fit. Transitional fits are referred to as "line-to-line" and are those where the size tolerances for mating parts always result in a combination of

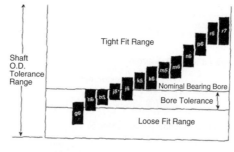

FIGURE 9-38

Shaft and housing fit symbols.

Courtesy of Timken/Torrington.

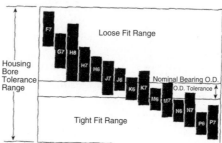

tight and loose fits. Loose fits, often called slip or clearance, are where no force is required to mate the parts. They are called on for light load or stationary inner rings. For example a bearing supporting an electric motor shaft will have a tight fit of the inner ring to shaft, but a loose fit on the outer ring to housing bore.

The fit tolerance selection process must take into account numerous factors: the character and magnitude of the load, bearing dimensions, type of bearing, temperature, relative load rotation, and running accuracy requirements (see Figure 9–39). Consideration also must be given to applications with a hollow shaft. Hollow shafts contract and deform when a bearing is pressed onto them. Once the proper fit designation is determined, the allowable fit tolerance deviations must be looked up in bearing manufacturers' tables.

The fit for both new assembly and repaired machines must be correct. It is crucial that measurements of the shaft and housing be taken prior to assembly (see Shafting chapter). This is especially true of bearings that are being replaced as a part of the machine rebuild process. All bearings must be seated properly to prevent failure.

Another important consideration when fitting a bearing onto a shaft is how it abuts a shaft shoulder. When the bearing is mounted on a shaft, it usually is located against a shaft shoulder to assist in supporting and locating the bearing. The shoulders or abutments should be machined squarely with the bearing seat and perpendicular to the shaft axis. Shaft fillets should be used, but in some instances the corner is undercut. Shaft fillets are a rounding off or curving of the corners that provides favorable stress distribution. The shaft fillet radius should clear the corner radius of the inner ring (see Figure 9–40). The shoulder height must be sufficient to support the bearing but must not contact any moving bearing parts. As a general rule, the height of the shaft shoulder should be one-half the height of the bearing inner ring. Correct shoulder dimensions, abutments, and corner radii can be obtained

Selection of Solid Steel Shaft Tolerance Classification for Metric *Radial* Ball and Roller Bearings of Tolerance Classes ABEC-1, RBEC-1 (Except Inch Dimensioned Tapered Roller Bearings)

Conditions	Shaft diameter, mm Ball bearings[1]	Cylindrical roller bearings, metric taper	Spherical roller bearings	Tolerance Symbol
Rotating inner ring load or direction of loading indeterminate				
Light loads	18≤100 100≤140	≤40 40≤100	— —	j6 k6
Normal loads	≤18 18≤100 100≤140 140≤200 200≤280 — — —	— ≤40 40≤100 100≤140 140≤200 200≤400 — —	— ≤40 40≤65 65≤100 100≤140 140≤280 280≤500 >500	j5 k5 (k6)[2] m5 (m6)[2] m6 n6 p6 r6 r7
Heavy loads	— — —	50≤140 140≤200 ≤200	50≤100 100≤140 >140	n6[3] p6[3] r6[3]
High demands on running accuracy with light loads	≤18 18≤100 100≤200 —	— ≤40 40≤140 140≤200	— — — —	h5[4] j5[4] k5[4] m5[4]
Stationary inner ring load				
Easy axial displacement of inner ring on shaft desirable	all	all	all	g6
Easy axial displacement of inner ring on shaft unnecessary	all	all	all	h6
Axial loads only				
	≤250 >250	≤250 >250	≤250 >250	j6 js6

[1]Shaft tolerances for Y-bearings (set-screw mounted) are available from SKF.
[2]The tolerances in brackets are generally used for metric taper roller and single row angular ball bearings used individually. They can also be used for other types of bearing where speeds are moderate and the effect of bearing internal clearance is not significant.
[3]Bearings with radial internal clearance greater than Normal are necessary.
[4]For ABCE 5 bearings, use Table 14; for higher precision bearings other recommendations apply. Consult SKF.
≤ less than or equal to.

FIGURE 9–39
Bearing and shaft fit selection table.
Courtesy of SKF.

from bearing manufacturers. Correctly machined shoulders and abutments contribute to the running accuracy of the bearing.

Bearing Radial Internal Clearances

Bearing internal radial clearance is defined as the total distance one ring can be moved relative to the other in a radial or axial direction (see Figure 9–41). It is the measurable play or looseness inside the bearing. This internal clearance is required

FIGURE 9–40
Example of a shaft
shoulder and bearing
placement.
Courtesy of SKF.

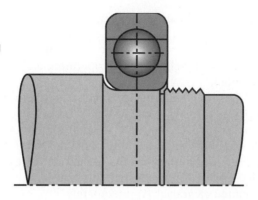

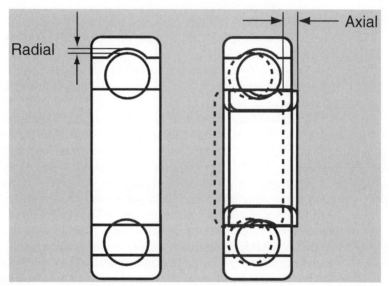

FIGURE 9–41
Radial and axial bearing internal clearances.
Courtesy of NSK.

for free rotation of rolling elements and to compensate for thermal growth of the
shaft and/or housing. Clearance also is in the bearing to absorb the effects of a
press fit. In fact excessive press fits can remove all of the internal clearance within a
bearing. Internal clearance allows space for minute amounts of lubricating film and
assists in obtaining optimum load distribution on the rolling elements and races.
Although both radial and axial clearance are important, radial clearance is accepted as the more significant factor with ball, cylindrical, and spherical roller bearings due to its direct relation to shaft/housing fits and the effects of temperature.

Taper roller bearings have a clearance referred to as "end-play." End-play is the total lateral or axial movement of a bearing assembly. Certain manufacturers of machinery will call out a "bench end-play" required for the bearing prior to mounting that will yield a specific operating end-play under load and speed after being mounted.

The radial internal clearance of a bearing essentially will have three stages that it passes through from the shelf to operation. Bearings will have an unmounted internal clearance built into them at the factory during assembly called unmounted clearance or bench clearance. This clearance will probably be reduced after mounting due to interference fits caused by the expansion or compression of the rings. That is why the proper shaft and housing fits are imperative to produce the desired mounted clearance. Approximately 50% of the amount of interference fit between the inner ring and the shaft will show up as a loss of internal clearance within the bearing. A reduction in clearance from mounting can also be caused by the bearing arrangement and mounting method used to install the bearings. When the bearing, under load, reaches its full running speed, it will then achieve its operational clearance. This operational clearance is attributed to factors such as load, speed, and temperature rise.

Satisfactory performance relies on the correct operational clearance for the bearing and application. As a general rule, ball bearings and taper roller bearings should have an operational clearance near zero or with a slight preload. Cylindrical and spherical bearings should function with a small amount of clearance. The clearance within the bearing influences many factors related to the performance and operation of the bearing. Noise, vibration, heat, load-carrying capacity, and ultimately the life of the bearing can be affected by the clearance within the bearing.

Clearance designations and amounts have been established by organizations such as the American Bearing Manufacturers Association (ABMA) and the International Standards Organization (ISO). The designations C2, C0, C3, C4, and C5 are in order from less to more clearance. Normal clearance, C0, has been selected so that a suitable operational clearance will occur when bearings are mounted and under normal operating conditions. Typically there will be no indication marked on the bearing if it is a normal clearance bearing. Those bearings with clearances other than normal will be designated in some manner, such as an etched mark on a ring face. C2 (snug) clearance bearings are used in applications where minimal radial and axial play is desirable. Printing press bearings and pump impeller bearings are examples where bearings with tighter internal clearance might be found. C3 (loose), C4 (extra loose), and C5 (extra-extra loose) are used in applications involving large press fits of the rings and when extreme temperature differentials occur. Examples would include oven, kiln, or dryer rolls on paper machines. Figure 9–42 illustrates the radial internal clearance limits for spherical roller bearings. Note that the maximum limit of one designation is the minimum limit of another. The size and type of the bearing will be a determining factor in the amount of unmounted clearance. Also observe that the recommended reduction and final mounted clearance is shown. Additional tables for various types of bearings can be obtained from manufacturers.

RADIAL INTERNAL CLEARANCE LIMITS

All data on this page, except Bore I.D. are in inches/millimeters

Each cell is given as *inch / mm*. For the clearance columns, adjacent classes share boundary values as printed in the staggered header (C2 min / C2 max = Normal min / Normal max = C3 min / C3 max = C4 min / C4 max = C5 min / C5 max).

Bore (nominal) mm over	incl	Cylindrical Bore — C2 min	C2 max / Normal min	Normal max / C3 min	C3 max / C4 min	C4 max / C5 min	C5 max	Tapered Bore — C2 min	C2 max / Normal min	Normal max / C3 min	C3 max / C4 min	C4 max / C5 min	C5 max	Recommended Reduction of RIC Due to Installation min	max	Recommended RIC after Installation[1] min
24	30	0.0006 / 0.015	0.0010 / 0.025	0.0016 / 0.040	0.0022 / 0.055	0.0030 / 0.075	0.0037 / 0.095	0.0008 / 0.020	0.0012 / 0.030	0.0016 / 0.040	0.0022 / 0.055	0.0030 / 0.075	0.0037 / 0.095	0.0006 / 0.015	0.0008 / 0.020	0.0006 / 0.015
30	40	0.0006 / 0.015	0.0012 / 0.030	0.0018 / 0.045	0.0024 / 0.060	0.0031 / 0.080	0.0039 / 0.100	0.0010 / 0.025	0.0014 / 0.035	0.0020 / 0.050	0.0026 / 0.065	0.0033 / 0.085	0.0041 / 0.105	0.0008 / 0.020	0.0010 / 0.025	0.0006 / 0.015
40	50	0.0008 / 0.020	0.0014 / 0.035	0.0022 / 0.055	0.0030 / 0.075	0.0039 / 0.100	0.0049 / 0.125	0.0012 / 0.030	0.0018 / 0.045	0.0024 / 0.060	0.0031 / 0.080	0.0039 / 0.100	0.0051 / 0.130	0.0010 / 0.025	0.0012 / 0.030	0.0008 / 0.020
50	65	0.0008 / 0.020	0.0016 / 0.040	0.0026 / 0.065	0.0035 / 0.090	0.0047 / 0.120	0.0059 / 0.150	0.0016 / 0.040	0.0022 / 0.055	0.0030 / 0.075	0.0037 / 0.095	0.0047 / 0.120	0.0063 / 0.160	0.0012 / 0.030	0.0015 / 0.038	0.0010 / 0.025
65	80	0.0012 / 0.030	0.0020 / 0.050	0.0031 / 0.080	0.0043 / 0.110	0.0057 / 0.145	0.0071 / 0.180	0.0020 / 0.050	0.0028 / 0.070	0.0037 / 0.095	0.0047 / 0.120	0.0059 / 0.150	0.0079 / 0.200	0.0015 / 0.038	0.0020 / 0.051	0.0010 / 0.025
80	100	0.0014 / 0.035	0.0024 / 0.060	0.0039 / 0.100	0.0053 / 0.135	0.0071 / 0.180	0.0089 / 0.225	0.0022 / 0.055	0.0030 / 0.080	0.0043 / 0.110	0.0055 / 0.140	0.0071 / 0.180	0.0091 / 0.230	0.0018 / 0.046	0.0025 / 0.064	0.0014 / 0.036
100	120	0.0016 / 0.040	0.0030 / 0.075	0.0047 / 0.120	0.0063 / 0.160	0.0083 / 0.210	0.0102 / 0.260	0.0026 / 0.065	0.0039 / 0.100	0.0053 / 0.135	0.0067 / 0.170	0.0087 / 0.220	0.0110 / 0.280	0.0020 / 0.051	0.0028 / 0.071	0.0020 / 0.051
120	140	0.0020 / 0.050	0.0037 / 0.095	0.0057 / 0.145	0.0075 / 0.190	0.0094 / 0.240	0.0118 / 0.300	0.0031 / 0.080	0.0047 / 0.120	0.0063 / 0.160	0.0079 / 0.200	0.0102 / 0.260	0.0130 / 0.330	0.0025 / 0.064	0.0035 / 0.089	0.0022 / 0.056
140	160	0.0024 / 0.060	0.0043 / 0.110	0.0067 / 0.170	0.0087 / 0.220	0.0110 / 0.280	0.0138 / 0.350	0.0035 / 0.090	0.0051 / 0.130	0.0071 / 0.180	0.0091 / 0.230	0.0118 / 0.300	0.0150 / 0.380	0.0030 / 0.076	0.0040 / 0.102	0.0022 / 0.056
160	180	0.0026 / 0.065	0.0047 / 0.120	0.0071 / 0.180	0.0094 / 0.240	0.0122 / 0.310	0.0154 / 0.390	0.0039 / 0.100	0.0055 / 0.140	0.0079 / 0.200	0.0102 / 0.260	0.0134 / 0.340	0.0169 / 0.430	0.0030 / 0.076	0.0045 / 0.114	0.0024 / 0.061
180	200	0.0028 / 0.070	0.0051 / 0.130	0.0079 / 0.200	0.0102 / 0.260	0.0134 / 0.340	0.0169 / 0.430	0.0043 / 0.110	0.0063 / 0.160	0.0087 / 0.220	0.0114 / 0.290	0.0146 / 0.370	0.0185 / 0.470	0.0035 / 0.089	0.0050 / 0.127	0.0028 / 0.071
200	225	0.0031 / 0.080	0.0055 / 0.140	0.0087 / 0.220	0.0114 / 0.290	0.0150 / 0.380	0.0185 / 0.470	0.0047 / 0.120	0.0071 / 0.180	0.0098 / 0.250	0.0126 / 0.320	0.0161 / 0.410	0.0205 / 0.520	0.0040 / 0.102	0.0055 / 0.140	0.0030 / 0.076
225	250	0.0035 / 0.090	0.0059 / 0.150	0.0094 / 0.240	0.0126 / 0.320	0.0165 / 0.420	0.0205 / 0.520	0.0055 / 0.140	0.0079 / 0.200	0.0106 / 0.270	0.0138 / 0.350	0.0177 / 0.450	0.0224 / 0.570	0.0045 / 0.114	0.0060 / 0.152	0.0035 / 0.089
250	280	0.0039 / 0.100	0.0067 / 0.170	0.0102 / 0.260	0.0138 / 0.350	0.0181 / 0.460	0.0224 / 0.570	0.0059 / 0.150	0.0087 / 0.220	0.0118 / 0.300	0.0154 / 0.390	0.0193 / 0.490	0.0244 / 0.620	0.0045 / 0.114	0.0065 / 0.165	0.0040 / 0.102
280	315	0.0043 / 0.110	0.0075 / 0.190	0.0110 / 0.280	0.0146 / 0.370	0.0197 / 0.500	0.0248 / 0.630	0.0067 / 0.170	0.0094 / 0.240	0.0130 / 0.330	0.0169 / 0.430	0.0213 / 0.540	0.0268 / 0.680	0.0050 / 0.127	0.0070 / 0.178	0.0040 / 0.102
315	355	0.0047 / 0.120	0.0079 / 0.200	0.0122 / 0.310	0.0161 / 0.410	0.0217 / 0.550	0.0272 / 0.690	0.0075 / 0.190	0.0106 / 0.270	0.0142 / 0.360	0.0185 / 0.470	0.0232 / 0.590	0.0291 / 0.740	0.055 / 0.140	0.0075 / 0.190	0.0045 / 0.114
355	400	0.0051 / 0.130	0.0087 / 0.220	0.0134 / 0.340	0.0177 / 0.450	0.0236 / 0.600	0.0295 / 0.750	0.0083 / 0.210	0.0118 / 0.300	0.0157 / 0.400	0.0205 / 0.520	0.0256 / 0.650	0.0323 / 0.820	0.0060 / 0.152	0.0080 / 0.203	0.0050 / 0.127

[1]For bearings with normal initial clearance

FIGURE 9–42

Spherical roller bearing internal clearances.

Courtesy of Timken/Torrington.

Measuring of the unmounted and mounted radial internal clearance of bearings is necessary under certain conditions. Measuring clearance on small ball bearings is difficult and requires special clamping fixtures and instruments. This is in part because the clearance is measured in tens-of-thousandths or microns. A radial deep groove ball bearing with a bore of 25mm (.9843″) that has a normal clearance

FIGURE 9–43
Measuring internal clearance with feeler gauges.
Courtesy of Timken/Torrington.

designation will only have 0.0002″ (5 micrometers) to 0.0008″ (20 micrometers) internal clearance. Larger spherical roller bearings can have their internal clearance measured by the use of feeler gauges (see Figure 9–43). The significant issue is that the final recommended mounted clearance be correctly established, based on the recommendations of the bearing clearance tables. If the clearance is too great or too little, failure can occur. On tapered journals and sleeves, the drive up of the bearing onto the taper in the axial direction will determine the mounted clearance. Overtightening of lock nuts can result in a greater reduction of clearance than specified, due to the expansion of the inner ring. On cylindrical bore bearings, such as those mounted with a press fit and abutting a shoulder, the amount of interference will determine the mounted clearance. In either case it may be necessary to measure and control the clearance during and after the installation process.

The following are a few general guidelines for measuring the radial internal clearance of a spherical roller bearing:

■ Check the bearing manufacturer's clearance tables for the specific bearing to be mounted against the part number and designation on the bearing. Determine the clearance values.

■ Obtain a clean feeler gauge in good condition, with a mixture of blade thicknesses.

■ Place the unmounted bearing in an upright position with inner and outer race faces parallel and gently oscillate the bearing while pressing down firmly to seat the bearing. As an alternative the bearing may be suspended from a journal.

- On both sides of the bearing, position a roller so that it is at the top of the bearing and press the rollers inward to assure proper contact with the races.

- With a thin blade of the feeler gauge, slide or gently saw the blade carefully between the top of the roller and the outer ring race surface. Some manufacturers recommend inserting the feeler gauge blade across the width of one roller. Others recommend inserting the blade across both sets of rollers. If the bearing is placed on a table, measure at the top of the bearing. If the bearing is suspended from a journal, measure at the bottom of the bearing between the roller and race surface.

- Repeat the process and measure both sides to achieve an average. Use progressively thicker blades until one is found that will not pass through. The blade thickness that preceded the "no-go" blade is the internal clearance of the unmounted bearing.

- During the mounting procedure on a tapered journal or sleeve using a lock nut to drive up the bearing, periodically check the clearance to track and achieve the desired final mounted clearance. On straight shaft, cylindrical bored bearings, a final clearance check after seating the bearing should take place.

- Do not exceed the recommended reduction and make sure the remaining mounted clearance matches or slightly exceeds the recommended minimum.

Bearing Preloads

A few special applications require a bearing to start or operate with a negative clearance. This negative clearance is commonly referred to as "preload." Negative clearance is somewhat of a misnomer. There cannot be negative clearance because this would require the metal of the races to literally penetrate the metal of the rolling elements. What actually happens is that the bearing housing walls will bend, compress, and deflect marginally to cause a metal-to-metal condition. There also may be minute amounts of deflection within the bearing. Preload is the opposite of end-play.

The reasons for preloading bearings vary. In some cases preload is required to enhance the stiffness of the housing and bearing arrangement, which can result in increased running accuracy and rigidity. Wear and settling of bearing arrangements can be compensated by preloading practices. Preloading can reduce running noise or chatter within the machine. With taper roller bearings it ensures full line contact along the length of the roller and distributes the load among the rolling elements, yielding long service life. Thermal growth of associated assembly components, such as the housing sidewalls, can be compensated by initial preloading of the bearing arrangement (cross-located bearings) through shim pack adjustment.

Determining the correct amount of preload force is done through a series of calculations and testing, and is based on application experience. Preload may be expressed as a force (pounds or Newtons) or lateral distance. The preloading of a bearing is indirectly related to frictional torque. In some cases the amount of preload can be determined by how much torque is required to turn the bearing arrangement. The optimum preload is the minimum preload required to achieve the desired results of stiffness, accuracy, quietness, reliability, and greatest life for the application. In some cases preloading of bearings is a trade-off between life versus rigidity and accuracy.

Preload adjustment is achieved by various means. Tightening nuts to specified torque specifications; adjusting the shims between housing side walls and cover plates, calibrated spacers, and springs; and offset machine dimensions and tolerances of associated components can all be used to preload a bearing. Gearbox manufacturers' manuals oftentimes will call out a specific amount of required preload or end-play when rebuilding the shaft and bearing assemblies. The amount of end-play in the shaft—once it is shimmed and bolted—can be measured with a dial indicator placed to read the end of a shaft as the shaft is barred back and forth. The total amount of lateral movement is the end-play. If a preload is required the measured end-play must be taken out, plus any specified preload, by adjusting the shims behind the bearing cover plate. The maintenance technician is advised to consult with the machine manufacturer, service bulletins, and bearing handbooks when installing and adjusting preloaded bearings.

HANDLING BEARINGS

The single most important word in the handling of bearings is "cleanliness." Bearings must be kept free of contamination from dirt and moisture. Even the smallest foreign particle or minute amount of a nonlubricating fluid can interrupt the lubricant film as well as damage the bearing surfaces. Improper storage and handling of bearings prior to installation is a major contributing factor to premature failure. The bearing must be treated as a precision piece of equipment that is crucial to the overall performance and life of the machine it is operating in.

The following guidelines, if observed, will help deliver adequate bearing service life:

- Store the bearings in a clean and dry environment, lying on their sides. Do not store them in an area where there is excessive ambient vibration present.
- Keep the bearing wrapped and boxed in its original container until just prior to installation. Do not use them as a gauge to check shaft fits or housing bore dimensions. If the bearing has been mounted into an assembly and placed in storage, make sure that it is wrapped with plastic, waxed paper, or a lint-free cloth to protect it from contaminants.
- Handle the bearings with gloves or clean dry hands. Never work in a dirty shop environment. Cover unfinished jobs.
- Properly sling, lift, and handle all bearings—especially large items—to prevent ring distortion, damage, or dropping. If the bearing being handled is dropped, the handler may be injured, and certainly the bearing should be considered damaged.

- Do not clean the original bearings of the factory-provided oil protection. It is compatible with most standard lubricating oils and greases.
- Never spin the bearing by hand or with compressed air. Damage to the bearing will occur and there is the potential for personal injury.
- Never attempt to modify, drill, alter, or re-machine the bearing rings, retainers, rolling elements, and seals/shields.
- Use only the proper recommended tools that are clean and in good working order.
- All related shafts and housings must be cleaned, inspected, measured, toleranced, and correctly sized. Mounting surfaces must be free of nicks, burrs, and any damage or wear.
- Use correct mounting and dismounting practices. Make sure all parts are aligned during assembly. Never transmit the mounting forces through the rolling elements during installation.
- Under no circumstances should a bearing ring or rolling element be struck by a hard surface hammer. The bearing may chip or even explode, resulting in personal injury.

BEARING DISMOUNTING AND MOUNTING PRACTICES

The methods used to mount and dismount bearings are as numerous as there are types of bearings. Mechanical forces, heat, hydraulics, sleeves, pullers, and a mixture of other means can be employed to remove and install bearings. The important issue for the maintenance mechanic to remember is that how the used bearing is removed and the new bearing is installed will effect the life and condition of the bearing and related components. Improper mounting and dismounting procedures account for a large portion of premature bearing failures. All of the issues pertaining to shaft/housing fits, clearances, precision, tolerances, proper measuring techniques, and record keeping should be addressed. Sloppy removal and installation practices will yield unsatisfactory bearing life and may result in an injury to the maintenance technician.

Bearing Dismounting/Removal

Bearings can be removed from shafts and housings by use of pullers, presses, heat, hydraulics, and special removal sleeves. It is important that the maintenance technician not damage any machine components during the removal process. Once the bearing is removed, the question of reuse arises. It is my opinion that care must be taken to not damage the bearing when it is being removed, but not for the express purpose of reuse. The primary purpose of removing a bearing without causing any further damage to the bearing is to identify the cause of failure and correct it. Inspection and identification of the bearing for its replacement and for corrective measures is more easily accomplished with an undamaged bearing.

Cleaning bearings with various solvents or compressed air involves issues of cleanliness and safety. Is the plant air dry and clean? I doubt it! Is the cleaning solvent and tank clean, safe, and approved? Probably not! What is the cost of man-hours spent to clean a bearing versus the cost of downtime and a new bearing? No comparison. New bearings are relatively inexpensive compared to the cost of downtime and man-hours! Is the person who is performing the cleaning and inspection qualified to identify future potential failure symptoms? Maybe! There are only two exceptions to this recommendation: 1) If there is no replacement bearing immediately available and/or 2) the bearing is of significant cost and size to warrant its rebuilding, it should be reused. Qualified bearing manufacturing engineers and service technicians should do the inspection and rebuilding of costly bearings. It should be done at their plant under controlled circumstances and conditions.

Pullers Conventional bearing pullers are the optimum choice for removing a bearing whenever practical because they are quick, safe, and clean. The potential for damage to the shaft and housing is minimal if pullers are correctly employed. Small to medium sized bearings, with low interference fits, may be removed cold by using a conventional bearing puller. An assortment of sizes and types of pullers are made to suit a particular design bearing in specific applications. Jaw pullers are one of the most common types. They have arms that reach behind and grab the bearing outer ring while a threaded rod pushes against the shaft (see Figure 9–44). Hydraulic assist jaw-type pullers deliver extreme, even pulling force for larger bearings (see Figure 9–45). Internal arm pullers are used to remove deep groove ball bearings from blind housing bores (see Figure 9–46). Slide hammer pullers are used for smaller bearings mounted in hard-to-reach locations (see Figure 9–47). Back-type pullers are used to get behind a bearing and pull evenly against a press fit inner ring without transmitting the dismounting forces through the bearing and damaging it (see Figure 9–48).

Presses Hydraulic and mechanical arbor presses may be used to dismount or mount small to medium bearings with light to moderate press fits (see Figure 9–49). Proper blocking of the ring that has the interference fit is required to prevent damage to the shaft, press, and bearing. Proper placement and alignment of the ram is

FIGURE 9–44
Three-arm jaw puller.
Courtesy of SKF.

FIGURE 9–45
Hydraulic assist jaw
puller.
Courtesy of SKF.

FIGURE 9–46
Internal arm puller.
Courtesy of SKF.

FIGURE 9–47
Slide hammer puller.
Courtesy of SKF.

FIGURE 9–48
Back puller used to pull
from behind the bearing.
Courtesy of SKF.

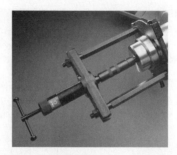

FIGURE 9–49
Small press used for
mounting or dismounting
a bearing.
Courtesy of SKF.

essential to produce an even press off of the bearing. When the shaft is pressed through the bearing, consideration must be given to prevent it from falling to the floor or onto the technician's toe.

Hydraulics Hydraulic pressure can be used with tapered journals and sleeves that have been machined for that purpose. The end of the tapered journal or sleeve can be drilled and tapped to accept a hydraulic fitting. This hole or port leads through the shaft/sleeve and up under the bearing inner ring, which has a specially machined groove. A metered hand pump, connected by hose to the port, is used to pump oil under pressure into the groove. When sufficient pressure is achieved between the mating journal surface and bearing inner ring inside diameter, the bearing will release or require minimal force to remove (see Figure 9–50). This process is commonly referred to as the oil injection method. A nut should be kept on the threaded end of the journal to prevent the bearing from sliding or popping off. This method is often used on bearings that operate on the rolls of papermaking machines.

Heat Heat from a torch is used frequently, but is a poor alternative to more acceptable methods. Care must be taken to not burn through to the shaft and damage it. The temperature of most cutting torches is high enough to change the temper and composition of any steel that it comes into contact with. If a torch must be used, do it quickly and cleanly. Do not raise the kindling temperature of the shaft or allow slag to drip onto the shaft. High-speed cutters may be used but the operator must be careful to control the depth of cut to not damage the shaft surface. Keep in mind that any time sparks fly around a bearing, safety is an issue.

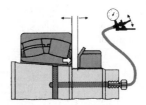

FIGURE 9–50
Oil injection hydraulic removal of a bearing.
Courtesy of SKF.

Excessive Force A hammer should never be used as a bearing removal tool. The potential of personal injury is too great and the shock forces will certainly be transmitted to any associated components. Bearings are under an inherent force known as "hoop stress." During the manufacturing, finishing, carburizing, and heat treating process, residual stresses are present within the bearing rings. These stresses want to expand the inner and outer rings. Additional hoop stress can be imposed into the bearing when it is press fitted onto the shaft or journal. When the bearing is heated for removal, it expands to an even greater amount. It does not require much to relieve these built-up stresses; a single blow of the hammer will cause the ring to fracture and potentially explode. At the very least a chip may fly off of the hard bearing steel and cause injury. Never strike a bearing with a hammer! Brass drifts are commonly used along with a hammer as a removal tool. The ends of the drift will mushroom and chip. These chips can damage surfaces or end up in areas where they should not be. If it is absolutely necessary to use a drift, make sure it is not cracked or damaged. Wear safety glasses when working with bearings!

Bearing Installation and Mounting Practices

The installation method chosen to install a bearing depends on factors such as the number of installations to be done, type of bearing, bearing arrangement, magnitude of the interference fit, the available tools, and the skill level of the technician. As in the removal of bearings there are different methods available for installing bearings, such as heat, mechanical force, presses, and hydraulics.

Cold Mounting Small diameter bearings with light interference fits can be cold mounted using direct force. The mounting forces required to push the bearing onto the shaft can be provided by a mechanical or hydraulic press. This can safely be done if the interference between the shaft and inner ring or the housing bore and outer ring is minimal. A general recommendation is a maximum bore of 50 mm on cylindrical bore bearings and 200 mm on taper bore bearings. The pressure must be applied evenly against the ring with the interference fit (see Figure 9–51). Never transmit the mounting forces through the rolling elements into the other ring because this will dent or brinnel the race surfaces (see Figure 9–52). A special fitting tool with sized impact rings is designed to mount small light press fit bearings (see Figure 9–53). This sleeve-type tool is used with a soft face dead-blow hammer. Make sure all mounting surfaces have been measured, cleaned, and are free of nicks and burrs. Clean the metal surfaces with light oil prior to assembly. Ensure that the bearing has been seated fully against the shoulder.

FIGURE 9–51
Proper press setup.
The inner ring of the
bearing is blocked
because it has a
press/tight fit to the
shaft.

Courtesy of Timken/Torrington.

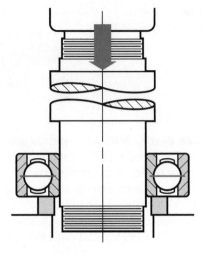

FIGURE 9–52
Improper press setup
that would result in the
mounting forces being
transmitted through the
rolling elements.

Courtesy of Timken/Torrington.

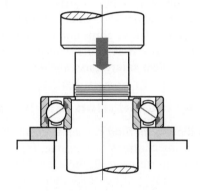

FIGURE 9–53
Mounting sleeve tool used
for cold mounting a bearing.

Courtesy of SKF.

Heat Controlled heating of the bearing is the preferred mounting method that should be used for medium to large bearings. A bearing inner ring temperature 150°F above shaft temperature, not to exceed 250°F (120°C) overall, is adequate to mount most bearings with normal amounts of interference fits. As the bearing cools, the ring contracts and a tight fit between shaft and ring results. The heat must be regulated and uniformly applied to prevent damage to the bearing and even expansion. A torch should never be used because it will damage the metallurgical properties of the bearing steel and change its temper. Heat methods include hot plates, oil baths, ovens, and electric induction bearing heaters. Oil baths are one of the prevalent traditional methods used, but decreasingly, for obvious reasons: The oil is messy and must be kept clean, its temperature is difficult to control, it must have a high enough flash point (minimum 300°F), and eventually it will need to be disposed of in a safe and environmentally correct manner. Heating plates and ovens offer their own problems: They must be kept clean, temperatures tend to be difficult to control, and they require relatively long times to bring the bearing up to temperature. The amount of time the bearing spends in the oven must be closely monitored.

The best tool for heating bearings for mounting is the induction heater because it is clean, fast, and safe (see Figure 9–54). An induction heater is a high current, low voltage coil transformer. The bearing acts as a short-circuited, single-turn secondary coil. The modern induction heater will have controls to allow heating by time or temperature. A contact probe is placed on the inner ring and temperature feedback is provided constantly to the heater controls. A cycle will occur in the process to demagnetize the bearing. Heat resistant gloves must be worn when handling a heated bearing.

Shafts or housings are sometimes cooled through the use of dry ice or a freezer in conjunction with heating the bearing. Condensation will form on the parts, so this should be avoided except in extreme press fit situations.

Tapered Journals and Tapered Sleeves A common method used to mount a tapered bore bearing is to position it onto a tapered sleeve or mount it directly onto a tapered journal. The taper of the sleeve or journal must match the taper of the bearing inner ring inside diameter. The tapered seat can be checked by using a tool known as a "sine bar" gauge. A sine bar gauge is a specially machined measuring

FIGURE 9–54
Bearing induction heater.
Courtesy of SKF.

block that mounts onto a tapered surface and allows the diameter, tolerance, and taper angle to be checked.

Split tapered adapter sleeves serve three purposes. First, they adapt shaft diameters from the American inch standard to the metric bore of the bearing (metric-to-metric sizes are available). Second, adapter sleeves lock and position the bearing assembly in place. And finally, they allow the internal clearance within the bearing to be set to the recommended specification. A split tapered adapter sleeve is slid on the shaft in the correct mounting position. The tapered bore bearing is placed over the sleeve. The sleeve has a threaded end for a lock nut. The mounting is achieved by turning the nut, which applies sufficient axial force to the bearing inner ring and causes it to move up the sleeve. As the bearing moves up the sleeve, the sleeve clamps down on the shaft. The bearing and adapter sleeve become one, and a tight fit is created between the bearing assembly and shaft. In some cases a special adapter sleeve can be used for oil injection removal. The setting of the radial internal clearance is accomplished simultaneously along with the locking of the adapter and bearing assembly to the shaft. Turning the nut is accomplished by using a spanner wrench and hammer (see Figure 9–55). The nut is kept in place with a lock-washer with tabs that are bent over after installation. If at all possible, install the lock-washer after mounting and adjusting the internal clearances. Damage may occur to the lock-washer; reuse is not recommended.

Measuring the axial displacement of the bearing inner ring from an initial starting point can set the radial internal clearance of the bearing. The approximate starting point is established by pushing the bearing onto the sleeve/journal with manual effort, inducing minimal hoop stress to the inner ring. Some manufacturers' tables will list a displacement value for a particular bearing and taper. As an alternative to the conventional adapter sleeve, a hydraulic tapered adapter sleeve assembly can be used to drive up the bearing ring without using a hammer and spanner wrench (see Figure 9–56). The axial displacement of the bearing ring must be measured and controlled. The required axial movement and pressures for the appropriate mounting arrangement and particular size bearing can be obtained by consulting

FIGURE 9–55
Spanner wrench used to tighten a lock-nut holding a bearing.
Courtesy of SKF.

FIGURE 9–56
Hydraulic nut used for installation of a spherical roller bearing.
Courtesy of SKF.

the manufacturer of the hydraulic nut. Once the bearing is mounted, the hydraulic nut is removed and replaced with a lock-nut.

The following are general guidelines for installing and mounting bearings.

- Lockout, tag-out, and verify that the machine that is being worked on is in a safe position.
- Wear protective gloves and glasses when working with bearings.
- Remove bearing assemblies to a clean, well-lighted, and organized workshop, if practical, for further disassembly and installation.
- Measure and inspect all shafts and housings for damage or wear.
- Properly prepare all mounting surfaces. Inspect all components (lock-nuts, washers, etc.). Replace damaged parts.
- Properly inspect and identify the replacement bearing. Cross-check part numbers.
- Use tools that are clean and in good working order.
- Consult the machine and bearing manuals for reference, procedures, and required adjustments.
- Ensure alignment of all joining and mating components. Make sure the bearing is fully seated against the supporting shoulder or abutment.
- Properly install the located and nonlocated bearing based on the arrangement and application considerations.
- Never mix bearing parts or housing halves. Observe match marks.
- Use proper mounting techniques. Never transmit mounting forces through the rolling elements. Never apply flame heat directly to a bearing. Bearing temperature should not exceed 250°F (120°C). Never strike a bearing with a hammer.
- Fully support the housing base of pillow block and flange bearings.

- Properly install any auxiliary bearing seals. Make sure they are not worn or damaged.
- Avoid using antiseize type compounds around bearings. Some lubricating agents are abrasive in nature. Coat parts with light oil.
- Lubricate the bearing with the correct lubricant and proper amounts prior to operation.

ROLLING BEARING LUBRICATION

Rolling bearings require adequate amounts of the correct type of lubrication properly delivered to achieve their calculated life expectancy. The key words and phrases in that sentence are "adequate," "correct type," and "properly delivered." Answering the questions of what is adequate and what type of lubricant should be used requires extensive study, knowledge, and experience. The method used to deliver the lubricant is just as important as the type and amount. Many factors such as speed, load, type of bearing, size, temperature, application, and delivery system must be considered when lubricating rolling bearings. Manuals on the lubrication of rolling bearings are available from bearing manufacturers and they should be consulted to assist the technician in the correct application of bearing lubricants. (Consult the chapter on Lubrication for more detailed information on lubrication terminology.) In this section the basics of "How much?" and "What kind?" of lubrication of rolling element bearings will be addressed.

Bearing Lubrication

How the lubricating agent reacts with the rolling elements under load and at operating speeds is a science in itself. Not only does the lubricant react or change when operating within a rolling bearing, but the bearing itself also undergoes some minute changes. Rolling element bearing lubrication differs from plain or sleeve bearing lubrication and serves the following primary functions:

- Minimizes friction between mating parts.
- Lubricates all rolling contacts elastohydrodynamically.
- Lubricates contact areas between the rolling element, raceways, and retainers that are not truly rolling, but sliding.
- Protects all surfaces from corrosion.
- Dissipates heat.
- Acts as a sealing barrier to any contaminants and removes them from the bearing.

One of the purposes of lubricating a rolling bearing is to reduce friction. The source of frictional resistance in ball bearings is attributed to sliding action between the balls, races, and retainers. Also, frictional resistance is created between the rolling parts (balls) and the lubricating agent. Fluid friction due to shearing within the lubricant itself is another friction factor and is a function of its chemical and physical composition.

Another contributing source to frictional resistance is the deformation of the rolling element and the race surface. Deformation of the ball and races occurs under load due to the elastic qualities of bearing steel. This deformation is minute but it does occur. The amount of deformation is a function of rolling element size, steel elasticity, geometry, and the magnitude and direction of the applied load. When the rolling element is in a static condition, the load is generally distributed in a symmetrical pattern on the rolling element and race. If the rolling element is a roller, the load contact area is roughly a line along the roller. The area of line contact is affected by the roller shape and its curvature. With ball bearings the contact area is referred to as a "point," but it is actually an ellipse. When a tangential load is applied to the rolling element and causes it to move, the steel in the race bulges in front of the rolling element and flattens behind it (see Figure 9–57). The ball or roller also undergoes shape changes as it moves. This elastic deformation causes resistance to movement. We have, therefore, resistance to movement from rolling, sliding, and deforming components, and due to the lubricant itself generating friction and heat. That is why lubrication is essential to separate rolling bearing elements and reduce frictional heat.

During the operation of a properly lubricated rolling element bearing under normal loads and speeds, a bearing lifesaving phenomenon occurs. The phenomenon is "elastohydrodynamic lubrication." Elastohydrodynamic lube film (EHL) is a condition in which surfaces of heavily loaded rolling element bearings are either completely separated, or separated in part, by a very thin lubricant film under extreme pressure. Extreme pressures at the point of contact cause micro deflections in the metal and the surfaces are momentarily pressed together and flattened slightly (elastic deformation). Elastic deformation of the contacting surfaces traps the lubricant, subjecting it to extremely high pressures that increase its viscosity and its load-carrying capacity. When the rolling elements roll on, the surfaces return

FIGURE 9–57
Ball bearing deformation during operation.
Courtesy of Timken/Torrington.

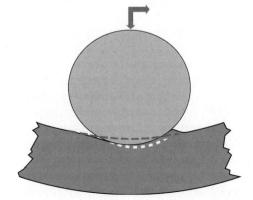

to their original shape and viscosity returns to its original condition. If a lubricating agent were not present in the contact ellipse, a metal-to-metal condition would be present and welding, seizing, and tearing away of surfaces would occur. One might think that the pressure from applied loads would squeeze out the lubricant from where it is needed, but this is not what happens in an elastohydrodynamic state. A crude analogy would be the hydroplaning of an automobile tire on a wet or icy surface. The film thickness is almost immeasurable, but it is present if the correct lubricant has been properly applied.

Pure hydrodynamic lubrication is the ideal condition for rolling element bearings. When the bearing is stationary or during initial start-up conditions, only a boundary layer lubrication state may be present. This is a critical time in the life of the bearing, and preflooding or wetting of rolling surfaces might be required under certain conditions of high loads.

Bearing Lubricant Types and Selection

The type and selection of the correct type of lubricant is based primarily on the operating conditions of the bearing. Factors including load, speed, temperature, type of application, environment, delivery systems, and many others must be considered when selecting the proper lubricant. (Further information regarding lubrication specifics can be found in the Lubrication chapter.)

There are two basic types of lubricants used with bearings: grease and oil. Each has distinct advantages and disadvantages. Oil is better for high-speed bearing applications. It will diffuse heat quicker and can be cooled in a circulation process. The amounts delivered to the bearing can be metered. There is an assortment of delivery systems that can be used with oil. Oil can be filtered for reuse in circulating systems. Oil does tend to leak and must have adequate sealing to prevent spillage. Grease is used in the vast majority of bearing applications operating under what are considered normal loads and speeds. Grease is an excellent choice for bearing lubrication if the correct type is properly dispensed. Grease stays put and clings to bearing surfaces. Very little grease is lost from the system from leakage and tends to stay in the housing. It acts as a more efficient barrier to contaminants than oil and can be purged to push foreign materials out and away. It is also generally less expensive than oil.

The single most important property of a lubricant is viscosity. Viscosity is a measure of the resistance to flow. Using the common unit of measurement of SUS (Saybolt Universal Second), a general recommendation of at least 100 SUS at normal bearing operating temperature for the application is adequate. Ideally, using oil that has the same viscosity at all temperatures would be best. This is not possible because of the effect that temperature changes have on viscosity. Oil becomes less viscous, or thins out as temperature rises. This effect can be inhibited somewhat with VI (Viscosity Index) additives. Temperature, speed, and the size of the bearing all play a role in selecting the correct oil viscosity for bearing lubrication. Figure 9–58 is a viscosity selection chart that may be used to approximate the proper oil viscosity for bearing applications.

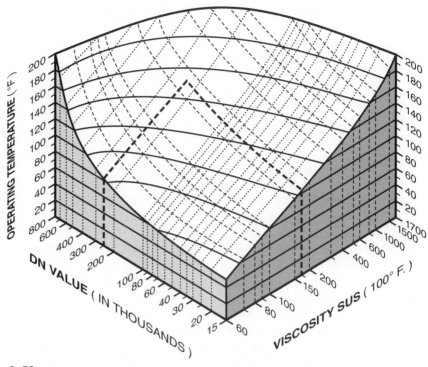

FIGURE 9–58
Oil viscosity selection.
Courtesy of Timken/Torrington.

As an example of the chart use:

1. Determine the DN value of the bearing. Multiply the bore diameter, measured in millimeters, by the speed of the shaft, measured in revolutions per minute.
2. Select the operating temperature of the bearing.
3. Enter the DN value in the DN scale on the chart.
4. Follow or parallel the dotted line to the point where it intersects the selected solid temperature line.
5. At that point follow or parallel the nearest dashed line downward and to the right to the viscosity scale.
6. Read the approximate viscosity value expressed in SUS at 100°F.

Bearing Lubrication Amounts and Intervals

Delivering the correct amount of lubricant at controlled intervals is key to the survival of the bearing. Just as many bearings fail from too much, too little, and too often as they do from the wrong kind of lubricant. Filling the bearing with lubricant initially is just as important as relubrication. Sealed for life bearings do not require relubrication for the rolling elements. Many machines will have an additional grease chamber as part of the machine housing; filling it allows the grease to act as a barrier to contaminants. Oftentimes when relubricating a sealed for life bearing, the technician is using the grease to purge the housing of contaminants—not to relubricate the bearing. Relubrication is necessary when the life of the lubricant is exceeded and its lubricating properties have been lost. The life expectancy of the lubricant is dependent on numerous variables. A decision for relubricating can be made considering loads, speeds, operating environment, type of bearing, temperature, etc., but most often experience is the key. As a general rule, dirty and hot applications require more frequent relubrication.

Initial lubrication with oil should be done with clean oil of the correct type and viscosity. The level of the oil within the sump must provide splash to the highest bearing within the machine during operation. For housed bearing blocks the oil level should be approximately at the center of the lowest rolling element within the bearing. Sight glasses should be positioned to indicate both static and dynamic levels. Drain and vent plugs must be located to prevent spillage and allow ventilation, if needed. All oil feed holes, along with return gutters and troughs, should be kept clear and primed with a supply of oil.

Grease lubricated bearings should be handpacked after mounting with clean, gloved hands and clean grease. The hand application of grease should be done after mounting whenever possible to avoid the potential for contamination. Usually it is recommended to fill the bearing completely with grease. The housing should have free space with only 1/3 to 1/2 of the void within it filled. In very slow moving applications, 100 to 200 rpm or less, the housing cavity may be filled 80% or more. In these cases, such as conveyor pulley applications, the grease is primarily acting as a contaminant barrier. In high-speed applications, such as machine tools and motor bearings, a small amount of lubricant is usually adequate.

The relubrication intervals for bearings operating in oil are dependent on the operating conditions, degree of oil cleanliness, and quantity of oil in the system. Oil bath lubricating systems should be changed at least once per year providing the bearing temperatures do not exceed 120°F (50°C). Higher temperatures require more frequent changes. Applications with bearing operating temperatures at 220°F (105°C) should have the oil changed every three months. Regularly scheduled testing of the oil to determine oil properties and the presence of contaminants should be done. Inspection should include observation of any obvious changes such as color, foaming, or oil levels.

Relubrication of bearings operating in grease is a function of some of the same factors as those lubricated with oil. Relubrication is a function of speed, size, type, and environment. Figure 9–59 illustrates suggested relubrication intervals for different types of bearings operating at various speeds. Relubrication should be done prior to bearing failure, not as a means to quiet the bearing. Because many bearings that operate in severe environments are grease lubricated, the prevention of water or solid contaminants penetrating the bearing assembly is of extreme importance.

The amount of lubricant necessary to provide an EHL film is small, but the amount required to keep out foreign substances can be large. Grease acts as a

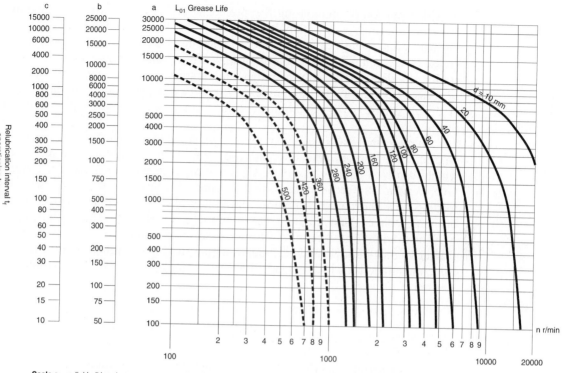

Scale a: radial ball bearings

Scale b: cylindrical roller bearings, needle roller bearings

Scale c: spherical roller bearings, CARB™ toroidal roller bearings, taper roller bearings, thrust ball bearings; cylindrical roller thrust bearings, needle roller thrust bearings, spherical roller thrust bearings ($0.5t_f$); crossed cylindrical roller bearings with cage ($0.3t_f$); full complement cylindrical roller bearings ($0.2t_f$)

d: bearing bore diameter (mm)

FIGURE 9–59

Grease relubrication chart. To use the chart determine the operating rpm of the bearing. Follow the line at that speed until it intersects with the bore size of the bearing in millimeters. Then read to the left in the appropriate column, depending on the type of bearing. This number is a suggested minimum lubrication interval in hours.

Courtesy of SKF.

barrier to foreign substances and when pumped under pressure it allows the bearing to be purged, pushing out contaminants. Replenishing a bearing with new grease serves the dual purpose of displacing lubricant that has deteriorated and flushing out contaminants. Purging can be accomplished through the sealing mechanisms. Care must be taken to apply grease at low, even pressures. Application can take place effectively by using a standard grease gun connected to a nipple/fitting that is placed in a drilled and tapped hole in the housing. Administering the grease at a low pressure will ensure that the seals will not get damaged or blown out. On larger bearing assemblies, purging may be done through a machined hole in the housing that has a "purge plug." A "purge plug" is a threaded plug that fits a hole in the lower portion of the housing. Some machines will have a relief valve that must be opened during relubrication. It is recommended that when relubricating a bearing with grease, the purge plug should be removed to allow old contaminated grease to vacate the bearing housing. The bearing should then be allowed to operate for a period of time with the purge plug removed to vacate any excess amounts of grease. Periodic inspection and cleaning of bearing housings is required to remove any grease or wax-like bases that have congealed or hardened in corners of the housing.

An oversupply of grease may cause churning and temperature rise within the bearing. Excessive amounts of grease usually mean more resistance to movement, which results in friction and heat. Heat can damage the bearing and the grease. Overgreasing of electric motors can cause the lubricant to flow into the windings, leading to a short circuit. Overgreasing may also cause contamination to the machine product. An overabundance of grease may result in a safety hazard.

The amount and frequency of grease supplied to a bearing should be based primarily on operator and technician experience in conjunction with the machine manufacturer's recommendations. As a guideline, the following formula may be used:

Grease quantity in ounces = outside diameter inches multiplied by the bearing width in inches, multiplied by .114,

or

Grease quantity in grams = outside diameter millimeters multiplied by the bearing width in millimeters, multiplied by .005.

ROLLING BEARING FAILURE ANALYSIS

Bearings are an integral part of any machine; when they fail, the operating performance of the machine and its life will be limited. The analysis of failed bearings is important for one primary reason. Bearings should be inspected and analyzed to determine the cause of failure so that the cause may be corrected. It is imperative that the technician examine the failed bearing and connected components thoroughly to determine the root cause of the premature failure. Premature failure can be considered anything less than its calculated and predicted L_{10} life. Oftentimes the cause and symptoms exhibited by the bearing failure become intertwined. It is a difficult practice to determine the exact primary cause of failure because it is typically obscured by consequential damage to related components. Primary failure causes are closely followed by secondary causes. For example, a bearing may exhibit signs of abrasive wearing of metal surfaces (secondary) in conjunction with foreign particle contamination (primary). Many times the failure is not determined until it has reached a catastrophic condition and determining the cause is near impossible without additional input. When a severely damaged bearing that has seized up is removed from a machine, it becomes a question of "which came first, the chicken or the egg?"

Determining the causes, understanding the contributing factors, and identifying symptoms of bearing failure is a complete study in itself. It is certainly beyond the limited scope of this book. Bearing manufacturers have published numerous handbooks on failures that should be consulted. This section will attempt to simplify the primary reasons for bearing failures and illustrate some of the most common.

Detecting the failure early by observing such obvious signs as wear, smoke, noise, heat, and vibration is an important aspect of the problem solving process. Early detection will result in accurate analysis of the root cause and prevent further damage to connected components. It also allows for planned replacement of the failing bearing within the regular maintenance schedule. We traditionally rely on our senses to determine the operating performance of a bearing. Our eyes, ears, nose, or a screwdriver placed between the ear-bone and bearing (not recommended!) have all been used to determine the condition of the bearing. We still rely on these senses but through modern technology we are able to quantify, isolate, and record (trend) these symptoms. Condition monitoring tools such as vibration analysis, thermography, and ultra-sound, to name a few, are valuable tools in the analyst's arsenal. Trained individuals should use these failure analysis methods in a planned and recorded program. One of the keys to proper failure analyses is determining what is a symptom, such as vibration, and what may be the cause, such as misalignment.

As part of the diagnostic process the failed bearing should be inspected and questions need to be asked. Careful attention should be paid to the load zone or contact pattern that will appear on the raceway surfaces of a used bearing (see Figure 9–60). These patterns will indicate the loading patterns within a bearing and are a valuable clue to the cause of failure. For example, a bearing that has been subjected to excessive thrust loading will have a pattern such as Figure 9–61. Misalignment will have a pattern similar to Figure 9–62. The maintenance technician should ask questions of the operator and service personnel. Those questions should include: How long did it operate? What type and method of lubrication was applied?

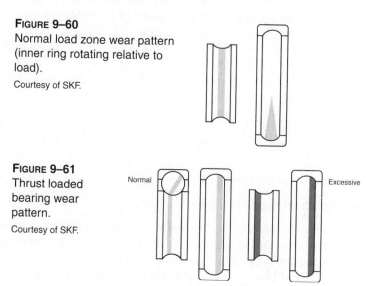

FIGURE 9–60
Normal load zone wear pattern (inner ring rotating relative to load).
Courtesy of SKF.

FIGURE 9–61
Thrust loaded bearing wear pattern.
Courtesy of SKF.

Normal

Excessive

FIGURE 9–62
Misaligned bearing wear pattern.
Courtesy of SKF.

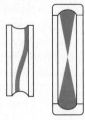

What installation methods were used? What symptoms were observed? Were oil, vibration, or temperature readings obtained prior to and during failure? These questions and many others will assist the technician in determining the cause of failure and the corrective steps required to eliminate future failures.

Many terms used in describing failure modes are used interchangeably among experts. Certain terms have definitive meanings and describe the symptom; others describe the cause. Terms such as "brinneled" (dented) may be used to describe the appearance of the race surface, but the root cause was a dropped bearing prior to installation. The technician should have a rudimentary understanding of these terms to help him define and understand causes and effects of bearing problems. Some of the most important terms used in bearing failure analysis will be explained in this section.

There are many reasons why bearings fail. Operating conditions, installation practices, environment, loads, etc., all contribute to the lack of adequate service life. Experience teaches that there are usually numerous contributing factors that lead to the premature failure of a particular bearing. In some cases, e.g., a failure attributed to corrosion, the corrosion may have occurred during storage, handling, from the environment, or even from the lubricant. There are hundreds of descriptive terms used to describe various failure types, modes, conditions, symptoms, and contributing causes. For the sake of clarity and discussion, most bearing failures can be placed into six categories of primary causes. These causes are listed below, not necessarily in order of importance or frequency.

- Manufacturing defect
- Fatigue
- Handling and installation
- Operational factors
- Environment
- Lubrication

Manufacturing Defect

Manufacturing defect as a cause of premature bearing failure is by far the least common. Because reputable manufacturers of precision rolling element bearings employ exacting inspection procedures and standards throughout the step-by-step production process, there is seldom if any error. Occasionally bearings are mismarked or improperly boxed, but the bearing that is made to incorrectly specified dimensions and tolerances is almost nonexistent. Occasional problems arise with cutting, stamping, grinding, and finishing machine tools, but these are quickly detected and corrected prior to shipment. Rolling bearing manufacturers warranty their product to be free of manufacturing defects. If a legitimate defect problem exists, bearing manufacturers want to be made aware of the issue so it may be corrected.

Fatigue

Fatigue failure is the right reason why bearings should fail. When a bearing meets or exceeds its calculated life expectancy (L10), it will probably fail from fatigue. Fatigue failure of a bearing occurs assuming that proper lubrication, installation, selection, sealing, and operational factors have been addressed. Fatigue is a result of cyclic stresses applied beneath the load-carrying surfaces of the bearing. Steel has a fatigue life and when exceeded, begins to fail. A crude analogy would be to take a clothes-hanger and bend it back and forth repeatedly until it cracks and snaps. With bearings the fatigue life is directly related to the quality and degree of cleanliness of the steel. Fortunately steel is cleaner now than ever before. Any micro-inclusions (impurities) within the steel that are located in the load zone area will act as a focal point for stress. Micro cracks originating from the stress risers will propagate and eventually make their way to the surface, creating what is known as "spall." This spalling of the race or rolling element surface appears as a small area of removed metal—a "pot-hole" of sorts.

The beginning of a fatigue spall will be subsurface and may appear initially as a crack in the steel surface. Early fatigue spalling is shown in Figure 9–63. Spalling is a progressive failure and the time from incipient to advanced spalling varies. As the spalled area propagates and advances, it will create a noisy bearing that has some associated vibration that can be detected with monitoring equipment. More

FIGURE 9–63
Initial fatigue spall.
Courtesy of SKF.

FIGURE 9–64
Advanced fatigue spall.
Courtesy of SKF.

FIGURE 9–65
Fragment denting.
Courtesy of SKF.

advanced spalling is shown in Figure 9–64. If caught early enough, the amount of metal removed from the spalled area will be limited and not cause additional damage to race surfaces. The metal that is removed can end up between the rolling elements and races, interrupting the lubricant film and denting the surfaces (see Figure 9–65). Eventually, in the case of advanced extreme spalling, catastrophic failure may occur and the entire ring or roller will crack. Ideally fatigue failure is what should be the desired cause. In reality, fatigue failure represents only a small percentage of bearing failures in industrial applications.

It is important to differentiate between spalls that have occurred as a natural result of the bearing being at the end of its useful life, and those that have been induced through artificially created stress risers. Chemical attack, inadequate lubrication, vibration, or physical damage from faulty mounting practices can also cause spalls to eventually occur on the race or rolling elements.

Handling and Installation

Improper handling of rolling element bearings is a major reason for premature failure. Poor fitting practices, contamination during assembly, improper installation techniques, and incorrect clearance/pre-load setting also fit in this group.

Many bearings are committed to premature failure prior to turning one revolution. If a bearing is stored improperly and exposed to ambient vibration or hostile environments, it will fail early. Exposure to moisture in storage will cause the bearing to be stained and corroded (see Figure 9–66). If the bearing is installed by applying the mounting forces through the rolling elements, dropped, or receives a severe shock load, brinnel marks will be evident at roller intervals (see Figure 9–67). Brinneling is the plastic flow of raceway material. "Scuffing" of the contact surfaces can also occur during installation from mishandling or the bearing not being properly

FIGURE 9–66
Corrosion staining on a
bearing roller from moisture
and acids.
Courtesy of SKF.

FIGURE 9–67
Magnified view of a
brinneled bearing
surface.
Courtesy of SKF.

FIGURE 9–68
Scuffing from excessive
mounting forces
magnified 100 times.
Courtesy of SKF.

orientated during installation (see Figure 9–68). Scuffing is a shallow marking or smearing of the steel at the contact points of race and rolling elements. Vibration forces that are transmitted through the rolling elements and races can cause "false brinneling." False brinneling is a gradual wearing away of material at metal-to-metal contact points from repeated vibrating stress and movement (see Figure 9–69). This can happen during shipping, storage, or while the machine is shut down.

How the bearing fits on the shaft and in the housing has an effect on its life. Relative movement of a ring to shaft or ring to housing is referred to as "creeping." Creeping of the rings can smear or score bearing ring surfaces (see Figure 9–70).

FIGURE 9–69
False brinneling from
vibration.
Courtesy of Timken/Torrington.

FIGURE 9–70
Scoring caused by inner
ring creep.
Courtesy of SKF.

FIGURE 9–71
Fretting corrosion of rings.
Courtesy of SKF.

FIGURE 9–72
Cracked inner ring from
excessive press fit.
Courtesy of SKF.

FIGURE 9–73
Pitting from electrical
arcing.
Courtesy of SKF.

Improper fitting of the bearing onto the shaft or into the bearing housing can cause fretting corrosion (see Figure 9–71). Fretting corrosion is the wearing away of the nonrolling component metal high points and their eventual oxidation. High interference fits can actually crack the bearing rings (see Figure 9–72). Bearing damage can also be caused by faulty shaft and housing seats that are concave, convex, or tapered.

Improper grounding of welding equipment can cause the current to pass through the bearing and arc at the contact points of rollers and races (see Figure 9–73). These arc points create micro-pits and will be a focal point for stress failure.

Operational Factors

Operational factors such as misalignment, vibration, excessive loads, or thrust loads will result in premature failure. Many of these failures will eventually cause spalling. Excessive speeds and loads are in this category. Misalignment of taper rollers creates edge loading and eventually localized spalling (see Figure 9–74). Excessive thrust is evident by observing an off-center roller path that in extreme cases can cause the metal to flow to one side (see Figure 9–75). Vibration during

FIGURE 9–74
Edge loading/spalls on
taper roller bearing.
Courtesy of SKF.

FIGURE 9–75
Failure of inner ring from
excessive thrust.
Courtesy of SKF.

operation can cause a wavy pattern on race surfaces (see Figure 9–76). This type of vibration failure is closely associated with abrasive wear.

Another failure that can be categorized as an operational failure is "fluting." Fluting is the pattern that appears on the bearing created by the passage of stray electrical current through the bearing while it is turning (see Figure 9–77). The current that is seeking ground can be generated by magnetic fields, short circuits, improper grounding of welders, or in rare cases static electricity. If the process continues over a prolonged period of time, numerous individual pits may be created. An orderly narrow wavy pattern of consistent depth and width will appear on the race or rollers. This phenomenon is not completely understood but is probably

FIGURE 9–76
Vibration pattern
associated with abrasive
wear.
Courtesy of Timken/Torrington.

FIGURE 9–77
Fluting pattern on roller
and ring.
Courtesy of SKF.

related to the cycle of current, amperage, vibration, and shock loading that is ongoing within the bearing. Once fluting has starting, it is self-perpetuating and will result in vibration. The secondary failure mode associated with fluting may be vibration and abrasive wear from removed metal.

Environment

The operating environment that a bearing is subjected to can determine the bearing's service life. Exposure to moisture, chemicals, dust, dirt, or excessive heat can result in a shortened life. Contaminating agents interrupt the all-important lubricant film and cause metal-to-metal contact, wear, and welding. Foreign objects can also dent and score the rolling elements and raceways. Smaller particles that are trapped and suspended within the lubricant will act as a lapping compound and wear away the surfaces of the steel and retainer materials. This is called "abrasive wear" (see Figure 9–78). In some extreme cases the geometry of the races will be altered and the clearances opened up to such an extent that vibration would occur due to the looseness within the bearing.

Moisture from external sources such as wash-down hoses or from the operating environment (rain) can penetrate into the bearing and create a problem with staining (see Figure 9–79) or severely etching the surfaces (see Figure 9–80). Water, combining with corrosive acids formed in oxidized lubricants, can cause numerous pits.

FIGURE 9–78
Abrasive wear and
vibration failure pattern.
Courtesy of SKF.

FIGURE 9–79
Moisture
staining/corrosion.
Courtesy of Timken/Torrington.

FIGURE 9–80
Severely etched bearing race
surface.
Courtesy of SKF.

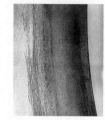

Lubrication

Lubrication is probably the primary cause of most bearing failures. This is because of an improper type or amount of lubricant (Handling & Installation), speeds are too high for grease (Operational), or if the grease/oil were contaminated (Environmental). The correct amount, type, application method, viscosity, and frequency of lubrication can have a positive or negative effect on the service life of the bearing. Lubrication failure is not just a case of not getting enough.

Bearings that receive inadequate lubrication will have a glazed appearance. This "glazing" is a result of metal being pulled from the surface after momentarily welding together. This "picking," or "frosting" as it is sometimes called, leaves a rough finish on the race and roller surfaces (see Figure 9–81). If the surfaces appear rough or discolored, it is probably due to problems with lubricant viscosity. In some cases, "smearing" will take place as the roller or ball skids into the load zone (see Figure 9–82). A lubricant that is too stiff or of an inadequate supply can yield smearing. Spherical roller ends may have a pattern known as "cycloidal smearing" from improper lubrication (see Figure 9–83). Smearing may also occur on any

FIGURE 9–81
Glazed or frosted bearing
surface from inadequate
lubricant film.
Courtesy of SKF.

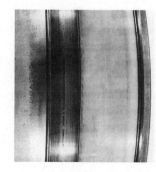

FIGURE 9–82
Skid smearing on
raceway in load zone
of raceway.
Courtesy of SKF.

FIGURE 9–83
Cycloidal smearing on roller
ends.
Courtesy of SKF.

FIGURE 9–84
Fine grain spalling
on rollers resulting
from inadequate
lubrication.
Courtesy of
Timken/Torrington.

thrust faces. Contact areas within the bearing that have a lack of adequate lubrication will produce a fine grain spall prior to catastrophic failure (see Figure 9–84).

When lubrication supply or viscosity is inadequate, the temperature will rise. This temperature increase, if high enough, will cause metallurgical changes and cause a discoloration of the bearing steel. Initially the bearing steel will turn a straw color, then blue, and eventually black. In cases of a total lack of lubricant, races and components will not only discolor but will change shape (see Figure 9–85) and possibly even weld together (see Figure 9–86).

FIGURE 9–85
Discoloration and shape change of rolling elements attributed to inadequate lubrication, heat, and pressure.
Courtesy of SKF.

FIGURE 9–86
Welding of bearing components from heat and no lubrication.
Courtesy of SKF.

Student Exercise

List the primary components of an antifriction-type ball bearing (Figure 9–87). Identify the various types of bearings. Fill out the exercise worksheet.

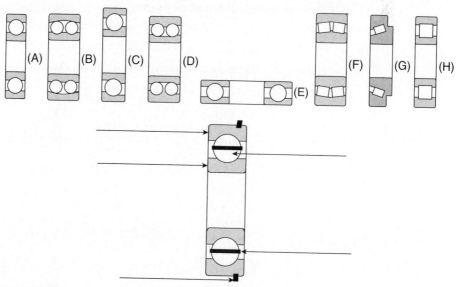

FIGURE 9–87
Student exercise worksheet.

Questions

1. What is a bearing and what is its purpose?
2. What is the difference between plain and antifriction bearings?
3. List three reasons a plain bearing would be used as opposed to an antifriction bearing.
4. List three reasons an antifriction bearing would be used as opposed to a plain bearing.
5. List four installation, application, or operating concerns of plain bearings.
6. Name and describe three materials used in plain bearings.
7. What is an oil groove cut on the inside diameter of a plain bearing used for?
8. Name five shapes of antifriction rolling elements.
9. What are the basic parts of an antifriction rolling bearing?
10. What three basic types of loads are rolling bearings required to handle?
11. What two groups of variables effect bearing life?
12. What is bearing radial internal clearance?
13. What is the difference between preload and end-play?
14. Name five types of rolling element bearings.
15. Why is it necessary to have a proper fit of the bearing onto the shaft and into the housing?
16. List three types of housed bearings.
17. Why is lubrication important to bearings?
18. Name the advantages of both oil and grease lubrication.
19. What is the recommended tool or method for removing bearings?
20. List three general methods for correctly installing bearings.
21. What is the preferred reason for bearing failure?

CHAPTER 10

V- and Synchronous Belt Drives

This chapter examines V-belt and synchronous drive systems. V-belt and synchronous drives are used to connect parallel shafts of separate machines transferring power and rotary motion. We will examine the numerous types of belts, sheaves/pulleys, bushings and numbering systems; how belt drives function and operate; and the various forces involved in the operation of drives. Selection, installation, and maintenance of belts will also be covered.

Objectives

Upon completion of this chapter, the student will be able to:

✔ Understand and explain the basic operating principles of V-belt and synchronous belt drives.
✔ Identify the various types of V-belts and synchronous belts.
✔ Identify the various types of drive sheaves/pulleys and their mounting methods.
✔ Understand and use the numbering system for various types of belt drives.
✔ Have a basic understanding of V-belt selection procedures.
✔ Demonstrate correct alignment and tensioning procedures for belt drives.

INTRODUCTION

The modern V-belt drive was developed in the early twentieth century as a means to transmit power safely and efficiently between parallel shafts (see Figure 10–1). Prior to its use in the late eighteenth and early nineteenth century the use of flat drive belting was common. These early belt systems typically had a single power source driving an overhead line shaft that turned pulleys with flat leather belting. Around that same time period grooves were cut into the flat pulleys to fit round wire, manila, or cotton ropes for multiple and endless drives. These grooved pulleys became known as sheaves.

In 1917, John Gates is credited with inventing the first rubber and textile V-belt. His V-belt was basically a V-shaped belt of rubber and fabric composition. It soon became the industry standard drive belt.

Two significant inventions of the early twentieth century, the automobile and the electric motor, drove the research and development of the V-belt. The electric

FIGURE 10–1
Typical V-belt drive.
Courtesy of Emerson Power
Transmission.

motor required an efficient and compact connecting belt for power transmission. From these single belt connections, multiple belt drives became a popular means to transmit large amounts of power. The multiple belt drive not only allowed more horsepower to be transmitted but it eliminated some of the jerking and slapping that is associated with a single belt drive. There are obvious benefits to connecting the prime mover in close proximity to the driven machine, which is made possible with V-belts.

The advantages of V-belt drives include low relative cost, quiet operation, no lubrication, energy efficiency, a broad range of available ratios and sizes, shock absorbancy, and easy installation. The disadvantages of the V-belt drive are more of an issue in applications that have adverse operating environments such as extreme temperature ranges, high moisture, or chemically filled atmospheres. On systems that must maintain positive speed, synchronous belts—rather than V-belt drives—should be used. A synchronous belt is a toothed drive belt that does not slip. Synchronous belts have their own advantages and disadvantages that are discussed later in this chapter.

There are numerous types of V-belts in use today on a wide range of machines. The V-belt of today is a product of considerable research and development. The combination of different types of woven synthetic fibers and rubbers incorporated into a modern V-belt make it a clean, efficient, and safe means of power transmission.

V-BELT PRINCIPLES OF OPERATION AND TERMINOLOGY

V-belts make use of the mechanical principle of the wedge and the belt being in tension between the driver and driven shafts. The V-belt depends upon frictional contact between the sidewalls of the sheave and belt to transmit power. The tension forces cause the belt to be wedged into the groove (see Figure 10–2a). The efficiency of the drive is directly related to the grip of the belt to the sheave. Although a V-belt drive is considered a relatively efficient means of power transmission, it loses effectiveness if the belt slips in the sheave. Other key factors such as the condition and shape of the contact surfaces and the installation tension applied also affect the belt drive system. The amount of contact between the belt and sheave depends on the sheave diameters, groove shape, the center distance between the

FIGURE 10–2a
V-belt tensions
and friction
forces.

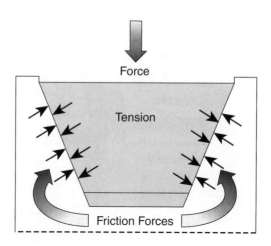

Force

Tension

Friction Forces

driver and driven machines, and the surface speed. The two most important princi-ples to successful V-belt operation are tension and friction.

V-Belt Working, Effective, and Peak Tensions

A force or tension is applied when a belt is installed. The tension wedges the belt into the groove by moving the driver and driven shafts apart. This is accomplished by mov-ing the prime mover, usually a motor, on an adjustable base away from the driven sheave. When the prime mover begins to turn, it causes the connected shaft to move. A V-belt does its work by pulling the driven sheave, which causes it to rotate. In a two-wheel drive (driver and driven) there are two spans or sides of the belt. Both sides must be under tension to operate correctly. When the drive is not operating (static condition), the tension on the belt is equal on both sides. In a dynamic or running con-dition, the side being pulled into the driver sheave by the rotation of the prime mover is referred to as the tight side of the drive and designated T1. A certain amount of ten-sion must be maintained on the other side to keep the belts in the groove and prevent slipping. This slack side, designated T2, is actually holding back (see Figure 10–2b).

The difference between the tight side and the slack side tension equals the net pull or effective tension. Design tension ratios—the difference between the tight side and the slack side (T1 − T2)—are generally around 5:1.

T1 − T2 = Net Pull or Effective Tension (Te).

For example: 50# (T1) − 10# (T2) = 40# Net Pull or Effective Tension

The transmission of power by the V-belt drive depends on effective tension or net pull, which is how hard the belt is pulling. The effective tension depends only on the horsepower being transmitted and the belt speed. V-belt speed is usually meas-ured in a velocity of feet per minute (fpm).

Power is usually measured in horsepower units with V-belts. The horsepower to be transmitted by a V-belt equals effective tension in pounds (Te) multiplied by the

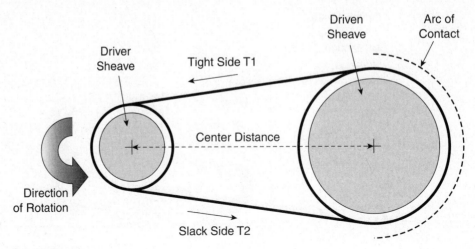

FIGURE 10–2b
V-belt drive terminology.

belt speed in feet per minute (velocity), the sum of which is divided by 33,000 pound-feet per minute.

$$\text{Te (T1} - \text{T2)} = 33{,}000 \times \frac{\text{HP}}{\text{Velocity}} \qquad \text{HP} = \frac{\text{Te} \times \text{Velocity}}{33{,}000}$$

Higher speed means less net pull. Higher horsepower load means more net pull is needed to operate properly. Working tensions in a V-belt drive are fairly predictable if one can determine the horsepower transmitted and the velocity.

Total tension within the drive needs to be considered. The tight side tension plus the slack side tension equals the total tension in the drive.

$$\text{T1 (tight side)} + \text{T2 (slack side)} = \text{Total Tension} = 50\# + 10\# = 60\#$$

The sum of the tight side (T1) and slack side (T2) is the tension the bearings and shafts must support. On a drive with two equal diameter sheaves, the total tension is the tension that the bearings and shafts feel. If the sheave diameters are unequal, they feel the vector sum of T1 and T2. If a drive were to have two belts, the tension per belt would be reduced by 1/2, and the total tension of the drive would not change. The shaft and bearings also are subjected to installation tensions that must be controlled to prevent failure of the supporting bearings from excessive overhung loads.

There are other tensions in an operating V-belt that affect the belt. There are centrifugal and bending tensions. As the belt travels around the sheaves it tries to pull away from the grooves due to centrifugal force. As it bends around the sheaves it resists bending and compression, creating more tensions within the belt. The smaller the sheaves, the greater the forces. Working, bending, centrifugal, and installation tensions all combine to form peak tensions. The V-belt is exposed to these tensions in a cyclic manner as it rotates around the drive. These repetitive stresses eventually cause the belt to fatigue. Tension is what helps make a belt work, but eventually it is also what makes it fail. Manufacturers of V-belts consider these variables in the design and construction of the belt. The life of the belt can be calculated when loads are known and power ratings can be assigned which will limit the peak stresses and allow the desired life.

V-Belt Drive Ratio

The difference in diameter between the driver and driven sheaves can be expressed in a ratio. For example, a 4″ diameter 1750 rpm driver (motor sheave) and a 12″ driven sheave have a ratio of 3:1. This would be a speed down or speed reduction drive. The resultant driven speed would be 583.3 rpm. Putting the larger sheave on the driver and the smaller on the driven would result in a speed up drive.

(Driver Sheave Diameter ÷ Driven Sheave Diameter) × Driver RPM
 = Driven RPM

Belt Drive Arc of Contact

Arc of contact is the amount of belt wrap around a sheave groove expressed in degrees. Two sheaves with the same diameter, a 1:1 ratio, would each have a 180° arc of contact. The power rating of a V-belt is based on an arc of contact of 180°. When two sheaves are of different diameters (for example, 3:1 ratio) there will be contact greater than 180° on the larger diameter sheave and less than that on the smaller sheave. The determining arc of contact is always the smaller sheave. When it is less than 180° the V-belt's horsepower rating must be adjusted to a value reflecting the amount of contact. This is done through the use of tables provided by belt manufacturers.

Belt Drive Center Distance

The center distance of the V-belt drive is the measured distance between the centerline of the driver and the driven shaft (see Figure 10–2b). On drives that have large ratios and short center distances, the amount of belt wrap can be inadequate to transmit the required power without substantially increasing the tension. As a guideline, the minimum center distance should be equal to the sum of the two sheave diameters. General recommendations for center distances are based on the sheaves and belt sizes that are available from manufacturers.

Belt Pull and Bearing Loads

The total bearing load induced by the belt drive is a combination of sheave or pulley weight and the belt pull. Belt pull must be vectorially added to the weight of the sheave and shaft to find the true bearing loads.

Sheave and pulley weights can be found by consulting the appropriate manufacturer tables or contacting the supplier. Belt pull can be calculated with the drive
data. It is a function of the following variables:

■ Horsepower transmitted—for the same drive, more horsepower requires more
belt pull.

■ Belt speed—for the same horsepower, higher belt speed (large sheave diameters) means less belt pull.

■ Arc of contact—for V-belt drives, reduced arc of contact requires greater tension
to prevent slip, resulting in increased belt pull for the same horsepower load.

■ Total drive installation tension—in a V-belt drive, this can vary depending upon
how it is installed.

It is important to note that the required belt pull is independent of the number
of V-belts used on a drive. The number of belts affects only the amount of overhang
from the center of the belt pull to the bearings.

When designing a new drive system, it is sometimes necessary to calculate the
belt pull to properly size the shafts and bearings. On an existing piece of equipment, it
is a common practice to assume that the driven equipment can tolerate as much belt
pull as the driving machine. It is usually required to investigate allowable belt pull for
the driver only. The driver is usually an electric motor or an engine. Motor manufacturers or NEMA publishes minimum sheave diameter tables for the purpose of limiting belt pull to acceptable amounts. Recommended minimum sheave diameters are
used to reduce bearing loads and provide longer belt life (see Figure 10–3).

For internal combustion engines the drive designer and the machine designer
should collaborate to determine belt pull. Typically the power takeoff manufacturer
can supply the appropriate formula to determine belt pull.

The resultant shaft pull of a conventional V-belt drive installed with the proper
installation tension may be calculated by applying the formulas provided in V-belt
manufacturers' manuals.

V-Belt Idlers

Idlers act as a belt tightening sheave or pulley on a drive to increase tension. Idlers
are occasionally used on V-belt and synchronous drives for the following reasons:

■ To provide take-up on fixed center drives.
■ To clear obstructions.
■ To eliminate belt whip.
■ To maintain tension by a spring or weighted idler.

If at all possible, idlers should be avoided because they have a tendency to shorten
belt life and reduce power ratings. Idlers may be situated either outside or inside
the drive. Outside idlers increase the arc of contact on the adjacent sheaves. It is

Frame No.	Shaft Dia. (In.)	Horsepower at Synchronous Speed, rpm				Super HC® V-Belts & PowerBand® Belts	Hi Power® II, PowerBand & Tri-Power® V-Belts
		3600 (3450)*	1800 (1750)*	1200 (1160)*	900 (870)*	Min. Outside Dia. (In.)	Min. Datum Dia. (In.)
143T	0.875	1-1/2	1	3/4	1/2	2.2	2.2
145T	0.875	2–3	1-1/2 – 2	1	3/4	2.4	2.4
182T	1.125	3	3	1-1/2	1	2.4	2.4
182T		5	—	—	—	2.4	2.6
184T		—	—	2	1-1/2	2.4	2.4
184T	1.125	5	—	—	—	2.4	2.6
184T		7-1/2	5	—	—	3.0	3.0
213T	1.375	7-1/2 –10	7-1/2	3	2	3.0	3.0
215T		10	—	5	3	3.0	3.0
215T	1.375	15	10	—	—	3.8	3.8
254T		15	—	7-1/2	5	3.8	3.8
254T	1.625	20	15	—	—	4.4	4.4
256T		20–25	—	10	7-1/2	4.4	4.4
256T	1.625	—	20	—	—	4.4	4.6
284T		—	—	15	10	4.4	4.6
284T	1.875	—	25	—	—	4.4	5.0
286T	1.875	—	30	20	15	5.2	5.4
324T	2.125	—	40	25	20	6.0	6.0
326T	2.125	—	50	30	25	6.8	6.8
364T		—	—	40	30	6.8	6.8
364T	2.375	—	60	—	—	7.4	7.4
365T		—	—	50	40	8.2	8.2
365T	2.375	—	75	—	—	8.6	9.0
404T		—	—	60	—	8.0	9.0
404T	2.875	—	—	—	50	8.4	9.0
404T		—	100	—	—	8.6	10.0
405T		—	—	75	60	10.0	10.0
405T	2.875	—	100	—	—	8.6	10.0
405T		—	125	—	—	10.5	11.5
444T		—	—	100	—	10.0	11.0
444T		—	—	—	75	9.5	10.5
444T	3.375	—	125	—	—	9.5	11.0
444T		—	150	—	—	10.5	—
445T		—	—	125	—	12.0	12.5
445T	3.375	—	—	—	100	12.0	12.5
445T		—	150	—	—	10.5	—
445T		—	200	—	—	13.2	—

*Approximate Full Load Speeds

FIGURE 10–3
Chart lists the NEMA minimum sheave diameters allowable for a particular motor along with the shaft diameters based on the motor frame sizes and motor rpm.
Courtesy of The Gates Rubber Co.

recommended that they be placed on the slack side of the drive. Outside idlers should be 1/3 larger than the smallest loaded sheave; inside idlers should be at least as large as the smallest loaded sheave. Mounting brackets for idlers should be sturdily constructed and properly aligned. Drive problems described as "belt stretch,"

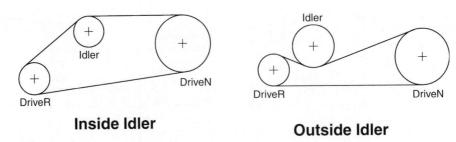

Inside Idler

Outside Idler

FIGURE 10–4
Typical belt idler placement. Note the distance of the idler with respect to the driving sheave.
Courtesy of The Gates Rubber Co.

"short life," and "belt vibration" are often caused by improperly applied and installed idlers. Idlers impose additional bending stresses on the belt resulting in less life. Figure 10–4 represents typical idler installation.

V-Belt Matching

Many applications consist of a multibelt drive where more than one belt is necessary to transmit the horsepower required. The Rubber Manufacturers Association (RMA) standards specify permissible length variations for a set of classical or narrow industrial belts. For example, all belts up to 63″ in a set must not vary more than .15″ from the longest to the shortest belt. If they do not meet this specification the load will not be evenly distributed and the belts will wear out faster. Some longer V-belts may appear to hang unevenly when installed. This "sag" is termed the "catenary effect." Field tests have proven that it has virtually no effect on drive performance. On longer span drives where the sag is noticeable, it is recommended that a "run-in" procedure be done. This "run-in" process consists of starting the drive, letting it run under full load and then stopping, checking, and retensioning it to the recommended values. Running the belt under load will allow "tension decay" to take place as a result of the belt elongating and seating in the grooves. Belt sag will be less noticeable after a proper run-in procedure. On longer center distance drives, the use of banded belts will help eliminate belt whipping.

Some industrial belt manufacturers still group their belts by the older match number system. It involves numbers printed on individual belts that represent the measured belt length range. The numbers are grouped in sequential order for matching by length. Matched sets of belts that have closer tolerances than RMA standards may be provided upon request from certain manufacturers.

V-Belt Construction

Standard V-belts are a composite construction of special rubbers and synthetics. The parts of a V-belt are the tensile members, cushion, compression section, top rubber, and cover (see Figure 10–5). The tensile cord members of either single or multiple

FIGURE 10–5
Typical V-belt construction. Notice the location of the tensile cords.

Courtesy of The Gates Rubber Co.

layers are made of wound synthetic fibers such as polyester, and are the load-carrying element of the belt. A cushion surrounds each tensile member and bonds the cords to the carcass of the belt. It helps to absorb the shock loads imparted to the tensile fibers. The V-belt's bottom or compression section, made from rubber or a rubberized fabric, supports the tensile cords. A top rubber section equalizes cord tension loads and assists in alignment. The flexible woven cover wraps all of these components together and affords some protection and containment.

V-Belt Types

The Rubber Manufacturers Association (RMA) and the Mechanical Power Transmission Association (MPTA) have established standards for most V-belts and sheaves. The standards cover nomenclature, dimensions, tolerances, and other technical features. Letter and number combinations typically identify V-belts. Some of the general classifications of V-belts are fractional horsepower, classical, narrow, banded, double angle, V-ribbed, variable speed, and endless types.

Fractional Horsepower V-Belts Fractional horsepower V-belts are principally used for light duty service on fractional horsepower drives. They are designed for single belt drives and to run on small diameter sheaves. Prefix numbers 2L(0), 3L(1), 4L(2), and 5L(3) designate the nominal cross-section dimensions. The last three digits of the part number designate the length of the belt measured on the outside circumference. For example, a 3L480 (1480) part-number means 3/8″ top width and 48″ outside length.

Classical V-Belts Classical V-belts, sometimes referred to as industrial or standard V-belts, are designed to be used in heavy-duty applications. They are one of the original series of industrial V-belts. The nominal cross-section size is designated as A, B, C, D, or E (see Figure 10–6). The numbers following the letter prefix signify the belt length measured on the inside surface or circumference of the belt. For example, an A42 belt indicates a belt width of 1/2″ across the top and 42″ on the inside of the belt's circumference.

Notched V-Belts Classical V-belts are available in a "notched" or "cogged" form (see Figure 10–7). These notches relieve bending stress during operation, which allows the belts to run on smaller diameter sheaves than comparable unnotched

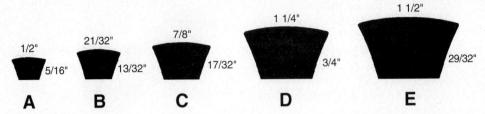

FIGURE 10–6
Classical V-belts—sizes A, B, C, D, and E—with dimensions.
Courtesy of The Gates Rubber Co.

FIGURE 10–7
Notched V-belt with
the tensile members
exposed on the side.
Courtesy of The Gates
Rubber Co.

belts. The notches or cogs can be cut into the belts after they are made, or they can be molded into the belt during the manufacturing process. An "X" is used in the belt number to designate a notched construction. For example, AX26 indicates a notched version of a 1/2″ top width and 26″ inside circumference belt.

Narrow Section V-Belts Narrow section V-belts are a relatively newer design with a narrower cross-section but comparable ratings to classical V-belts. Their compact form allows them to do more work in less space, reducing costs and space requirements. The standardized cross sections numbers are 3V, 5V, and 8V (see Figure 10–8). The top width code precedes the numerical nominal length designation. The length of these belts is measured by the approximate outside circumference or "effective length." Most of these belts are available in the notched or cogged form. For example, a 3VX750 indicates a 3/8″ top width and 75″ outside circumference notched narrow series belt.

Banded Belts Banded or joined V-belts are a matched set of V-belts joined together at the top with a permanent tie band of a high strength rubber material (see Figure 10–9). The backing provides lateral rigidity and prevents belt "rollover." They were developed for use in applications that are subject to extreme vibrations and pulsating and shock loads. These belts are available in the classical or narrow series.

FIGURE 10–8
Narrow series V-belts showing dimensions. The "X" series is a notched version of the belt.
Courtesy of The Gates Rubber Co.

FIGURE 10–9
Banded V-belt with five belts joined as one.
Courtesy of The Gates Rubber Co.

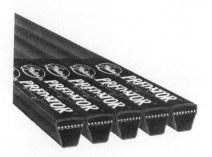

FIGURE 10–10
Double angle V-belt used on serpentine drives.
Courtesy of The Gates Rubber Co.

Double Angle V-Belts   Double angle V-belts are a special version of classical A, B, or C cross-section belts (see Figure 10–10). Their shape permits them to transmit power from both the top and bottom half of the belt. They are used on serpentine drives with multiple shafts rotating in different directions. Double angle V-belts are identified by a duplicate letter code prior to the length number; for example, AA, BB, CC.

Open End Belts   Open end or spliced belts are manufactured in the same cross-section as classical V-belts. They were designed for emergency replacement of V-belts on equipment with no tension adjustment, and on equipment with internal drives that have limited access. Open end belting is available in continuous lengths from a reel. Certain types require fastener clips and a special assembly tool to join the ends. Adjustable V-link belting is made of individual links joined by a rivet or by internal locking links (see Figure 10–11).

V-Ribbed Belt   V-ribbed or "Poly-V" (Micro-V®) belts are single endless rubber and fiber belts with a series of V-ribs molded lengthwise around the inside circumference.

FIGURE 10–11
A belt drive with adjustable
V-link belting.
Courtesy of The Gates Rubber Co.

FIGURE 10–12
V-ribbed or Micro-V belt.
Courtesy of The Gates Rubber Co.

These belts are smooth running and are used on high-speed applications. They are made in three cross-sections designated J, L, and M (see Figure 10–12). The part number code designates the length, pitch (distance between ribs), and number of ribs. For example, a 655L6 code has an effective pitch length of 65.5″ and six ribs spaced 3/16″ apart.

Metric V-Belts There are several different standards for metric V-belts. The numbering system differs between U.S. metric standards, and within the metric system between ISO and DIN standards. An ISO number indicating datum width and datum length in millimeters identifies classical metric belts manufactured to ISO standards. For classical belts made to DIN standards, the numbers indicate nominal top width and inside circumference in millimeters. Typically belts in the metric system are marked with the letters' width and length in millimeters. For example: an SPA1500 belt has a 1500 mm datum length. Metric belts are also available in the "notched" form.

Variable Speed Belts Variable speed belts (see Figure 10–13) are used on drives that incorporate the use of adjustable pitch sheaves. They are V-shaped but considerably wider than standard V-belts. The operating principles of the variable speed belt are similar to the V-belt. The variable pitch sheave design allows for the adjustment of one or more flanges of the sheave to be moved axially, causing a radial movement of the belt as it rides in the adjusted pitch diameter. Changing the diameter of the sheave permits drive ratio variations. The numbering system employs a combination of numbers and a letter to designate width, groove angle, and top width. For example, a 1422V450 variable speed belt has a 14/16″ top width, a 22° groove angle, and a 45″ pitch length.

FIGURE 10–13
Variable speed belt.

V-BELT SHEAVES

V-belt sheaves are grooved wheels that are designed to run with the correct corresponding belt. Standards have been established for most of the boundary angles and dimensions (see Figure 10–14a and 10–14b). The critical dimension used in identifying and sizing the sheave is the pitch, or datum, diameters. By definition, the diameter through which the pitch line passes should be the pitch diameter. It is the dimension that reflects the diameter of the tensile members in the belt form as it wraps around the sheave. The "datum system" is a more accurate reflection of this and is an established standard now used by the International Standards Organization (ISO) and the Rubber Manufacturers Association (RMA) as the means of designating sheave size. There are some differences between the "pitch system" and "datum system" used by manufacturers and design engineers. These differences can

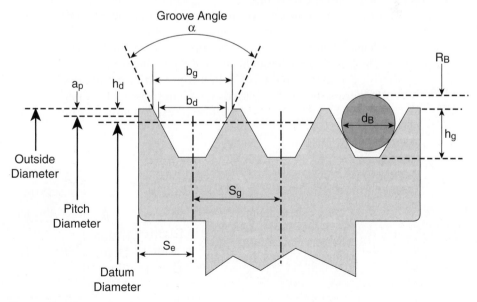

FIGURE 10–14a
Standard V-belt sheave groove dimensions, angles, and terminology.

Gates Hi-Power™ II Sheave Groove Dimensions

Cross Section	Datum Diameter Range	α Groove Angle ±0.33°	b_d Ref.	b_g		h_g Min.	2h_d Ref.	R_B Min.	d_B ±0.0005	S_g ±0.025	S_e	Minimum Recommended Datum Diameter	2a_p
A, AX	Up through 5.4 / Over 5.4	34 / 38	0.418	0.494 / 0.504	±0.005	0.460	0.250	0.148 / 0.149	0.4375 (7/16)	0.625	0.375	+0.090 / -0.062	A 3.0 / AX 2.2	0
B, BX	Up through 7.0 / Over 7.0	34 / 38	0.530	0.637 / 0.650	±0.006	0.550	0.350	0.189 / 0.190	0.5625 (9/16)	0.750	0.500	+0.120 / -0.065	B 5.4 / BX 4.0	0

A-B Combination

Cross Section	Datum Diameter Range	α Groove Angle ±0.33°	b_d Ref.	b_g		h_g Min.	2h_d Ref.	R_B Min.	d_B ±0.0005	S_g ±0.025	S_e	Minimum Recommended Datum Diameter	2a_p
A, AX Belt	Up through 7.4(1) / Over 7.4	34 / 38	(2) 0.508	0.612 / 0.625	±0.006	0.612	0.634 (3) / 0.602	0.230 / 0.226	0.5625 (9/16)	0.750	0.500	+0.120	A 3.6(1) / AX 2.8	0.37
B, BX Belt	Up through 7.4(1) / Over 7.4	34 / 38		0.612 / 0.625	±0.006		0.268 (3) / 0.276	0.230 / 0.226				-0.065	B 5.7(1) / BX 4.3	-0.08
C, CX	Up through 7.99 / Over 7.99 to and including 12.0 / Over 12.0	34 / 36 / 38	0.757	0.879 / 0.887 / 0.895	±0.007	0.750	0.400	0.274 / 0.276 / 0.277	0.7812 (25/32)	1.000	0.688	+0.160 / -0.070	C 9.0 / CX 6.8	0
D	Up through 12.99 / Over 12.99 to and including 17.0 / Over 17.0	34 / 36 / 38	1.076	1.259 / 1.271 / 1.283	±0.008	1.020	0.600	0.410 / 0.410 / 0.411	1.1250 (1 1/8)	1.438	0.875	+0.220 / -0.080	13.0	0

1) Diameters shown for combination grooves are outside diameters. A specific datum diameter does not exist for either A or B belts in combination grooves.
2) The b_d value shown for combination grooves is the "constant width" point but does not represent a datum width for either A or B belts (2h_d = 0.340 reference).
3) 2h_d values for combination groove are calculated based on b_d for A and B grooves.

Machined Surface Area	Maximum Surface Roughness Height, R_a (Arithmetic Avg.) (Microin.)
Sheave Groove Sidewalls	125
Sheave O.D.'s and Rim Edges	250
Rim I.D.'s Hub Ends, Hub O.D.'s	250
Straight Bores	125
Taper Bores	175
Cast Surface Area	As Cast

Face Width of Standard and Deep Groove Sheaves

Face Width = $S_g (N_g - 1) + 2S_e$

Where: N_g = Number of Grooves

FIGURE 10–14b
Standard V-belt sheave groove specifications and dimensions.

Courtesy of The Gates Rubber Co.

be important when designing V-belt drive systems. From a practical standpoint, the maintenance mechanic can consider the two measurement terms the same. For classical V-belts (A, B, C, D, and E), the pitch or datum diameter of a given sheave is .4″ less than the outside diameter (O.D.). On sheaves that run with 3V, 5V, and 8V belts, the outside diameter is .1″ greater than the pitch diameter.

The identification of sheaves also includes the belt cross-section, the number of grooves, and the bore or mounting bushing style. For example, a single groove of 4″ diameter and a section sheave with a 1″ bore would have a part number of 1A4.0 × 1″. The part number is usually marked on the sheave.

Certain sheave types called combination or "dual duty" allow more than one belt cross-section to be used (see Figure 10–15). Running an A section V-belt in the groove of a dual duty sheave would result in a smaller pitch diameter than that of a

FIGURE 10–15
Datum/pitch diameter comparison for dual duty sheaves.

Courtesy of The Gates Rubber Co.

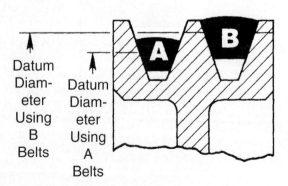

Datum Diameter Using B Belts

Datum Diameter Using A Belts

B section V-belt. For example, the part number of a "dual duty" sheave would read 1A4.0 B4.4 × 1″.

VARIABLE SPEED/PITCH SHEAVES

Variable speed sheaves (pulleys) are used in applications where speed changes are required. A significant number of these pulleys are variable pitch in construction. The adjustable pitch of the sheave allows for speed adjustment without changing the sheave. The basic operating principle of the sheave is that the angular faced discs forming the V-shaped groove where the belt rides are movable. When the discs are moved apart, the belt rides lower and the pitch diameter is smaller. When the discs are moved together, the belt rides higher in the groove and the pitch diameter is greater. Variable speed sheaves are frequently used on air handling and pump applications. In the case of fans, it is sometimes necessary to adjust speeds to balance multiple blower systems. Once the system is balanced, no additional adjustment is required. These sheaves should not be used on applications that need frequent adjustment. There are two basic forms of variable pitch sheaves for V-belts: manually adjusted sheaves (see Figure 10–16) and spring-loaded sheaves (see Figure 10–17).

Variable speed belts have their own form of variable pitch sheaves. Variable pitch sheaves used with variable speed belts have wider grooves than standard V-belt variable pitch sheaves. Most variable speed belts will operate with one

FIGURE 10–16
Manually adjusted variable pitch, single groove V-belt sheave.

Courtesy of Emerson Power Transmission.

FIGURE 10–17
Spring-loaded variable pitch V-belt sheave.

Courtesy of Emerson Power Transmission.

FIGURE 10–18
Spring-loaded variable pitch pulley for use with a variable speed belt.

Courtesy of T. B. Woods Incorporated.

adjustable pulley/sheave and one nonadjustable companion pulley/sheave, or two adjustable types (see Figure 10–18).

There are two common forms of variable speed drives: single and dual variable sheaves. The simplest form is a single variable pitch sheave used with a fixed pulley. The drive consists of one spring-loaded variable pitch pulley mounted on the driver shaft. An adjustable motor base is employed to vary the center distance of the drive, which results in a speed ratio change. The spring mechanism in the sheave keeps the flanges against the sides of the belt and automatically provides the required tension and friction. The ratio of these single variable pulley drives is generally in the 3:1 range.

Dual variable drives will incorporate two variable pitch pulleys. With a variable pitch pulley mounted on the driver and driven shafts, higher ratio ranges are achieved in the 9:1 range. The center distance on these drives will be fixed and one of the pulleys, usually the driver, will have an adjustable mechanism. Actuating the adjustment device of the sheave will open or close the flanges, which changes the pitch diameter that the belt runs at.

Sheave Mounting Methods

Sheaves are mounted to the shaft either by the use of bushing systems bored to fit, or with special mounting flanges. A press fit or bored-to-size sheave has a straight hole bored in the center for the shaft. The bore in the sheave hub can be either a clearance or interference fit to the shaft. In addition, a setscrew and keyway help lock and hold the sheave in place. Loose fit shaft-to-sheave hub fits are generally used on lighter duty applications with fractional horsepower belts.

Bushing mounted sheaves are available in a variety of different types of shaft locking mechanisms. The taper bushing, or taper-lock, actuates the principle of the wedge to hold the bushing and sheave in place on the shaft. The combination of friction, clamping forces, and keys transfers the torque from shaft to bushing/hub. There are several advantages to using bushings with sheaves. Tapered bushing mounting styles are relatively easy to install and remove. They have gripping forces equal to or greater than bored-to-size sheaves. They can be used on shafts that are slightly over- or undersized. Inventory costs can be reduced because bushings for different shaft sizes can be stocked to fit many different sheaves.

Preparation of all tapered mating surfaces involves cleaning and checking for nicks and burrs that should carefully be removed. All surfaces should be free of grease and contaminants. Torque wrenches should be used when assembling tapered bushings to prevent uneven mounting and overtightening. If the bushing is drawn up unevenly to the sheave hub, excessive axial run-out will be created.

V-BELT DRIVE SELECTION

Proper selection of V-belt drives must take into consideration the nature of the application, speed, horsepower, cost, required life, service factor, and space limitations. Manufacturers of V-belts can provide the design engineer with the proper manual or selection software. If the drive is a speed up, nonstandard motor speed, serpentine, or special in any shape or form, it is recommended that qualified manufacturers' design engineers be consulted. Basic standard drive design requires the following information:

- Type of driver (prime mover), rpm, and shaft size.
- Type of driven equipment (pump, conveyor, fan, etc.), rpm, and shaft size.
- Service factor from V-belt manufacturers' tables.
- Center distance between driver and driven shaft.

Once this information is determined, a drive design manual should be consulted and the following basic steps followed:

1. Select the belt cross-section from appropriate charts.
2. Determine the "design horsepower" (horsepower of prime mover × service factor).
3. Calculate the drive ratio and driven speed, then locate it under the appropriate column in the drive selection tables.

   (Driver speed ÷ Driven speed = Ratio)
4. Determine the sheave combination from the choices listed based on desired ratio, minimum and maximum diameter constraints, and any space limitations. When selecting the sheave combination consider these points:

   - If sheave space width must be kept to a minimum, select the largest diameter pair.
   - Drive tension and shaft pull will be less with larger diameter sheaves.
   - Multiple belt drives are more dependable.
   - Space limitations might require smaller diameter sheaves.
   - NEMA charts on minimum diameters for driver sheaves must be consulted.
   - Choose standard stock sheaves.

5. Determine belt length. In the column of the chosen sheaves, find the closest center distance listed to the actual centers, which will determine the belt length size to be used.

$$\text{Belt length} = 2C + 1.57(D + d) + \frac{(D - d)^2}{4C}$$

Where: C = center distance

D = diameter of large sheave

d = diameter of small sheave

6. Note arc of correction factor shown based on center distance and wrap of belt on sheaves. This number corrects the listed rating of the belt.

7. In columns listing rpm of small sheaves, determine rated horsepower per belt. Use the correction factor multiplied by the rated horsepower to determine the actual rating.

8. Determine the number of grooves (belts) required for each sheave to meet or exceed the design horsepower.

9. Convert the selection to part numbers for sheaves and bushings to confirm availability.

NOTE: This selection procedure should be used in conjunction with a V-belt manufacturer's drive selection catalog.

SYNCHRONOUS BELT DRIVES

Synchronous belt drives are referred to by a variety of names: "timing belt," "gear belt," and "positive drive belt" are all common terms used to identify the family of nonslip toothed belt drives. The drive system was originally developed in the early 1940s for the garment industry. It operates in many applications where an efficient and constant speed ratio drive is required.

Synchronous drive belts have teeth that are accurately molded to conform with and engage the teeth of the mating pulleys (see Figure 10–19a). Power is transmitted by the positive engagement of belt teeth with pulley teeth—not by friction forces, as in V-belts. The design allows virtually no slip to occur between belt and pulley. This feature results in the drive being extremely efficient. The nonslip characteristics of the synchronous belt drive permit it to be used in place of gear and chain drives when no backlash is required in the system. If the drive is applied in an application that is subject to jamming, the teeth may shear off the belt. A V-belt will slip; a synchronous belt will not.

Synchronous belts are molded, endless flat belts with uniformly spaced teeth of varied size and shape. They are composed of tensile members, a backing, teeth, and a face. The tensile members are wound cords of a steel, fiberglass, Kevlar, or other

FIGURE 10–19a
Synchronous belt
and pulley.

Courtesy of The Gates
Rubber Co.

synthetic material. The tensile members provide the belt its strength and flexibility. The backing, teeth, and face are made from a neoprene or polyurethane material. The backing is the body of the belt and protects the tensile members from contaminants and frictional wear. The teeth are molded integrally with the backing and are made of a shear-resistant compound. The drive is designed to have at least six or more teeth engaged in the driver pulley. This amount of engagement produces a tooth shear strength that exceeds the tensile strength of the belt.

Synchronous belt drives have certain advantages over V-belt, gear, and roller chain drives. There is no creep of the belt that would allow speed variation and inefficiency. They require no lubrication as in chain and gear drives. The tooth configuration engages and clears the sprocket grooves in a continuous smooth transmission of power. This design characteristic of constant angular velocity causes no chordal action to occur, as in chain drives. The result is less vibration in the system. Because synchronous drives do not rely completely on friction forces for power transfer, the bearing loads are less than V-belt drives. Slack side tension is minimal and the tight side tensions are less than V-belt drives, resulting in reduced overhung loads on the bearings and shafts.

Synchronous Belt Types

Synchronous belts are available in numerous types and sizes. The basic classes are the original trapezoidal tooth shape in inch dimensions, and the newer design forms that have rounded tooth profiles and metric dimensions. Two of the terms used in the identification of synchronous belts are "pitch" and "pitch line." Pitch is defined as the distance between corresponding points on adjacent teeth as measured on the pitch line of the belt. The pitch line refers to a theoretical line in the belt that is midway between the areas of tension and compression when the belt is wrapped around the pulley, which is the approximate location of the tensile members (see Figure 10–19b). The belt pitch line and the pulley pitch diameter are the same dimension when the belt is in mesh on the pulley. Because of the thickness of the belt's backing, the pulley pitch diameter is always greater than the pulley outside diameter.

The original standard synchronous belt and pulley sizes, pitch, and other dimensions are in inch units of measurement and have a trapezoidal shape

FIGURE 10–19b
Synchronous belt and
pulley terminology.

Courtesy of The Gates
Rubber Co.

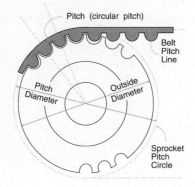

FIGURE 10–20
Traditional
synchronous belt pitch
letter designation
system showing
relative sizes.

Courtesy of The Gates
Rubber Co.

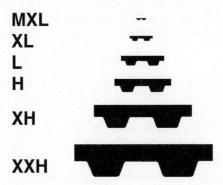

(see Figure 10–20). These original standard inch sizes are still in use on older drive systems. The five stock pitches found in industry are:

Pitch	Code	Standard Widths	Description
1/5″	XL	1/4″, 5/16″, 3/8″	Extra Light
3/8″	L	1/2″, 3/4″, 1″	Light
1/2″	H	3/4″, 1″, 1 1/2″, 2″, 3″	Heavy
7/8″	XH	2″, 3″, 4″	Extra Heavy
1-1/5″	XXH	2″, 3″, 4″, 5″	Double Extra Heavy

The belts have a code number that reflects the length, pitch, and width of the belt. The prefix is the belt pitch length in inches and tenths of inches. The middle letter designates the tooth pitch code. The last part of the number designates the belt width in inches multiplied by 100. For example, a synchronous belt with a part number of 500L 100 has a pitch length of 50″, a pitch of 3/8″, and a width of 1″.

The newer designs of synchronous drive belts have rounded tooth profiles and metric dimensions. They are capable of transmitting higher loads than traditional

FIGURE 10–21
Common metric synchronous belt pitches.

Courtesy of The Gates Rubber Co.

8mm
Pitch

14mm
Pitch

timing belts, and are less prone to tooth loss from shearing forces because of better distribution of the imposed stresses along the tooth form and improved composition. These belts are made from specially formulated materials such as Kevlar, polyesters, and polyurethane that have the ability to transmit high torque loads. Design engineers are recommending and installing the newer metric pitch sizes with rounded tooth forms in place of inch series synchronous belts. Some manufacturers produce a unique tooth shape that must be run with the correct corresponding pulley, while others are interchangeable. When installing synchronous belt drive components it is imperative that the belt tooth profile not only have the same pitch, but it also must match the shape of the mating pulley. Standard pitches are 2 mm, 3 mm, 5 mm, 8 mm, 14 mm, and 20 mm (see Figure 10–21).

The metric synchronous belts also have a part number system that designates length, pitch, and width. For example, a belt with a part number of 1490-14M-55 would have a 1,490 millimeter pitch diameter, with a 14 millimeter pitch, and a width of 55 millimeters.

Miniature pitch synchronous belts are produced in both inch and metric pitch and length dimensions. Belts can be constructed from nylon covered, fiberglass-reinforced neoprene materials. Typical miniature pitches are 2 and 3 millimeters. Applications include computers, office equipment and other low torque, high-speed conditions.

Double-sided synchronous belts have teeth on both sides (see Figure 10–22). This feature makes possible drive designs that are both serpentine and reverse shaft direction. The composition, dimensions, and part numbers of these belts are similar to standard synchronous belts, with the exception of a special prefix letter (TP) to designate that they are dual sided.

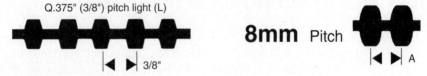

Q.375" (3/8") pitch light (L)

3/8"

8mm Pitch

A

FIGURE 10–22
Double-sided synchronous belts.

Courtesy of The Gates Rubber Co.

Synchronous Drive Pulleys

Synchronous drive pulleys, sometimes referred to as sprockets, are designed and produced to properly fit the matching belt. This precise fit between belt and pulley teeth is imperative to produce minimal backlash and optimum efficiency. Pulleys for synchronous drives are usually made of iron or steel. The pulleys can be supplied with taper bushings or bored to fit a variety of shaft sizes with keyways and setscrews. Most manufacturers produce a pulley that is statically balanced with a maximum rim speed rating of 6000 feet per minute. Dynamically balanced pulleys of a special material are available from manufacturers that are capable of higher rim speeds. It is a recommended practice to have flanges on at least one of the pulleys, preferably the driver. The flanges will help prevent the belt from walking over the side due to inherent axial thrust forces created by the spiral lay of the tensile members and inevitable installation misalignment. The pulley diameter should be greater than the face width to prevent belt overriding of the flanges.

Synchronous Drive Selection

The selection for synchronous drives is similar to V-Belt drives. Components must be determined by considering the nature of the application, speed, horsepower, cost, required life, service factor, operating environment, and space limitations. The design of a synchronous belt drive makes it a positive drive. Therefore it must be capable of carrying the full load of the system and handling shock loads and starting torques produced by the prime mover and driven equipment. Careful consideration must be given to selecting a synchronous drive for pulsing loads and machines that have high inertia loads. These types of applications can cause the belt to jump off or teeth to shear. Because the capacity of the belt is determined by the shear strength of the belt teeth while engaged with the sprocket (pulley), it is best to design drives with large pulleys. Following this general instruction allows for more teeth in mesh. It also will prevent fatigue of the belt from excessive flexing. On high-speed drives, install smaller pitches and pulleys as large as reasonably possible. The noise levels generated by certain synchronous drive types can be above acceptable OSHA limits. Consult the drive belt manufacturer for noise ratings of the belt selected at the operating speed. Manufacturers of synchronous drives can provide the design engineer with the proper manual or selection software.

V-BELT AND SYNCHRONOUS DRIVE MAINTENANCE

V-belt and synchronous drives will achieve optimum performance and maximum service life if correctly installed and maintained. Proper inspection, tensioning, alignment, and installation procedures are common to both drive types. All aspects of the inspection, installation, and maintenance process should be measured and documented. These records should be kept in an organized manner and accessible to all maintenance personnel. A comprehensive belt drive preventative maintenance program that is effectively implemented will reduce downtime, save costs, and create a safe working environment.

Effective belt preventative maintenance programs should include the following:

- Correct drive selection.
- A safe environment.
- Regular belt drive inspections.
- Clean and proper component storage.
- Correct belt and sheave installation.
- Evaluation and troubleshooting.

Prior to proceeding to the drive, all data pertinent to the machine should be collected. Tensioning specifications and alignment tolerances should be part of the work order. Original equipment specifications or subsequent recorded changes on the belt and sheave sizes must be used as a reference to ensure that the replacement parts will be correct. Unintentional changes in the components can lead to inefficiency or, worse, failure. For example, installing increasingly smaller diameter driver sheaves on the motor can lead to bearing failures from excessive overhung loads.

Following a logical sequence of maintenance steps will assist in determining drive problems and corrective actions. A comprehensive inspection of the drive should take a multifacetted approach. Visual, audible, and instrument inspection procedures should all be drawn on. Common sense questions can be asked of operating personnel, such as: Is the drive noisy? Is it vibrating? Does it smell? Are the drive belts off of the sheaves? The answers to these questions can help determine the root cause of drive problems and possible corrective steps.

The frequency of drive inspection is determined by such factors as speed, temperature, environment, operating cycle, critical nature of the equipment, and experience. Normal drives should be given a basic sight and sound inspection once per month. Critical drives should be checked weekly. A complete shutdown and thorough system check should be done approximately every six months.

Belt Preventative Maintenance Checklist and Troubleshooting

The following procedures are general guidelines for the operation and maintenance of belt drives. It should be supplemented with technical literature supplied by the manufacturers. It is imperative that the maintenance staff be educated and trained in all aspects of belt drive maintenance to yield the desired results from the system operating in a safe manner.

Safety All safety procedures, both general and specific to your plant, must be followed. Always shut off the power source. Lock it out and follow tagout policies (see Figure 10–23). Verify that the machine is electrically cut off. Make sure the drive

Figure 10–23
Lockout tag. Follow all
lockout, tagout, and
verification procedures
before working on the drive.
Courtesy of The Gates Rubber Co.

is mechanically neutralized so it will not turn because of back pressure from air,
fluids, and uneven loads.

Safety Gear Never wear loose fitting or bulky clothes around belt drives. Wear
gloves, glasses, and any other mandatory safety equipment when working on a drive
(see Figure 10–24).

Safe Working Environment Maintain a clean and safe access area around the
drive. Keep the floor and connected machines free of dirt, oil, obstructions, and
debris (see Figure 10–25). Avoid using any combustible cleaners.

Drive Foundation Solid foundations and base plates are absolutely required to
prevent vibration and misalignment. Inspect the mounting surfaces for any unwanted looseness and allow free movement of any adjustable plates required for
belt tensioning. Replace all worn or stressed fasteners.

Figure 10–24
CAUTION! Do not
wear unsafe
clothing such as
shown during an
inspection.
Courtesy of The Gates
Rubber Co.

Figure 10–25
An unsafe and
cluttered drive area
is an accident
waiting to happen.
Courtesy of The Gates
Rubber Co.

Drive Guards Always keep drives guarded when operating. The guard should be designed and installed to OSHA standards. It is preferable to have a guard that is constructed to breathe, reducing heat build-up. Grills and vents facilitate ventilation. An approximate 36°F ambient drive temperature rise will halve belt life. Make sure the guard is clean and not contacting any moving parts. Remove the guard for complete inspection.

Belt Removal Loosen the adjustable driver base by turning the jacking screw to remove the belts. If the driver base plate is the nonadjustable type, loosen the motor base bolts in their slotted holes and pry the motor toward the driven equipment. This will slacken the belts and facilitate removal. Do not pry the belts off because this might damage the sheaves and cause injury. On drives using an idler, loosen and swing the idler arm out of the way. Avoid placing fingers and hands in any pinch points.

Belt Inspection Inspect the belt, observing any worn or damaged surfaces (see Figure 10–26). Check for fraying, cracking, cuts, or obvious wear patterns. Glazing, which appears as a slick, shiny surface on the sides of the belts, is an indication that the belt needs to be replaced. A glazed belt has reduced friction and will not function as it is intended. It can be caused from foreign substances rubbing between the belt and sheaves, improper tensions, and worn sheaves. On synchronous belts, the teeth should be checked to see if they have been sheared off. Identify the belt's size and part number for correct replacement by checking for labeled numbers and taking measurements.

Sheave Inspection Inspect the sheaves and clean them prior to inspection. Use a noncombustible solvent, avoiding contact with the belt. Identify the part number and size of the sheaves and record the information. Check the hubs and bushings for cracks, which could cause catastrophic failure while operating. Replace all threaded fasteners. Thoroughly inspect the sheaves for wear, damage, or highly polished surfaces. Nicks and scratches on the groove walls will damage the belt when it is operating. Dirt and oil build-up in the grooves can cause the belt surfaces to be

FIGURE 10–26
The technician should periodically visibly inspect the drive belt.

Courtesy of The Gates Rubber Co.

FIGURE 10–27
Checking a sheave groove for wear with a sheave groove template.

Courtesy of The Gates Rubber Co.

abraded. A worn or "dished out" sheave groove profile will cause the belt to bottom out, slip, or roll over. To compensate for the loss of proper contact between sheave sidewalls and the belt due to wear, tension must be increased to prevent slipping. This causes increased loading on the bearings and shorter belt life. A sheave groove gauge can be obtained and used as a "go-no-go" tool (see Figure 10–27). Select the correct sheave groove gauge based on belt type, size, and sheave diameter. Insert the gauge into the groove and observe the shape of the sidewalls. If more than 1/32″ of wear can be seen, replace the sheave.

System Inspection Inspect related equipment. Inspect other drive components such as shafts and bearings for wear and looseness. If the drive uses a static conductive system, check it to make sure it is functioning properly. Replace or repair parts as needed.

Sheave Replacement Replace all worn or damaged sheaves. Press fit sheaves might require the use of a bearing removal tool. On taper-lock and quick-disconnect bushing types, follow correct removal procedures. Once the previous sheaves have been carefully removed, install the correct replacement sheaves. If they are the "bored-to-fit" type make sure all setscrews are loosened during removal and retightened when mounting. All keyways should be inspected and repaired if necessary. Proper fitting keys that do not require extreme force to insert into the slot must be used. If a press fit is required to mount the sheave, heat it in an oven or on a bearing heater. Do not heat the sheave to more than 350°F (177°C) or the metal may be weakened. Employ soft face or "dead blow" hammers if force is required. Do not strike the surface of the sheave with a metal object. Consider the positioning of the component to facilitate alignment of both sheaves. Attempt to place the sheaves as close to the supporting bearings as possible within alignment and installation constraints. This will help reduce the overhung load on the bearings and shafts.

Check the TIR The TIR (total indicator run-out) of the driver, driven sheaves, and shafts can be determined using a magnetic base dial indicator. Excessive radial and axial (rim face) run-out of components can cause vibration in the system. Bent shafts, improper mounting, and poor fastener tightening methods can cause this problem to occur. At speeds of 1,750 rpm and above, this can be a critical issue. As a general rule the following tolerances can be used:

Shaft radial run-out = .004″ or less.

Sheave radial run-out up to and through 10.0″ outside diameter = 0.010″ or less. (For each additional inch of outside diameter, add 0.0005″.)

Sheave axial run-out up to and through 5.0″ outside diameter = 0.005″ or less. (For each additional inch of outside diameter, add 0.001″.)

Sheave Alignment Alignment of the sheaves is a critical requirement for the drive to run trouble-free. Misalignment causes vibration, component wear, belts to run out of grooves, and premature bearing failure. Misalignment can occur in the angular and offset (parallel) planes, or as a combination of both (see Figure 10–28). The driver and driven shafts must be in the same plane.

Precision alignment of sheaves and pulleys can be achieved by several different methods. The use of levels, straight edges, feeler gauges, and lasers are all approved tools for the job.

A preliminary step to alignment is to check all motor and driven machine bases for "soft foot." This can be done using a feeler gauge to measure any air gaps between machine mounting pads and the foundation. If a gap is greater than .003″ it must be corrected by shimming. If a machine is tightened down without correcting the "soft foot," a twist will be induced into the machine's frame and cause misalignment and distortion of the machine housings.

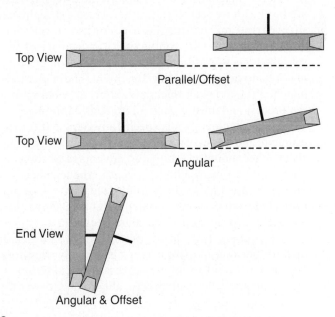

FIGURE 10–28
The three types of sheave/pulley misalignment: parallel/offset, angular, and a combination of both.

FIGURE 10–29
The maintenance technician can use a straightedge to check sheave alignment. Gaps between the straightedge and the side of the sheave/pulley indicate that correction is required.

Courtesy of The Gates Rubber Co.

Place a machinist's level on both the driver and driven shafts to make sure they are level with respect to each other. Place a straight edge across the faces of the two sheaves and make sure there is contact at all four points (see Figure 10–29). A string or piano wire can be tightly drawn across the face in place of a machined straight edge. Reposition the sheaves to accomplish parallelism. Adjust motor jacking bolts to achieve angular alignment. These methods work for sheaves that are a matched set. Sidewall thickness differences between manufacturers can aggravate the process and must be taken into consideration.

Laser alignment tools specifically designed for sheave alignment are the most efficient and exact. Two of the types currently in use are groove mounted and sheave-face mounted. The great advantage of these lasers is that all forms of misalignment become readily apparent to the technician. The lasers shoot a wide beam of light that is picked up on reference surfaces or magnetic targets. The laser line projected from the tool onto the target face allows the user to quickly ascertain and correct angular and offset misalignment.

As a general rule, sheave misalignment on V-belt drives should be less than $1/2°$ or $1/10''$ per foot of drive center distance. Synchronous drives should be aligned to $1/4°$ or $1/16''$ per foot of center distance. These limits are for the belt. The equipment runs smoother with precise alignment. Strive for perfection.

Belt Replacement Replacement belts must be stored in a cool, dry, and clean storage area. They should not be exposed to direct sunlight, oils, or solvents. If belts are hung in an excessively warm environment they will dry out, become brittle, and then crack. If they are kept in a tight coil for an extended period of time, they will take a "set" and not perform satisfactorily. Proper handling and storage will ensure that the belts will see their maximum service life. Previously run belts should not be run with new belts. All belts should be changed out as a set. It is not recommended to run V-belts of different manufacturers on the same drive because of possible differences in the cross-sectional shape, included angle between side-walls, and length.

When installing belts never use a pry bar or screwdriver to roll the belt over the sheave edge. Forcibly prying the belts into the grooves will damage the internal tensile members of the belt. Broken or damaged tensile cords will cause the belt to turn over in the grooves while operating. Once the tensile cords are broken, the belt will stretch excessively and begin to slip. Repetitive stopping and retensioning becomes the only way to get the drive to transmit power. Pry bars may chip the sheave edge leaving sharp surfaces to cut the belt or installer.

Under no circumstance should the motor be jogged to roll on the belts. Slacken the drive, narrowing the centers, so the belts may be placed on by hand without forcing them. Rotate the drive by hand after installing. This will help seat the belts in the pulley. With synchronous belts make sure the belts are not installed so that they rub against the pulley side flanges. If the synchronous belt walks over quickly to one side when manually being rotated, it is probably due to misalignment. It is imperative that minimal force is required to install all belts.

Belt Tensioning Tension is a vital issue to all belt drives. If too little tension is applied, the belt will slip. A loose belt will be obvious because of the squealing noise associated with it. The use of belt dressings to stop the noise and slipping should be avoided. The chemicals in the dressing cause the rubber to become soft, which results in a temporary change in friction, that ultimately accelerates the demise of the belt. In addition, abrasive contaminants in the environment can stick to the material and belt, causing abrasive wear of components. If the belt is overtightened, the excessive tension can reduce belt and bearing life. The correct tension is the lowest tension needed to run the drive without the belts slipping under full load.

There are various methods used to tension drive belts. One method starts with calculating static tension, taking into consideration several factors pertaining to speeds, arc of contact, sheave size, belt type, various constants, and formulas. These calculations are in turn used to determine recommended deflection forces. Another method, for banded belts, involves determining the amount of elongation the belt should be subjected to upon installation by referring to published charts available from belt manufacturers. The material of the tensile member must also be taken into consideration when using the elongation method. Polyester cord has enough elongation to measure with some degree of accuracy. Most of these procedures are for tensioning new belts. In some cases special measuring equipment (load cells) is required to accurately determine the numbers. It is recommended that on critical or special drives, these methods should be applied and belt design engineers consulted.

One of the newer methods involves vibration and acoustical measurement. The application of these procedures works best on positive drives. By using vibration equipment, transducers are fixed on the driver unit sheave end bearing in the plane of shaft centerlines. The belt tensioning mechanism is loosened until a maximum amount of vibration is observed. This point is marked. The belt then is tightened by the technician observing and recording the maximum degree of vibration, marking the adjuster, and setting it to a median point. Special acoustical devices called "sonic tension meters" accurately measure tension based on the theory that synchronous belts vibrate at a particular frequency determined by mass and span length. The belt is strummed to cause it to vibrate and the resulting oscillating sound wave is recorded. The belt is "tuned" into tension. Carrying out these procedures and implementing them in the field is often impractical, expensive, and difficult with the time and tools available. On the simple side, using your thumb to determine correct tension is unreliable.

The most practical method to tension belt drives is to use a belt tensioning tool (see Figure 10–30). The tool measures both the amount of deflection and the applied force at the center of the belt span. It is to be used in conjunction with manufacturer tables that take into account belt cross-sections, speeds, ratios, and sheave sizes. It is a simple process and draws on principles that reflect average static tensions for drives. The following is a belt tensioning procedure using a tension tester.

1. Measure or calculate the span length.

$$\text{Span length} = C\left[1 - 0.125\left(\frac{D-d}{C}\right)^2\right]$$

Where: C = center distance in inches

D = large sheave diameter in inches

d = small sheave diameter in inches

2. Adjust the lower of the two o-rings on the tool on the scale reading "Deflection Inches." Set the o-ring to show a deflection equal to 1/64" per inch of span length. Or, on the scale reading "Inches of Span Length" set the o-ring to show a deflection equal to the inches of measured span length. Read the scales at the bottom of the o-ring. Leave the upper o-ring in the maximum up position.

3. At the center of the span apply force with the tester perpendicular to the span. The force must be enough to deflect one belt until the lower edge of the bottom o-ring is even with the tops of the remaining belts on the drive. The adjacent belt can be used as a reference. A straightedge placed across the belt tops will ensure accurate positioning. On banded belts, lay a bar across the belt so all strands are deflected evenly. Divide the tension value by the number of strands in the banded belt.

4. Measure the amount of force required to deflect one belt on the drive 1/64" per inch of span length from its normal position by reading the upper scale on the tester. The o-ring will slide down the scale as the tool compresses and stay in position, marking the pounds pressure.

5. Compare the actual deflection force with the recommended forces in the tables based on cross section, diameters, ratios, and speeds (see Figure 10–30).

How to Tension Belt Drives With Your Gates Tension Testers

(Up to 30 lbs.)

1. Measure span length **(t)**.

2. Position the lower of the two O-Rings using either of these methods:

   a. On the scale reading "Deflection Inches", set O-Ring to show a deflection equal to 1/64" per inch of span length **(t)**.

   b. On the scale reading "Inches of Span Length", set O-Ring to show a deflection equal to the inches of measured span length **(t)**.

3. At the center of span **(t)**, apply force with Gates Tension Tester perpendicular to the span, large enough to deflect one belt of a multiple belt set on drive until the bottom edge of the lower O-Ring is even with tops of remaining belts. For drives with only one belt, a straightedge across pulleys will assure accuracy of positioning.

4. Find the amount of deflection force on upper scale of Tension Tester. The Sliding Rubber O-Ring slides up the scale as tool compresses—and stays up for accurate reading of pounds force. Read at the bottom edge of ring (slide ring down before reusing).

5. Compare deflection force with range of forces recommended. If less than minimum recommended deflection force, belts should be tightened. If more than maximum recommended deflection force, drive is tighter than necessary.

NOTE: There normally will be a rapid drop in tension during the "run-in period" for V-belt drives. Check tension frequently during the first day of operation.

Read the scales at the bottom edge of the O-Ring. Leave the upper O-Ring in maximum "down" position.

Belts, like string, vibrate at a particular natural frequency based on mass and span length. Gates unique Sonic Tension Meter simply converts this frequency into a measurement of tension. Here's how it works:

First, enter belt width, span length and unit weight into meter using built-in keypad. Next, hold meter sensor to belt span, then lightly strum belt to make it vibrate. Press "measure" button and that's it. Meter instantly converts vibrations into belt tension. Readings are displayed on a liquid-crystal screen. (detailed instructions accompany tester)

Recommended Deflection Force Per Belt For Super HC® V-Belts, Super HC PowerBand® Belts, Super HC Molded Notch V-Belts or Super HC Molded Notch PowerBand Belts*

V-Belt Cross Section	Small Sheave Diameter Range (In.)	Small Sheave RPM Range	Speed Ratio Range	Recommended Deflection Force (Lbs.) Minimum	Maximum
3V	2.65 - 2.80	1200-3600		3.0	4.3
	3.00 - 3.15	1200-3600	2.00	3.3	4.8
	3.35 - 3.65	1200-3600	to	3.7	5.4
	4.12 - 5.00	900-3600	4.00	4.4	6.4
	5.30 - 6.90	900-3600		4.8	7.1
3VX	2.20	1200-3600		2.8	4.1
	2.35 - 2.50	1200-3600		3.2	4.7
	2.65 - 2.80	1200-3600	2.00	3.5	5.1
	3.00 - 3.15	1200-3600	to	3.8	5.5
	3.35 - 3.65	1200-3600	4.00	4.1	6.0
	4.12 - 5.00	900-3600		4.8	7.1
	5.30 - 6.90	900-3600		5.8	8.6
5VX	4.40 - 4.65	1200-3600		9.0	13.0
	4.90 - 5.50	1200-3600	2.00	10.0	15.0
	5.90 - 6.70	1200-3600	to	11.0	17.0
	7.10 - 8.00	600-1800	4.00	13.0	19.0
	8.50 - 10.90	600-1800		14.0	20.0
	11.80 - 16.00	400-1200		15.0	23.0
5V	7.10 - 8.00	600-1800	2.00	11.0	16.0
	8.50 - 10.90	600-1800	to	13.0	18.0
	11.80 - 16.00	400-1200	4.00	14.0	21.0
8V	12.50 - 17.00	600-1200	2.00 to 4.00	28.0	41.0
	18.00 - 24.00	400- 900		32.0	48.0

Recommended Deflection Force Per Belt For Hi-Power II™ V-Belts, Hi Power II PowerBand Belts or Tri-Power® Molded Notch V-Belts*

V-Belt Cross Section	Small Sheave Diameter Range (In.)	Small Sheave RPM Range	Speed Ratio Range	Hi-Power II Minimum	Hi-Power II Maximum	Tri-Power Molded Notch Minimum	Tri-Power Molded Notch Maximum
A AX	3.0			2.7	3.8	3.8	5.4
	3.2	1750	2.00	2.9	4.2	3.9	5.6
	3.4 - 3.6	to	to	3.3	4.8	4.1	5.9
	3.8 - 4.2	3600	4.00	3.8	5.5	4.3	6.3
	4.6 - 7.0			4.9	7.1	4.9	7.1
B BX	4.6			5.1	7.4	7.1	10.0
	5.0 - 5.2	1160	2.00	5.8	8.5	7.3	11.0
	5.4 - 5.6	to	to	6.2	9.1	7.4	11.0
	6.0 - 6.8	1800	4.00	7.1	10.0	7.7	11.0
	7.4 - 9.4			8.1	12.0	7.9	12.0
C CX	7.0			9.1	13.0	12.0	18.0
	7.5	870	2.00	9.7	14.0	12.0	18.0
	8.0 - 8.5	to	to	11.0	16.0	13.0	18.0
	9.0 - 10.5	1800	4.00	12.0	18.0	13.0	19.0
	11.0 - 16.0			14.0	21.0	13.0	19.0
D	12.0 - 13.0	690	2.00	19.0	27.0	19.0	28.0
	13.5 - 15.5	to	to	21.0	30.0	21.0	31.0
	16.0 - 22.0	1200	4.00	24.0	36.0	25.0	36.0

*Note: This information is for Horsepower Ratings which are mentioned in this manual only. Use with older drives could result in overtensioning.

FIGURE 10–30

Instructions and tables for belt tensioning. A spring-type tensioning tool and sonic tester are pictured.

Courtesy of The Gates Rubber Co.

These tables reflect an average. If the reading is less than the recommended forces, the belts should be tightened. If the reading is more than the maximum, the drive should be slackened. Manually rotate the drive and repeat the process. Keep in mind that because of "tension decay" there will be a rapid drop in the initial tension of new belts. Tension decay is the phenomenon of the belt initially stretching a significant amount during operation and then gradually less as time progresses. For this reason, adjust the tension to the recommended maximum on new belts. Most belt drive manufacturers suggest that the drive be shut down and retensioned after an initial run-in period.

Final Drive Check Proceed with restart and final check. Follow the proper plant restart and notification procedures. Clear away all debris and replaced parts from the drive area and dispose of them according to established standards. Make sure the guard is properly repositioned. Start the machine and allow the belts to seat themselves in the grooves. Observe and note any unusual noises or smells. If a strobe light is available, tune it to the rotational speed of the component and adjust the flash rate so the sheave appears to rotate slowly. This procedure will let you observe various faults such as belt slip, tension, sheave condition, run-out, and shaft problems. Painting a thin line across the belts' top surfaces at 90° prior to start-up and then lighting the belts with the strobe while running will show belt slip. The strobe should be tuned to the operating frequency of the belts. Some initial slipping will naturally occur. The belts that are slipping will be moving away from the line at various speeds according to their degree of looseness. Vibration frequency data should be obtained on the machine, bearings, and belts. The frequencies for the belt drive can be calculated using the sheave diameters, motor speeds, and belt length.

$$\text{Belt frequency} = \frac{3.142 \times \text{Sheave rpm} \times \text{Sheave pitch diameter}}{\text{Belt length}}$$

A tachometer can be used to determine sheave speeds. During the operating mode, the belt drive system should not exceed a temperature of 140°F. Amp readings should be taken on the driving motor initially and at regular intervals to see if the load on the system is increasing. If possible, shut the system off (follow lock-out procedures) after 24 to 48 hours and retension the belts.

Following the above maintenance procedures will help ensure that the drive runs trouble-free throughout its service life.

V-Belt Drive Troubleshooting Quick Guide

Problem	Probable Cause	Corrective Action
Short Belt Life	Tensile members damaged during installation	Replace with a new set of belts
	Worn or damaged sheave grooves	Check groove profile with groove gauge, replace if needed
	Underdesigned drive	Redesign drive
	Foreign material between belt & sheave groove	Clean the drive & properly shield

Table continued

Problem	Probable Cause	Corrective Action
Belt Sidewalls Soft & Sticky	Oil, grease, or chemicals on belt	Remove contaminant source, clean belts & sheaves
Belt Dry, Hard, & Cracked	High temperatures	Remove source of heat & clean guard, ventilate
Deterioration of Belt Rubber Compounds	Belt dressing or chemicals	Do not use belt dressing
Extreme Cover Wear	Belt rubbing against guard or obstruction	Inspect and enlarge guard, remove obstructions
Belt Glazed, Spin Burns	Belt slipping under starting or stalling load	Properly tension drive, replace worn belts
Belt Noise	Belts slipping	Properly tension drive, inspect belts for damage
Bottom of Belt Cracked	High Temperatures	Remove source of heat & clean guard, ventilate
	Sheave diameter too small	Redesign drive, use notched belts
Belt Turnover	Misalignment	Properly align drive to recommended specifications
	Belt whipping	Properly tension, use banded belts
	Foreign material in drive	Clean the drive & properly guard
	Worn sheave grooves	Inspect sheaves with groove gauge, replace worn sheaves
	Incorrectly placed idler	Reposition idler
	Belt tensile member broken during installation	Replace belt set
Excessive Belt Stretch	Belt tensile member broken during installation	Replace belt set

Synchronous-Belt Drive Troubleshooting Quick Guide

Problem	Probable Cause	Corrective Action
Excessive Edge Wear (Exposed Tensiles)	Misalignment or nonrigid base	Align to recommended tolerances Inspect and make base rigid
	Bent pulley/sprocket flange	Straighten flange
Excessive Wear on Tooth Face	Overloads and/or overtension	Reduce drive tension and/or redesign drive

Synchronous-Belt Drive Troubleshooting Quick Guide *continued*

Problem	Probable Cause	Corrective Action
Softening or Cracking of Belt Backing	Exposure to extreme high & low temperatures	Eliminate extreme temperatures Ventilate or shield as needed
Tensile or Tooth Shear	Subminimum diameter pulley/sprocket	Increase diameters
	Excessive loads	Redesign drive, increase capacity
	Shock loads or jamming	Remove obstructions, possibly install clutching mechanism, increase drive size
Excessive Noise	Misalignment	Align to recommended tolerances
	Excessive tension	Properly tension drive
	Subminimum diameter	Increase diameters

Student Exercise

Calculate and determine the belt drive ratio, belt length, effective tension, sheave velocity (FPM), belt deflection distance, and the driven sheave speed of the example drive (Figure 10–31). Fill out the worksheet.

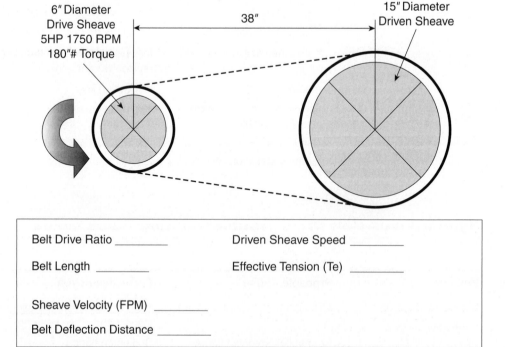

FIGURE 10–31
Student exercise worksheet.

Questions

1. How does a V-belt function?
2. Is tension vital to a V-belt?
3. V-belts replaced what type of belt and why?
4. Why is the V-belt arc of contact important?
5. How do you calculate the ratio of a V-belt drive?
6. What is the V-belt matching system and is it still relevant today?
7. How is a V-belt size determined?
8. Describe the "classical" V-belt nomenclature.
9. Name three types of mounting systems for V-belt sheaves.
10. What is the purpose of a V-belt idler?
11. What is the advantage of notching a V-belt?
12. What is a synchronous belt?
13. What advantage does a synchronous belt have over a V-belt?
14. How should sheave wear be checked?
15. How do you tension a V-belt drive?
16. Synchronous belt pitch is usually measured in what increments?
17. How are belt drives aligned?
18. How are V-belt sheaves measured?
19. List twelve safety, inspection, and maintenance steps for belt drives.
20. Name five specific applications for V-belts.
21. What are the three most important considerations pertaining to belt drives?

CHAPTER 11

Drive Chain and Sprockets

This chapter is devoted to the types of drive and conveying chains and how they function. Chain and sprocket drives are used to connect parallel shafts of equipment to transfer power and rotary movement. Industry uses drive chains not only for power transmission, but to move materials. We also will look at the different kinds of sprockets that engage and connect the drive and conveying chains. Numbering systems of some of the more common types of roller drive chains will be addressed, as well as installation, maintenance, lubrication, and safety procedures for drive chains.

Objectives

Upon completion of this chapter, the student will be able to:

✔ Understand the basic principles of drive chain function and operation.
✔ Identify the various components and types of chain.
✔ Identify the various types of chain sprockets.
✔ Use the numbering system for chain and sprockets correctly.
✔ Have a basic understanding of drive chain selection procedures.
✔ Disassemble and reassemble roller chain safely and correctly.
✔ Demonstrate correct alignment and tensioning procedures for drive roller chain.
✔ Have knowledge of the various lubrication methods for drive chain.

INTRODUCTION

Chains of one form or another have been driving machines and conveying materials for more than a thousand years. Leonardo da Vinci sketched a variety of different types of chain. During the 1800s, chain was installed on the bicycle. With the growth of agribusiness and the automotive industry, the demand for chain increased. The chain manufacturing industry met those demands with the development of numerous types and sizes of drive chains. Although drive chain is a mature form of power transmission, there are still many applications in industry that drive chain is suited for. It is economical, resistant to shock loads, easy to install, has the ability to transmit high torque, operates in hostile environments, and is efficient. Although it appears to be a simple component, it is highly engineered and manufactured within close tolerances. Chains are available in an assortment of styles, sizes, and materials.

Some of the different types of drive chain include roller, silent, detachable, tabletop, leaf, mill, pintle, and drag chain.

DRIVE CHAIN FUNCTION AND OPERATION

Chain can be grouped into two functions: material handling and power transmission. Some chains are used for both purposes. If the chain is transferring or conveying raw material or finished products, it is considered a material handling chain. If the chain's primary purpose is to transfer power from one shaft to another, it is for power transmission. In some applications, special attachments for moving materials are part of a power-transmitting chain's construction.

The chain drive system consists of a driving sprocket, one or more driven sprockets, and an endless chain consisting of individual links connected to form a strand. A sprocket is a wheel that has evenly spaced, uniformly shaped teeth on the outside diameter that provide positive engagement with the chain. Because the chain is hinged at every link, it is able to wrap around the sprocket's teeth, connecting the driver and driven units. Another portion of the drive chain and sprocket system is the chain lubrication system. Chain systems require lubrication.

Power, originating as torque, is transmitted from the drive sprocket mounted on the shaft of the prime mover and connected to one or more driven sprockets by the chain. As the chain drive system operates, every link in the chain undergoes "cycle loading." The working or tight side of the chain is under full tension, while the slack side is under minimum tension. The sprocket's ability to absorb this tension is determined by the pressure angle of the teeth, the number of teeth in mesh, and the resultant tension in the slack side of the chain. The sprockets, connected by a chain and turning on their respective shafts, deliver the power or transfer the load to accomplish work.

During operation a phenomenon known as "chordal action" occurs in a chain drive system because the chain link is a straight-line segment trying to follow the circular path of the sprocket pitch diameter. Because the line of approach of the chain is not tangent to the pitch circle, it is lifted to the top of the circle and then dropped down. There is a surge of force in the chain caused by the change in speed as it makes this rise and fall. An increase in chain velocity can aggravate this event, resulting in vibration and pulse loading in the system.

Chain Construction and Dimensions

Roller drive chain is essentially a connected series of bearings articulating (revolving) around an axis point. Typically, chain can be composed of side plates, pins, rollers, and bushings. With some chains, the side plates are an integral part of the barrels (bushings), and other chains are composed of different assembled parts. Roller chain is assembled from a variety of different parts all working together to transfer power and torque (see Figure 11–1). Roller chain has rollers that rotate on bushings as they come into contact with the sprocket teeth. The rolling action reduces friction, making the drive efficient and helping to increase its life. The bushings provide a bearing surface for pin rotation. The pins connect the side plates. The side plates carry the majority of

FIGURE 11–1
A section of roller drive chain showing the components.
Courtesy of Diamond Chain.

the working load and must endure static and dynamic tensile forces. Pins, rollers, and bushings are subject to shearing and bending forces. All of these parts work together to provide an efficient means to transmit power or move materials.

Drive and conveying chain dimensions can vary widely between manufacturers. The tolerances that are held during the production process can also be different from brand to brand. Most companies that produce roller drive chains hold similar width and length dimensions. The internal width dimension of the chain is critical, so that it properly engages the sprocket without binding. The primary length dimension to be concerned about is the "pitch" of the chain. The center distance from one hinge point to the next is defined as the pitch. The pitch is one of the critical chain dimensions and is important in identifying the part number of the chain. The design and material used in the construction of the chain, as well as the size of the chain pitch, determine its ability to transmit power. Larger pitch chains will have larger components that allow it to transmit greater horsepower. The pitch of the chain also determines the maximum operating speed of the chain. The larger the pitch, the lower the speed limit.

DRIVE CHAIN TYPES

Roller Chain

Standard roller chain is the most common type of drive chain used in industry. Standard roller chain is made up of alternating roller links and pin links. The roller links are constructed by placing the rollers onto bushings and pressing roller link plates onto the bushings. The pin links are constructed of a pin link side plate into which two pins are pressed and another side plate is pressed or slipped over the other end of the side plate. The assembly is secured by riveting the ends of the pins or passing various types of cotter pins through holes in the pins.

The American National Standards Institute (ANSI) has established standards for pitch and part numbers for roller drive chain (B29.1). This allows for interchanging of chain made by one company with chain produced by another. Typically a part number will be stamped on the side plate to identify the chain. The first number indicates the pitch of the chain in eighths of an inch.

EXAMPLE

A number 50 indicates that the pitch is 5/8″.

Figure 11–2 lists ANSI standard chain specifications and dimensions. Number 50 ANSI standard roller chain and smaller are furnished with riveted pins. Number 60 ANSI standard roller chain and larger are furnished with either riveted or cottered pins.

Connecting links are used to join the ends of a strand of roller chain to make it a continuous strand with an even number of pitches (see Figure 11–3). The connecting link or "master link" consists of two pins press fitted and riveted in one link plate. The holes in the free link plate are sized for either a slip fit or a light press fit on the exposed pins. The plate is secured by either cotter pins or spring clips. Connecting links can be slip fit or drive fit.

An offset link is used to join the ends of a strand of roller chain to make it a continuous strand with an odd number of pitches (see Figure 11–4). The offset or "half-link" is made up of a pin and roller with two offset side plates. The bent side plates tend to cause the chain to have a lesser strength rating so its use should be limited.

An offset section is an assembly of a standard roller link and an offset link together (see Figure 11–5). The press fit construction of this assembly increases its structural rigidity and durability, making it preferable to the single offset link.

Roller chains that are used for material handling are available with numerous types of attachments for moving different kinds of products. Figure 11–6 illustrates bent attachments on every link. Various kinds of stickers, plates, lugs, rollers, etc., can be connected to these attachments to assist in the movement of the product or material.

FIGURE 11–2

Standard ANSI roller chain specification and dimension table.

Courtesy of Diamond Chain.

Dimensions in Inches

ANSI & Diamond Number	ISO Number	Pitch Inches	Roller Width	Roller Diam.	Pin Diam.	Linkplate Thickness	Linkplate E Height	Linkplate H Height	C	R	K	Bearing Area Sq. inch	Weight Per Foot Pounds	Average Tensile Strength
25	04C-1	1/4	1/8	*.130	.090	.030	.205	.237	.37	.34	....	.017	.084	875
25-2	04C-2	1/4	1/8	*.130	.090	.030	.205	.237	.63	.59	.252	.034	.163	1750
25-3	04C-3	1/4	1/8	*.130	.090	.030	.205	.237	.88	.84	.252	.051	.246	2625
35	06C-1	3/8	3/16	*.200	.141	.050	.307	.356	.56	.50	....	.041	.21	2100
35-2	06C-2	3/8	3/16	*.200	.141	.050	.307	.356	.96	.90	.399	.082	.45	4200
35-3	06C-3	3/8	3/16	*.200	.141	.050	.307	.356	1.36	1.31	.399	.123	.68	6300
35-4	06C-4	3/8	3/16	*.200	.141	.050	.307	.356	1.76	1.70	.399	.164	.91	8400
35-5	06C-5	3/8	3/16	*.200	.141	.050	.307	.356	2.16	2.11	.399	.205	1.14	10500
35-6	06C-6	3/8	3/16	*.200	.141	.050	.307	.356	2.57	2.51	.399	.246	1.37	12600
40	08A-1	1/2	5/16	.312	.156	.060	.410	.475	.72	.67	....	.067	.41	4000
40-2	08A-2	1/2	5/16	.312	.156	.060	.410	.475	1.29	1.24	.566	.134	.80	8000
40-3	08A-3	1/2	5/16	.312	.156	.060	.410	.475	1.85	1.80	.566	.201	1.20	12000
40-4	08A-4	1/2	5/16	.312	.156	.060	.410	.475	2.42	2.37	.566	.268	1.60	16000
40-6	08A-6	1/2	5/16	.312	.156	.060	.410	.475	3.56	3.51	.566	.402	2.42	24000
41	085	1/2	1/4	.306	.141	.050	.310	.383	.65	.57	....	.049	.26	2400
50	10A-1	5/8	3/8	.400	.200	.080	.512	.594	.89	.83	....	.106	.68	6600
50-2	10A-2	5/8	3/8	.400	.200	.080	.512	.594	1.60	1.55	.713	.212	1.32	13200
50-3	10A-3	5/8	3/8	.400	.200	.080	.512	.594	2.31	2.26	.713	.318	1.98	19800
50-4	10A-4	5/8	3/8	.400	.200	.080	.512	.594	3.03	2.97	.713	.424	2.64	26400
50-5	10A-5	5/8	3/8	.400	.200	.080	.512	.594	3.75	3.69	.713	.530	3.30	33000
50-6	10A-6	5/8	3/8	.400	.200	.080	.512	.594	4.46	4.40	.713	.636	3.96	39600
50-8	10A-8	5/8	3/8	.400	.200	.080	.512	.594	5.89	5.83	.713	.848	5.30	52800
50-10	10A-10	5/8	3/8	.400	.200	.080	.512	.594	7.32	7.26	.713	1.060	6.62	66000
60	12A-1	3/4	1/2	.469	.234	.094	.615	.712	1.11	1.04	....	.161	.99	8500
60-2	12A-2	3/4	1/2	.469	.234	.094	.615	.712	2.01	1.94	.897	.322	1.95	17000
60-3	12A-3	3/4	1/2	.469	.234	.094	.615	.712	2.91	2.84	.897	.483	2.88	25500
60-4	12A-4	3/4	1/2	.469	.234	.094	.615	.712	3.81	3.74	.897	.644	3.90	34000
60-5	12A-5	3/4	1/2	.469	.234	.094	.615	.712	4.71	4.64	.897	.805	4.97	42500
60-6	12A-6	3/4	1/2	.469	.234	.094	.615	.712	5.60	5.53	.897	.966	5.96	51000
60-8	12A-8	3/4	1/2	.469	.234	.094	.615	.712	7.40	7.33	.897	1.288	7.94	68000
60-10	12A-10	3/4	1/2	.469	.234	.094	.615	.712	9.19	9.12	.897	1.610	9.92	85000

Dimensions in Inches

ANSI & Diamond Number	ISO Number	Pitch Inches	Roller Width	Roller Diam.	Pin Diam.	Linkplate Thickness	E Height	H Height	C	R	K	Bearing Area Sq. Inch	Weight Per Foot Pounds	Average Tensile Strength
Standard Series Chains														
80	16A-1	1	⅝	.625	.312	.125	.820	.950	1.44	1.32	....	.276	1.73	14500
80-2	16A-2	1	⅝	.625	.312	.125	.820	.950	2.59	2.47	1.153	.552	3.37	29000
80-3	16A-3	1	⅝	.625	.312	.125	.820	.950	3.74	3.62	1.153	.828	5.02	43500
80-4	16A-4	1	⅝	.625	.312	.125	.820	.950	4.90	4.79	1.153	1.104	6.73	58000
80-5	16A-5	1	⅝	.625	.312	.125	.820	.950	6.06	5.94	1.153	1.380	8.40	72500
80-6	16A-8	1	⅝	.625	.312	.125	.820	.950	7.22	7.10	1.153	1.656	10.07	87000
80-8	16A-8	1	⅝	.625	.312	.125	.820	.950	9.53	9.40	1.153	2.208	13.41	116000
100	20-A1	1¼	¾	.750	.375	.156	1.025	1.187	1.73	1.61	....	.402	2.51	24000
100-2	20A-2	1¼	¾	.750	.375	.156	1.025	1.187	3.14	3.02	1.408	.804	4.91	48000
100-3	20A-3	1¼	¾	.750	.375	.156	1.025	1.187	4.56	4.43	1.408	1.206	7.40	72000
100-4	20A-4	1¼	¾	.750	.375	.156	1.025	1.187	5.97	5.84	1.408	1.608	9.80	96000
100-5	20A-5	1¼	¾	.750	.375	.156	1.025	1.187	7.38	7.25	1.408	2.010	12.20	120000
100-6	20A-6	1¼	¾	.750	.375	.156	1.025	1.187	8.78	8.66	1.408	2.412	14.60	144000
100-8	20A-8	1¼	¾	.750	.375	.156	1.025	1.187	11.60	11.48	1.408	3.216	19.40	192000
120	24A-1	1½	1	.875	.437	.187	1.230	1.425	2.14	2.00	....	.605	3.69	34000
120-2	24A-2	1½	1	.875	.437	.187	1.230	1.425	3.93	3.79	1.789	1.210	7.35	68000
120-3	24A-3	1½	1	.875	.437	.187	1.230	1.425	5.72	5.58	1.789	1.815	11.10	102000
120-4	24A-4	1½	1	.875	.437	.187	1.230	1.425	7.52	7.38	1.789	2.420	14.70	136000
120-5	24A-5	1½	1	.875	.437	.187	1.230	1.425	9.31	9.17	1.789	3.025	18.43	170000
120-6	24A-6	1½	1	.875	.437	.187	1.230	1.425	11.10	10.96	1.789	3.630	22.11	204000
120-8	24A-8	1½	1	.875	.437	.187	1.230	1.425	14.68	14.54	1.789	4.840	29.47	272000
120-10	24A-10	1½	1	.875	.437	.187	1.230	1.425	18.26	18.12	1.789	6.050	36.83	340000
140	28A-1	1¾	1	1.000	.500	.219	1.435	1.662	2.31	2.14	....	.725	5.00	46000
140-2	28A-2	1¾	1	1.000	.500	.219	1.435	1.662	4.24	4.07	1.924	1.450	9.65	92000
140-3	28A-3	1¾	1	1.000	.500	.219	1.435	1.662	6.16	6.00	1.924	2.175	14.30	138000
140-4	28A-4	1¾	1	1.000	.500	.219	1.435	1.662	8.09	7.93	1.924	2.900	18.95	184000
140-6	28A-5	1¾	1	1.000	.500	.219	1.435	1.662	11.94	11.78	1.924	4.350	28.25	276000
160	32A-1	2	1¼	1.125	.562	.250	1.640	1.900	2.73	2.54	....	.987	6.53	58000
160-2	32A-2	2	1¼	1.125	.562	.250	1.640	1.900	5.04	4.85	2.305	1.974	12.83	116000
160-3	32A-3	2	1¼	1.125	.562	.250	1.640	1.900	7.35	7.16	2.305	2.961	19.03	174000
160-4	32A-4	2	1¼	1.125	.562	.250	1.640	1.900	9.66	9.47	2.305	3.948	25.60	232000
160-6	32A-6	2	1¼	1.125	.562	.250	1.640	1.900	14.27	14.09	2.305	5.922	37.78	348000
180	....	2¼	1¹³⁄₃₂	1.406	.687	.281	1.845	2.137	3.15	2.88	....	1.365	9.06	76000
180-2	....	2¼	1¹³⁄₃₂	1.406	.687	.281	1.845	2.137	5.75	5.48	2.592	2.730	17.67	152000
180-3	....	2¼	1¹³⁄₃₂	1.406	.687	.281	1.845	2.137	8.34	8.07	2.592	4.095	26.20	228000
200	40A-1	2½	1½	1.562	.781	.312	2.050	2.375	3.44	3.12	....	1.671	10.65	95000
200-2	40A-2	2½	1½	1.562	.781	.312	2.050	2.375	6.26	5.94	2.817	3.342	21.5	190000
200-3	40A-3	2½	1½	1.562	.781	.312	2.050	2.375	9.08	8.76	2.817	5.013	32.3	285000
200-4	40A-4	2½	1½	1.562	.781	.312	2.050	2.375	11.90	11.58	2.817	6.684	42.9	380000
200-6	40A-6	2½	1½	1.562	.781	.312	2.050	2.375	17.52	17.21	2.817	10.026	64.5	570000

FIGURE 11–2 (*Continued*)

Miscellaneous Roller Chains

Multiple Strand Roller Chain Multiple strand chain, which is made up of two or more strands, is constructed with common pins across the full width (see Figure 11–7). Multiple strand chains are used to provide increased power capacity without increasing the chain pitch. They are available in standard ANSI pitch lengths.

Heavy Series Roller Chain Heavy series roller chains have pitches and part numbers similar to standard chain, but they have thicker side plates. The thicker

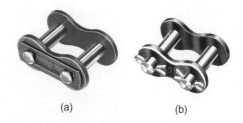

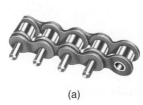

(a)   (b)

FIGURE 11–3
Roller chain connecting links: (a) cotter and (b) spring clip type.
Courtesy of Emerson Power Transmission.

FIGURE 11–4
Roller chain offset link.
Courtesy of Emerson Power
Transmission.

FIGURE 11–5
Roller chain offset section.
Courtesy of Emerson Power
Transmission.

(a)   (b)

FIGURE 11–6
(a) Pin and (b) plate roller chain attachments.
Courtesy of Emerson Power Transmission.

side plates provide greater resistance to shock loads, but no significant increase in tensile strength or wear life. The heavy series chains are often used to withstand the shock loads on mining, construction, and oil field machinery. The letter H following the part number designates the chain as a heavy series; they are available in ANSI number 60 (3/4 pitch) and larger.

FIGURE 11–7
Multiple strand roller chain.

Courtesy of Diamond Chain.

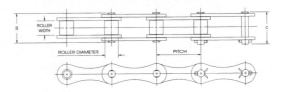

Dimensions in Inches and Pounds

ANSI Number	Pitch Inches	Roller Width	Roller Diameter	Pin Diameter	Link Pate Thickness	C	R	Weight Per Foot	Average Tensile Strength
2040	1	5/16	.312	.156	.060	.76	.68	.28	3700
2050	1¼	⅜	.400	.200	.080	.92	.84	.52	6100
2060	1½	½	.469	.234	.094	1.11	1.05	.72	8500
2080	2	⅝	.625	.312	.125	1.44	1.32	1.13	14500

FIGURE 11–8
Double pitch roller chain size specifications.

Courtesy of Diamond Chain.

Double Pitch Roller Chain Double pitch chain is similar to standard roller chain, except that the pitch is twice as long (see Figure 11–8). It is an example of a chain that can be used for both power transmission and conveying purposes. The link plates are available in two forms. One type has figure-eight side plates and is used for the transmission of power. The other form has straight-edged plates and is known as "conveyor series" chain; it is used in material handling. The rollers for conveyor series double pitch chain can be made oversized to allow the chain to roll on a bed. These chains are used extensively in agriculture and in conveying materials.

Self-Lubricating Roller Chain Self-lubricating chain has oil impregnated bushings. The chain is constructed with a one-piece powdered metal bushing/roller combination that has lubricant drawn in under extreme vacuum. The lubricant is released in service and provides supplemental lubrication to the pin/joint. They are capable of handling the same loads as standard roller chain but at slower speeds. When adequate or regular lubrication of the chain drive is difficult, self-lubricated chains can be used in place of standard ANSI chains. The ambient temperature at which the chain operates should not exceed 120°F.

O-ring Chain O-ring chain is also designed for applications that do not permit adequate lubrication. The chain is constructed with o-rings that seal lubricant into the

Figure 11–9
O-ring chain construction.
Courtesy of Diamond Chain.

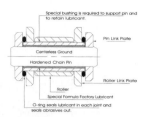

joints (see Figure 11–9). This prevents contaminants such as dirt, abrasives, and moisture from entering the chain components and causing wear. It is available in standard ANSI pitches and can operate in ambient temperatures up to 150°F unless special high temperature (400°F) o-rings are substituted.

Engineering, Conveying, and Miscellaneous Chains

There is a variety of different types of chain employed for both the purposes of transmitting power and conveying. Some of these chains fall into the catch-all category of "engineered class chain." There is limited standardization among the manufacturers of these types of chain and it is recommended that the specifications be checked before interchanging whole or parts of chains. This class of chains is used for the purpose of moving and holding loads. Generally they are of a durable and rugged construction. The major applications for these chains are mining, construction, food processing, material handling, and wood industry equipment.

Offset Sidebar Chain Offset sidebar chain is designed for power transmission or conveying service. On offset sidebar chain, each link is identical (see Figure 11–10). A link consists of two sidebars, one bushing, and one pin if a rollerless type. Sometimes this chain is called "welded steel chain." Offset sidebar chain is also available with rollers. These chains are economical, efficient, and have the ability to operate in rugged environments.

Pintle, Mill, and Drag Chains Wherever chain is needed to transfer power under shock loads, and raw materials need to be transferred in a hostile environment you will find these chains. These "engineered class" chains are generally of offset-type construction with integral side plates and barrels. They are constructed of malleable iron and steel. Malleable iron is resistant to wear and functions well in dirty applications. A hole has been drilled through the barrel for the connecting pin. This pin is locked in place by being riveted on one end and secured by a cotter key on the other.

Figure 11–10
Offset sidebar/
welded steel chain.
Courtesy of Webster
Industries.

FIGURE 11–11
Pintle chain.

Courtesy of Webster
Industries.

There are two basic classes of pintle chain, 400 and 700 (see Figure 11–11). Class 400 is used primarily for conveying and elevating. Class 700 has a longer pitch length, heavier construction, and is used extensively in wastewater treatment. Pintle chains are usually made of a cast metal, except for the steel pin.

Mill chain is generally heavier than pintle chain and is used in the paper and lumber industries for material handling service (see Figure 11–12). The sidebars can have flanged wear shoes to facilitate dragging or sliding on a surface. They are used extensively to move abrasive product in the wood, paper, and mining industries. It is sometimes referred to as "H-type mill chain."

Drag conveyor chain is available in three types: H-type drag chain, steel bar drag chain, and combination drag chain (see Figure 11–13). Large components made from steel and cast materials make it an excellent choice for handling heavy abrasive products. It is also referred to as "wide conveyor chain."

FIGURE 11–12
Mill chain.

Courtesy of Webster
Industries.

FIGURE 11–13
Drag chain.

Courtesy of Webster Industries.

FIGURE 11–14
Silent chain
and sprockets.

Courtesy Emerson Power
Transmission.

Silent Chain Silent chain consists of a series of toothed link plates assembled on joint or hinge components (see Figure 11–14). The accepted standard for silent chains has been established by ANSI (B29.2). Silent or "inverted tooth" chains normally are used with prime movers of less than 50 horsepower. They are often found in variable or high-speed applications. Silent chains are smooth running and very flexible. Standard pitches include 3/16″, 3/8″, 1/2″, 5/8″, 3/4″, and 1″. Silent chains are designated using the following system: Two prefix letters, SC, mean ANSI standard silent chain; one numeral indicates pitch in eighths of inches, and two or three numerals indicate width in quarter inches. For example, SC816 is a chain with 1″ pitch × 4″ wide. Special sprockets designed to operate with silent chain will have a circumferential groove or side plates to assist in keeping the chain on the sprockets.

Leaf Chain Leaf chains, sometimes called lift chains, are used for transmitting reciprocating motion of lift (see Figure 11–15). It is often incorporated into the lifting mechanism on lift trucks because of its high strength rating. It is constructed of interlaced plates and held together with riveted pins.

Polymeric or Plastic Chains Polymeric or plastic chains made from nonmetallic thermoplastic material come in many shapes and forms (see Figure 11–16). The pins are usually made from stainless steel. They are applied in applications requiring corrosion resistance. The low friction and lightweight characteristics reduce energy consumption and noise. It is recommended to use mating sprockets made from a similar material as the chain. The food and beverage industry makes extensive use of this

FIGURE 11–15
Leaf chain.

Courtesy Emerson Power
Transmission.

FIGURE 11–16
Plastic engineering
chain.

Courtesy of Rexnord.

type of chain because of its low weight and ability to be washed down without any detrimental effects.

Tabletop Chain Tabletop chains are constructed by placing joined plastic table tops onto a base roller chain to form a two-piece construction (see Figure 11–17). Single-piece construction is available that has each table piece hinged together. Tabletop, or flat-top, chains are used to convey materials such as bottles or cans in

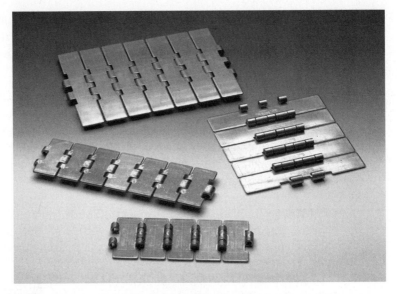

FIGURE 11–17
Tabletop chain used for conveying product such as canned goods.

Courtesy Emerson Power Transmission.

FIGURE 11–18
Plastic wash-down duty mat-top chain used to transfer loose product such as vegetables.

Courtesy of Rexnord.

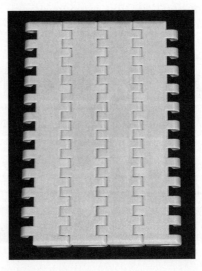

the brewing, bottling, dairy, pharmaceutical, and food industries. They are available in both the "straight running" and "sideflexing" types. They can be made of stainless steel or of some type of thermoplastic material that tolerates periodic wash-down and sanitizing. A version of tabletop chain is the "mat-top" chain which is made of wide-hinged plastic sections joined together (see Figure 11–18).

There are numerous other types of chain, such as detachable, combination, transfer, rooftop, flexible, tabletop, and conveyor chains. Chain manufacturers can be consulted for more detailed information on various special chains.

SPROCKETS

A sprocket is a toothed wheel whose uniform-sized teeth are evenly spaced and shaped to mesh with the appropriate chain (see Figure 11–19). The teeth provide positive engagement with a chain transferring power from one sprocket, mounted

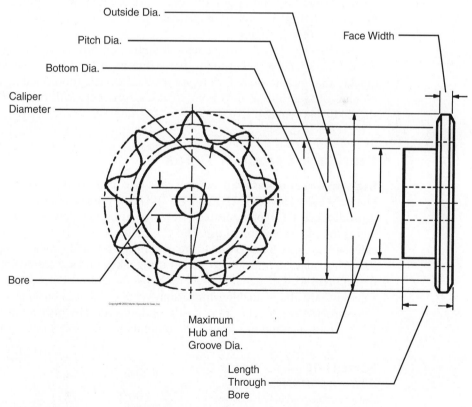

FIGURE 11–19
Sprocket terminology showing critical dimensional specifications used to identify the type and size of sprocket.

Courtesy of Martin Sprocket & Gear.

on a shaft, to another. It is imperative that the pitch of the sprocket be the same as the chain so that it will properly mesh to provide acceptable chain and sprocket life. The pitch diameter of a sprocket is the diameter of a circle followed by the centers of the chain pins as the sprocket revolves in mesh with the chain. The pitch diameter of a sprocket is in part determined by the chain pitch and the number of teeth in the sprocket. The hub diameter is the distance across the hub from one side to the other. The bottom diameter is the diameter of a circle tangent to the bottoms of the teeth spaces. The caliper diameter is the measurement across the teeth spaces nearly opposite. Length through bore (LTB) is the overall width of the sprocket.

Sprocket Construction

Sprockets usually consist of the body (plate), rim (teeth), and a hub that is attached to a shaft by some means. The body of the sprocket is the area between the circumference of the teeth and the hub. The body can be a solid, webbed, or spoked design. Generally the larger sprockets are spoked or webbed to reduce weight. Materials that sprockets are made from are iron, steel, sintered metals, and plastics of various compositions.

Hardening of the teeth through an induction process or by flame results in extended life of the sprocket teeth. Teeth that have been hardened tend to wear less quickly than nonhardened tooth sprockets. Because tooth loading increases as the number of teeth in the sprocket decreases, sprockets with twenty-five teeth or less should be hardened.

Standard Sprocket Types

Standard Sprockets Sprocket types are classified as A, B, C, and D (see Figures 11–20, 11–21, 11–22, and 11–23). The A type is a plate sprocket with no hub projections. The B type indicates a hub projection from one side of the sprocket face and is the most common in use. The C type sprocket has an integrated hub on both sides of the plate. Types B and C are generally provided with a straight finished bore in the hub and are attached to a shaft using a key and setscrews. Sprockets may also be supplied with detachable bolts on hubs; these are referred to as D types. Many sprockets are also available with various types of tapered bushings.

Sprocket nomenclature is easily understood. The part number of a sprocket will typically include the mating chain number, hub type, number of teeth, and bore

FIGURE 11–20
A-type or plate sprocket.
Courtesy of Emerson Power Transmission.

FIGURE 11–21
B-type sprocket with a single hub welded to the plate sprocket.

Courtesy of Emerson Power Transmission.

FIGURE 11–22
C-type sprocket with hubs on both sides of the plate sprocket.

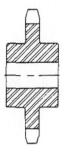

FIGURE 11–23
D-type or taper lock type bushing mount sprocket.

Courtesy of Emerson Power Transmission.

size. For example, a sprocket with a part number reading 50B21-7/8″ will be for a 50 ANSI standard roller chain, B-style hub, 21 teeth, with a bore of 7/8″. Those sprockets that are a D (tapered bushed) type might have a part number that reads 50SDS21. This sprocket would also be used with number 50 chain and have 21 teeth, but use an SDS "QD" style tapered bushing.

Special Sprocket Types

Split Sprockets Split sprockets are available for applications that require mounting between bearings, or in a "trapped" location (see Figure 11–24). Because it is split, it permits installation of the halves around the shaft without disturbing the bearings and other connected equipment.

FIGURE 11–24
Split sprocket.
Courtesy of Martin Sprocket & Gear.

FIGURE 11–25
Split and sectional rim sprockets.
Courtesy of Webster Industries.

(a)   (b)

Sectional Rim Sprockets Sectional rim sprockets have a body that is split or solid, with the rim (toothed sections) segmented into three or more sections (see Figure 11–25). In addition to the advantages of being a split-type sprocket, the rim sections can be replaced as they wear without disturbing the chain. This is advantageous in long conveyor applications or vertical drives. These types of sprockets are often used with conveyor chains.

Double Pitch Sprockets Double pitch chain sprockets come in two basic types: double cut and single cut (see Figure 11–26). Because of pitch diameter variations,

FIGURE 11–26
Double pitch sprockets.
Courtesy of Martin Sprocket & Gear.

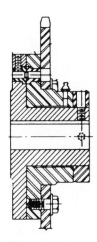

Figure 11–27
Shear pin sprockets.
Courtesy of Martin Sprocket & Gear.

standard roller chain sprockets should not be used for double pitch chain. Single cut sprockets have a special tooth form and are to be run with double pitch chain, some of which have rollers. Double cut sprockets are made to be run with long pitch engineering class chains (6″ or greater).

Shear Pin Sprockets Shear pin sprockets consist of a hub keyed to the shaft, and a sprocket that is free to rotate either on the shaft or on the hub when the shear pin breaks (see Figure 11–27). The pin is designed to break under a specified load. The material that the shear pin is made from has a known shearing strength, and a breaking point can be established by machining a groove with a calculated diameter at the shear plane. When a severe shock load or jam-up occurs, the pin breaks and protects the mating machinery. Replacing the broken pin will quickly put the assembly back into service. Under no circumstances should the shear pin be replaced with a bolt or rod. Replacing a shear pin with something other than the correct pin will result in damage to the equipment or injury to the operator.

Selection and Application of Roller Chain Drives

Proper selection of roller chain must take into consideration such factors as the nature of the application, speed, horsepower, cost, required life, service factor, and space limitations. Most manufacturers of high-quality chain can provide the design engineer with the proper manual or selection software that eliminates guesswork. The best chain drive system for a given horsepower, speed, and ratio will be one that transmits the required load using the smallest pitch chain and least number of strands, operating over the largest sprockets, and employing an adequate lube system. There are a number of general rules of good design to yield the best results from the chain drive.

Drive Ratio The ratio of the sprocket sizes is determined by the required speed ratio of the drive. The recommended maximum ratio under normal operating conditions is 7:1 for a single reduction drive. The minimum size of the small sprocket and maximum size of the large sprocket, along with the required adequate wrap on the small sprocket, affect the practical single reduction ratio limit.

Sprocket Size The selected sprockets must be large enough to accommodate the shaft and key that it is mounted on. Chain speed variation and impact between the rollers and sprocket caused by chordal action decrease as the number of teeth in the sprocket increases. For smooth operation, there is a recommended minimum number of teeth on the small sprocket: In a slow-speed application, 12 teeth; for a medium speed, 17 teeth; and for high-speed applications, 25 teeth. The number of teeth in the large sprocket will have an effect on allowable chain wear. As the chain wears, it becomes elongated. In percent of elongation, it is equal to approximately 200 divided by the number of teeth on the large sprocket. The lack of available sprockets over 100 teeth and space limitations usually determine the maximum tooth size.

Center Distance The center distance must be greater than one-half the sum of the sprocket's outside diameter to avoid tooth interference. A center distance of 30 to 50 pitches is good design practice, and a maximum of 80 pitches is appropriate for most applications. The shortest practical center distance is recommended for high-speed drives to avoid chain pulsing and vibration. For the best results the center distance should be in an arc of chain engagement of not less than 120°.

Idler Sprockets Chain tightening devices (idlers) are used to control tension and to prevent whipping on vertical drives and long spans. The most common type is an idler or free-wheeling sprocket mounted on an adjustable bracket. The bracket arm is either spring-tensioned or manually adjusted. Rollers, shoes, and wear plates are also used to tension chain. Idler sprockets should be no smaller than the driver sprocket. They should be located on the slack span of the chain. The idler sprocket should have a minimum of three teeth engaging the chain. Where adjacent sprockets mesh with opposite sides of a chain, there should be at least three pitches of free chain between mesh points.

ROLLER CHAIN SAFETY, INSTALLATION, AND MAINTENANCE

To attain maximum life from a chain drive system, proper handling, installation, lubrication, and maintenance of all parts are required. It is recommended that a regular maintenance schedule be established for all chain drives. Drives should be inspected after the initial 100 hours of operation, followed by 500-hour interval checks. Drives that are subject to shock loads or operate in a severe environment should be inspected more often. Chain and sprocket systems must be handled, installed, and maintained safely.

Chain Safety

Prior to and during maintenance checks, it is imperative that the following guidelines be adhered to in order to prevent personal injury:

- Always lockout and tag-out the equipment power switch followed by verification prior to working on the drive.
- Always wear safety glasses and other recommended safety equipment, such as gloves and steel-toed shoes, to prevent personal injury.
- Support the chain or mechanically neutralize it to prevent uncontrolled movement of the chain and parts.
- Properly use chain installation tools that are in good condition.
- Avoid lacing body parts in pinch points.
- Know the chain construction and the correct directions for chain assembly or disassembly.
- Do not substitute bolts, pins, etc., for the chain manufacturer's original parts.
- Damaged chain and sprockets should never be used.

Chain Installation, Maintenance, and Troubleshooting

The following installation and maintenance checks are general recommendations. Implementing these procedures will yield longer chain service life.

Chain Storage Storage of chain and sprockets in a clean, dry, and sheltered location is imperative. They should be kept in the original container until just prior to use. The lubrication applied at the factory is an excellent preservative and will be sufficient for most reasonable storage periods. If the chain is to be idle for an extended time on the drive it should be removed, cleaned with a nonflammable solvent, lubricated, marked, and stored.

Drive Interference Check for interference between the drive and other parts of the equipment. Contact between adjacent objects must not occur or it will cause abnormal and destructive wear of the system. Clearances must be allowed for normal pulses under load. Check for and eliminate any loose and abrasive materials that may have built up between the chain and sprockets. A small amount of debris in the sprocket roll seat can cause tensile loads to be extreme enough to break the chain if forced through the drive. Cleanliness of the system will extend its life.

Chain Case Chain casings, guards, or housings are to facilitate lubrication, isolate the drive from contaminants, and protect plant personnel (see Figure 11–28). These casings are usually made of sheet metal and stiffened by steel angles. Access doors are useful for inspection and maintenance. Proper clearances must be provided to prevent any contact with moving parts. They should be constructed to meet OSHA specifications. Periodic cleaning of the guards will benefit the life of the drive.

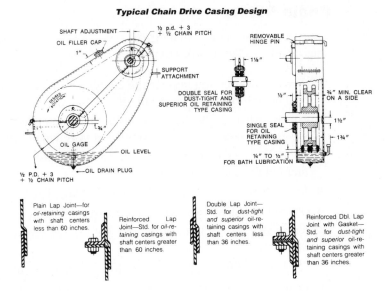

Typical Chain Drive Casing Design

FIGURE 11–28
Typical chain case.
Courtesy of Diamond Chain.

Chain and Sprocket Inspection Check the condition of the drive components. Some chain and sprocket wear is inevitable and normal as the drive operates. The articulation of chain as it enters and leaves the sprockets causes the pins and bushings to wear and the chain will gradually elongate. This is sometimes referred to as "chain stretch." Wear may be minimized by proper lubrication and maintenance procedures.

Measure the chain and if elongation exceeds functional limits or is greater than 3% (.36″ in one foot), replace the entire chain. Chain that is run beyond the 3% elongation maximum recommendation will not engage the sprockets properly and may cause damage to the components or jump off the sprockets. Do not join a new section of chain to a worn section because it may not operate correctly. Check the side plates for wear due to misalignment or cracks from fatigue. Make sure all chain joints are free to articulate and not stiff or frozen. Carefully inspect the components for signs of corrosion.

Sprockets should be closely examined for wear. Loose or wobbly sprockets on the shaft can indicate a worn bore, key, or shaft, and this improper fit could result in catastrophic failure. Light interference fits are recommended for most normally loaded applications. Wear on one side of the sprocket plate usually indicates misalignment. Wear on the working faces of the sprocket teeth can indicate inadequate lubrication. A hooked or shark's-fin-shaped tooth sprocket should be replaced. Watch for scratches, galls, grooves, or visible changes in the tooth form. Replace all worn and broken parts. Running a new chain on old sprockets will result in early failure of the chain.

FIGURE 11–29
Sprocket alignment. Angularity can be checked by using two feeler bars placed on both sides of the sprocket against the shafts.
Courtesy of Diamond Chain.

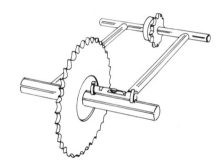

Chain Drive Alignment Drive alignment involves parallel shaft and axial sprocket alignment (see Figure 11–29). A misaligned drive results in uneven loading and will cause premature chain and sprocket wear. As part of the alignment process a check of bearings, shafting, and support structures is required. Excessive clearance in the bearings, bent shafts, and loose hold-down bolts—along with uneven foundations—will make alignment difficult. These issues should be examined and corrected if at all possible. Shafts should be parallel and level. Shaft parallel alignment should incorporate the use of a feeler bar, vernier calipers, and a machinist level. Apply the following formula to determine the allowable out-of-parallelism of shafts for most applications:

$$\text{Tolerance inches / foot 5 .0180 3} \frac{\text{Center distance (feet)}}{\text{Chain width (inches)}}$$

Sprocket axial alignment can be checked with a straightedge placed across the faces of the sprocket plates (see Figure 11–30). For long center distances, a taut cord or wire to extend beyond both sprockets can be employed. Use a piano wire (.029″ to .052″) stretched between the sprockets with a pull lead of approximately 20-lb. tension. Measure the gap between the reference edge and sprocket

FIGURE 11–30
Both offset and angularity can be checked with a straightedge.
Courtesy of Diamond Chain.

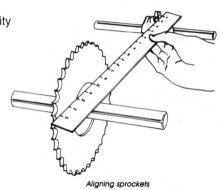

Aligning sprockets

face. Ideally, sprocket offset should be zero. Practically, apply the following specifications:

Chain Number	Offset-Inches
25 & 35	1/64
40, 50, 60	1/32
80	3/64
100	1/16
120 & 140	5/64
160	3/32

Mount and align the sprockets as close to the bearings as possible to lessen the overhung load. Allow for, but attempt to minimize, all float due to axial movement of the shaft. In some instances, consideration must be given to the difference between the starting and operating positions because of thermal growth of the shaft. Dial indicators can also be used to obtain an axial run-out reading on the sprockets. Exact alignment will extend the life of the drive system.

Calculating Chain Length Chain length can be calculated several ways. Counting pitches and using a tape measure together is advised. For simplicity, compute the chain length in terms of pitches, and then multiply the result by the chain pitch to obtain the length in inches. Most roller chain comes in 10′ sections. The following formula can be used to determine chain length in feet.

$$\text{Chain length in feet} = \frac{1.57(\text{PD1} + \text{PD2}) + 2\text{CD}}{12}$$

Where: PD1 = Pitch diameter small sprocket

PD2 = Pitch diameter large sprocket

CD = Center distance between shafts

Chain Removal and Installation Prior to actually removing or installing the chain, make sure all safety procedures are followed. Loosen the driver base to provide sufficient working slack. Make sure the chain is blocked and secure to prevent it from moving about during the process. Keep the chain straight when coupling or uncoupling. A link that is coupled while the chain is crooked may cause the chain to twist in operation.

To break the chain, remove any cotters or pins on the side plates where the break is to be made. Grind the heads off the pins so they are flush with the side plates. A roller chain pin extractor tool is recommended for ease in disassembly (see Figure 11–31). Drive out the pins with a drift or the proper removal tool. Care must be taken to remove the two pins alternately to avoid distortion of the roller link plates and adjacent links.

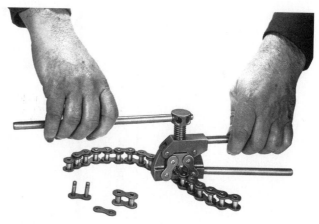

FIGURE 11–31
Chain pin extractor tool used to disassemble roller chain.
Courtesy of Dodge-Rockwell Automation.

FIGURE 11–32
Chain assembly tool used to bring the ends of the roller chain together to allow easy installation of a connecting link.
Courtesy Emerson Power Transmission.

To couple the chain, bring the free ends together on one sprocket while employing the teeth to hold the chain. When it is necessary to join the chain on long strands or between two sprockets, a chain-connecting tool may be used (see Figure 11–32). This tool has jaws that hook into each end of the chain and uses a screw assembly to draw the ends together. On heavy or conveyor chains, a come-along or block and tackle may be employed to draw the chain ends together. On engineering class or conveyor chains, the orientation of the barrels relative to the direction of travel and the primary purpose of the chain are important (see Figure 11–33).

Insert the pin or pin link to form an endless strand. Install the opposite cover plate. Side or cover plates are generally supplied in either slip fit or press fit. Press fit plates require some patience and tools to assemble. Never modify or drill out holes as this will compromise the integrity of the chain and could result in catastrophic failure.

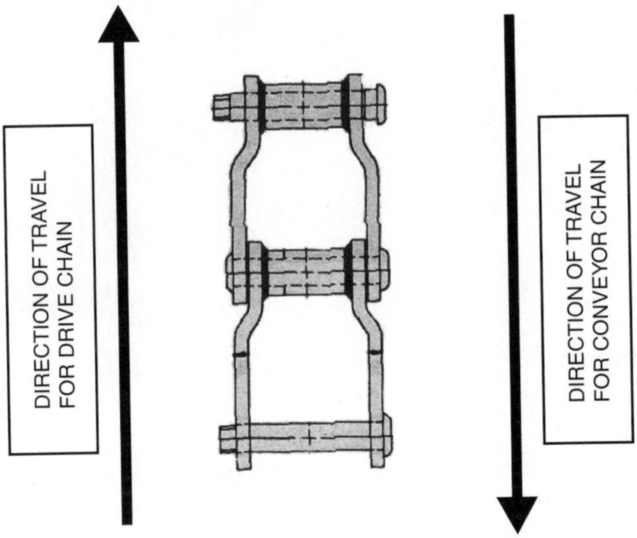

FIGURE 11–33
Chain installation direction
relative to narrow end.

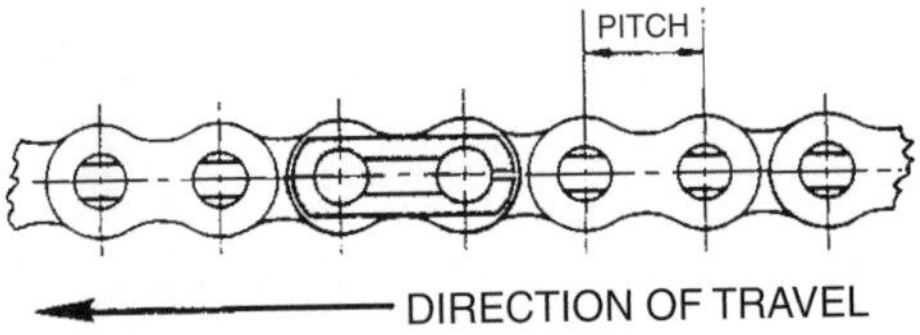

DIRECTION OF TRAVEL
FOR DRIVE CHAIN

DIRECTION OF TRAVEL
FOR CONVEYOR CHAIN

PITCH

DIRECTION OF TRAVEL

FIGURE 11–34
Proper spring clip assembly relative to direction of chain travel.
Courtesy of Diamond Chain.

After side plates have been installed, insert the cotters into pinholes or attach spring clips. Spring clips are normally supplied for chain pitches of 5/8″ and smaller. The clips are spring steel and fit into grooves in the link pins to retain the side plates. When installing clips by springing one leg of the clip over the pin, care should be taken not to overbend the leg. Spring clips should always be installed with the continuous or solid end pointing in the direction of travel (see Figure 11–34).

The cotter pins provided with cottered links are a special heat-treated design. After insertion in the pinhole, the prongs should not be spread to more than a 90° included angle (see Figure 11–35). The cotter pins should not be reused. Never

FIGURE 11–35
Cotter pin used on connecting links
and the proper amount of bend.
Courtesy of Diamond Chain.

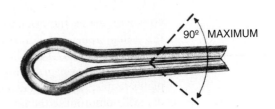

90° MAXIMUM

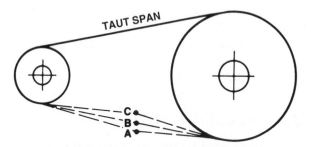

Recommended Possible Mid-Span Movement AC									
Drive Center-Line	Tangent Length Between Sprockets								
	5″	10″	15″	20″	30″	40″	60″	80″	100″
Horizontal to 45°	.25″	.5″	.75″	1″	1.5″	2″	3″	4″	5″
Vertical to 45°	.12	.25	.38	.5	.75	1	1.5	2	2.5

**AC = Total Possible Mid-Span Movement
Depth of Free Sag = .866 AB, approximately.**

FIGURE 11–36
Proper chain tension and midspan movement chart. Allowable midspan movement is
determined by the center distance (tangent length) and the mounting position of the drive
(vertical versus horizontal).

Courtesy of Diamond Chain.

substitute wire, bolts, rods, or anything else in place of the proper chain components. Always lubricate the connecting link upon assembly.

Chain Tension Proper tension for drive chain is extremely important. When chain
is too tight, the additional load results in excessive wear on the chain joints and
sprockets. Extreme tension also imposes additional loads on the bearings and shaft.
When chain is too slack, the system experiences vibration, noise, wear, and shock
loading. If the chain is excessively loose, it may jump off the sprockets.
    Check the chain tension so that the slack side span has 4 to 6% midspan move-
ment in horizontal drives and 2 to 3% in vertical drives. The following table in
Figure 11–36 lists the recommended midspan allowable movement.

CHAIN LUBRICATION

Suitable lubrication is one of the most crucial elements in chain operation. Roller
chain consists of a series of connected traveling metallic bearings that must be
lubricated to obtain the maximum service life. The adequacy of lubrication is one of
the prime influences on the life of the chain, providing the chain has been properly
selected, installed, aligned, tensioned, and maintained. There are a few exceptions
to lubrication, such as self-lubricating chains or those of a plastic type of material
with a low coefficient of friction.

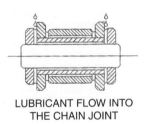

LUBRICANT FLOW INTO
THE CHAIN JOINT

FIGURE 11–37
Chain lubricant flow. (Space between mating parts has been exaggerated to show lube path.)
Courtesy of Diamond Chain.

Chain Lubricant Purpose

The purpose of lubrication is to reduce wear between moving parts. In addition, lubricant cushions impact loads, dissipates heat, retards corrosion, flushes away foreign materials, and lubricates between sprocket and chain contact surfaces. Most chains are made with minimum clearances between pins, rollers, and bushings to allow for a lube film. Proper lubrication penetrates these spaces and forms a separating wedge between the pins and bushings to minimize metal-to-metal contact (see Figure 11–37). The lubricant viscosity should be low enough to flow into the internal mating surfaces, yet have sufficient body to maintain the lubricating film under the bearing pressures. The lubricant must be free of contaminants. The highest viscosity clean oil that will flow in sufficient volume to cool and wash away contaminants will yield the best service life.

Chain Lubricant Types

The type of lubricant applied should be a good grade of clean petroleum oil with few additives, and free flowing at the prevailing temperatures. Certain additives will leave a varnish or gum deposit that prevents the oil from entering chain joints. Heavy oils and greases are generally too stiff to enter the chain joints and are not to be used except under unusual conditions. The proper lubricant viscosity for various operating temperatures is shown in Figure 11–38.

Ambient Temperature Degrees F	Recommended Lubricants				
	SUS Viscosity 100 F	SAE Engine Oil	SAE Gear Oil	ISO	AGMA
20-40	200-400	20	80W	46 or 68	1 or 2
40-100	400-650	30	85W	100	3
100-120	650-950	40	90	150	4
120-140	950-1450	50	90	220	5

FIGURE 11–38
Chain lubricant viscosity chart. Lubricant choice is based on operating temperature.
Courtesy of Diamond Chain.

Chain Lubricant Application

There are three types or methods of chain lube application (A, B and C). The method of application is determined by the power ratings tables provided by the chain manufacturer, the speed, and the conditions of the application. The lubricant should be applied to the upper edges of the link plates on the lower strand shortly before the chain engages a sprocket. As the chain travels around the sprocket, the lubricant is carried by centrifugal forces into the clearances between the pins and bushings. Spillage over the plates supplies lube to the interior and end surfaces of the rollers.

Manual and Drip Feed Lubrication (Type A) Oil is applied periodically with a brush, spout, or drip mechanism (see Figure 11–39). Volume and frequency should be sufficient to prevent discoloration in the chain joints. This method has the least degree of control. Delivery rates and quantities must be regulated to assure an adequate supply without flooding. It is recommended for slower speeds and lower horsepower drives. A general guideline is to lubricate at 8-hour intervals. Drip lubrication should be used in a clean environment.

Bath, Disc, or Slinger Lubrication (Type B) Bath lubrication requires that the lower strand of the chain run through a sump of oil in the drive housing or case (see Figure 11–40). The oil level should reach the pitch line of the chain at its lowest point while operating. Only a short span of chain should be submerged in the oil. With disc lubrication, the chain operates above the oil level. The disc or slinger picks up the oil from the sump and deposits it onto the chain, usually by means of a trough or plate. The diameter of the disc should produce rim speeds between 600 feet per minute minimum and 8000 feet per minute maximum. In the case of

FIGURE 11–39
Type A, drip lubrication method.
Courtesy of Diamond Chain.

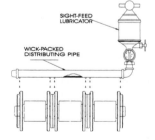

FIGURE 11–40
Type B, bath lubrication method.
Courtesy of Diamond Chain.

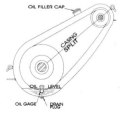

Figure 11–41
Type C, forced lubrication
method.
Courtesy of Diamond Chain.

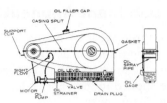

either a bath or disc method, the temperature of the lubricant and chain should not
exceed 180°.

Oil Stream Lubrication (Type C) This method of forced lubrication is required
for large horsepower, high-speed drives. A pump delivers oil under pressure to nozzles across the lower span of chain in a continuous stream just prior to the chain's
entry into one of the sprockets (see Figure 11–41). Nozzle orifices should be placed
to apply the oil across each strand of chain.

Troubleshooting Quick Guide

Problem	Probable Cause	Corrective Action
Excessive noise	Misalignment	Align to recommended tolerances for angularity & parallel/offset
	Inadequate lubrication	Check type of lubricant & delivery method is adequate for application conditions
	Loose support structure	Check mounting base and fasteners, tighten or support as needed
	Worn sprockets & chain	Check sprocket teeth & chain for excessive wear, replace if necessary
	Interference with associated components	Check to ensure there is adequate clearance between all supporting framework & the chain
	Improper tension	Check tension of system, adjust centers if needed
Excessive vibration	Misalignment	Align to recommended tolerances for angularity & parallel/offset
	Broken or missing chain parts	Inspect chain for missing rollers, replace or repair
	Bent shaft	Inspect both driver & driven shafts for run-out
	Failed bearings	Inspect supporting bearings for looseness
Wear on sprocket or chain sides	Misalignment	Align to recommended tolerances for angularity & parallel/offset

Troubleshooting Quick Guide *Continued*

Problem	Probable Cause	Corrective Action
Wear on sprocket tooth tips	Improper tension	Check tension of system, adjust centers if needed
	Chain elongation	Replace or shorten chain
	Soft teeth	Replace with hardened tooth sprocket
Broken sprocket teeth	Obstructions or material build-up	Check drive clearances & remove foreign material
	Excessive shock loads	Reduce loading, install clutching mechanism
Chain climbs sprocket	Improper chain fit	Check that chain & sprocket pitch match
	Chain elongated	Check if chain is worn or if it needs to be shortened
	Improper tension	Check tension of system, adjust centers if needed
Premature chain wear	Contamination	Clean & remove contaminants from drive and prevent foreign material intrusion
	Inadequate lubrication	Check type of lubricant & whether delivery method is adequate for application conditions
Broken chain	Drive overload or jam-up	Reduce loading, redesign drive
		Prevent foreign materials from interfering with the drive
		Install clutching mechanism
	Inadequate lubrication	Ensure proper lubrication
	Numerous link repairs	Replace chains that have numerous half-links & offset links installed

Questions

1. How do roller chain and sprockets function?
2. Name the parts of a roller chain section.
3. How is chain pitch measured?
4. What is the difference between standard roller chain and heavy series roller chain?
5. Name the types of links used to join the ends of chain.
6. What is "chordal action" in a roller chain drive?
7. What two methods are used to attach the sideplates of roller chain?
8. What has ANSI to do with drive chain?
9. What is the purpose of roller chain attachments?

10. Name four types of "engineering class" chain.

11. What is the proper method of roller chain assembly?

12. Name six safety concerns when working with drive chain.

13. What is a sprocket?

14. What are three methods for mounting a sprocket to a shaft?

15. Describe the difference between an A, B, and C type sprocket.

16. List a general part number for a sprocket with one hub projection, 21 teeth, used with #80 ANSI chain, and a bore of 1″.

17. How tight should chain be tensioned?

18. Is alignment important to chain, and why?

19. Describe three types of chain lubrication?

20. Why does chain require lubricant?

21. What is a common effective chain lubricant?

Student Exercise

Determine the roller chain drive ratio, driven speed, chain length, and the proper amount of midspan movement to tension the chain (see Figure 11–42).

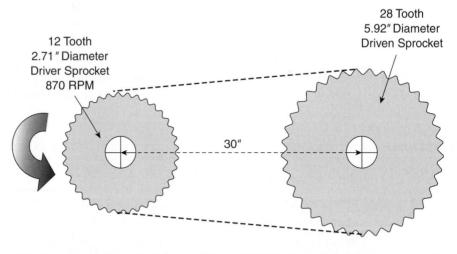

28 Tooth
5.92″ Diameter
Driven Sprocket

12 Tooth
2.71″ Diameter
Driver Sprocket
870 RPM

30″

Drive Ratio _____	Driven Speed _____	Chain Length _____
Midspan Movement for Proper Tension _____		

Figure 11–42

CHAPTER 12

Clutches, Brakes, and Backstops

This chapter covers some of the most common types of clutches, brakes, and backstop devices. The function of these devices is primarily to stop, slow, or prevent reversal of the load in a mechanical system. The various types of these mechanisms and how they are actuated will be discussed, as well as the basics of selection and maintenance.

Objectives

Upon completion of this chapter, the student will be able to:
✔ Understand and describe the basic types of industrial clutches and brakes.
✔ Understand the purpose of industrial clutches, brakes, and backstops.
✔ Define how clutches, brakes, and backstops function.
✔ Describe the considerations for selection of the mechanisms.
✔ Explain the methods used to actuate the various types of clutches and brakes.
✔ Detail the advantages and disadvantages of the various types.
✔ Comprehend and describe safety and maintenance concerns for clutches, brakes, and backstops.

INTRODUCTION

Clutches, brakes, and backstops are mechanical or electrical-mechanical devices that control the energy transfer and movement of rotating shafts on power transmission equipment. Typically clutches are used to engage and disengage a load to or from a prime mover without having to stop the prime mover. A major characteristic of a clutch is to connect two shafts rotating at different speeds and to engage them smoothly until the speeds are uniform. Brakes are designed to slow, stop, or hold a load. Backstops or any holdback device are intended to hold the load and prevent reverse rotation. These devices also act as a form of overload protection by limiting and controlling the torque and loads. Clutch and brake mechanisms can be purely mechanical in their function or they can incorporate the use of air, hydraulics, or electricity. Centrifugal, mechanical, electrical, pneumatic, or hydraulic forces can actuate them. In many modern units, the clutch and brake have been combined into a single apparatus.

In any mechanical system where control of the torque and rotation of the shafts are required, you will usually find some form of a clutch, brake, or holdback. Typically they are mounted on or near the prime mover or between the prime mover and the driven machine. Controlling torque, disconnecting loads, starting, stopping, and preventing load reversal are all functions of clutches and brakes.

Clutches and brakes are often used in tension-control systems where strip or web-fed products are being wound. Plastics, paper, metal strip, and wire are products that are fed or wound under a specific tension that must be maintained to prevent breakage or quality errors. Factors such as product elasticity and roll diameters need to be considered and compensated for during the operation. Control of the tension and operating speeds of the system is accomplished by using load cells or proximity sensors. Load cells sense the tension in a system and compare it to a set point or tension level. Feedback is provided to the controller and clutch/brake, automatically adjusting air pressure or electrically triggering the unit. Proximity sensors measure the diameter of the windup or unwind roll, providing feedback to a controller that modifies the speed to compensate for the changes. Speed and web tension control can be maintained by incorporating a clutch or brake into the system.

Clutches and brakes are integrated into the overall electrical and mechanical systems of most manufacturing processes. Electrical soft-starts, servo-drives, variable speed drives, and positioning systems—used in conjunction with other electrical and mechanical apparatuses—can be found on the low to mid range of the power and torque spectrum, acting as clutch/brake-like devices. Clutches and brakes—combined with drives, controllers, and sensors—provide accurate speed and torque control, and controlled starting and stopping.

CLUTCH/BRAKE SELECTION

To properly select a clutch or brake mechanism, many design, selection, and operating factors must be considered. Information regarding the application should be provided to the engineers to safely and correctly size and select the right unit for the job. The following information is typically required to select a clutch/brake unit.

- The type and nature of the application will determine the choice. What is the chief function of the unit: stopping, holding, braking, or slowing the speed?
- The speed of the rotating shafts. With mechanical applications, as speed increases torque decreases.
- Horsepower of the prime mover.
- Torque being transmitted through the system. All clutches, brakes, and holdbacks will have a torque capacity rating.
- The load size will affect the physical size of the unit, as well as the time required to accelerate and stop.
- Cycle rate. How frequently does the unit have to engage or disengage in a specified time period? Because these mechanisms often involve the use of

mating surfaces subjected to friction forces, heat is a by-product that must be considered.

- Environmental conditions. The operating environment will have an effect on the service life of the unit. Temperature, moisture, dirt, and exposure to chemicals all must be taken into account.

- The shaft size of the machine that it is being mounted on. This is often the shaft size of the prime mover and/or the driven equipment.

- Service factor. The class of service or the type of duty the unit will see should be considered in the selection procedure.

- Maintenance and service considerations are important. The accessibility, repair, and component part availability all vary from unit to unit and machine to machine.

CLUTCH, BRAKE, AND HOLDBACK TYPES

There are numerous variations and types of clutches, brakes, and holdbacks used on industrial machines. They are generally categorized by their operating principles. Listed below are some of the most common forms:

- Positive mechanical engagement clutches.
- Sliding or friction clutch/brakes.
- Electrical clutch/brakes.
- Pneumatic and hydraulic units.
- Torque limiting clutch/brake.
- Centrifugal fluid clutches.
- Single direction overrunning, indexing, and holdback.

Positive Mechanical Clutches

Positive mechanical clutches depend on interlocking mechanical parts to transfer power from one shaft to another. It is one of the earliest forms of a clutching mechanism used in industry. Jawed hubs, mounted on two separate shafts, rely on positive engagement to transfer power. The assembly is actuated or engaged by a connected lever arm (see Figure 12–1). One of the jaw hubs and shaft is movable; the other is stationary. There are square, tapered, and spiral shaped jaw types. Square jaws are engaged only when the unit is not rotating. Tapered designs will allow slow to moderate speed engagement. This type of mechanism is limited in its capacity to handle the speeds, torque, and power of most modern applications. By their design, shock loads that occur during engagement are directly transferred between shafts and can result in equipment breakage. It should be noted that purely mechanical clutch/brake devices increasingly are being supplemented or replaced with electrical, pneumatic, and hydraulic equipment that serves some of

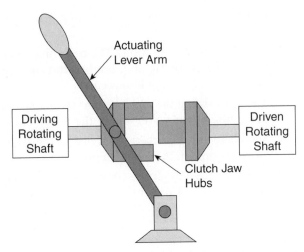

FIGURE 12–1
Side view of a square-jaw clutch actuated by a hand-operated lever.

the same functions. Some of the mechanical units are still in use but most are being replaced with better and more efficient electromagnetic or pneumatic designs.

Friction Clutches and Brakes

Friction clutches and brakes rely on friction forces to function. Electromagnetic clutch/brakes are a form of friction clutch/brakes. Typically there will be two closely spaced mating surfaces—one rotating and one stationary—that are part of the mechanism. Pressure or force is applied to one of the parts to engage or disengage the components. The pressure can be provided mechanically through a lever or from a source of air or fluids. At least one of these friction surfaces will be covered with or have attached plates of a friction material. This surface is known as the shoes or linings and will be made of a material that is resistant to heat and wear. The original material used in the construction of these shoes was made from asbestos. Asbestos is no longer an acceptable friction material because of health reasons and has been replaced with organic and metallic compounds.

The effectiveness of the unit is dependent on several factors. The coefficient of friction of the mating materials should be different—and preferably high—to offer more resistance to movement. The amount of surface area of the shoe or lining will increase its effectiveness proportionately and allow better heat dissipation. Finally, the amount of force that is applied through mechanical, electrical, pneumatic, or hydraulic means will cause more or less contact pressure. These devices that utilize external forces or pressures to assist are known as a power-assisted clutch/brakes.

There are two basic forms of friction clutch/brakes: radial and axial. The radial type uses contact pressure on the peripheral of a drum or rim. The most basic friction radial clutch or brake is a wagon wheel brake. It relies on the contact pressure between the wheel and the surface or face of the brake shoe. Another radial form is a band or strap brake. A band that wraps around the rotating wheel connected to a

FIGURE 12–2

(a) Caliper brake, and (b) drum brake.

Courtesy of Nexen Group Inc.

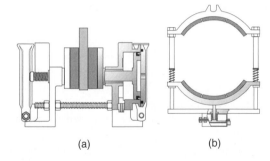

(a)   (b)

linkage and lever is used to apply pressure and stop rotation. Caliper brakes and drum brakes (see Figure 12–2) are two common forms of brakes. Both of these types generate significant amounts of heat and are subject to wear. Centrifugal clutches are radial types that rely on springs and centrifugal forces to engage and disengage the clutch mechanism. The axial type will use contact pressure that is applied perpendicular to the rotating shaft. Axial types generally provide a larger friction surface area than radial types, which translates to higher ratings. Disc brakes and clutches are examples of axial types.

Electrical Clutches and Brakes

Electrical clutch/brakes incorporate the use of electricity in the activation or control of the mechanism. Eddy current, magnetic particle, and electrically activated friction disc clutch/brakes are all mechanisms that use electricity to activate the unit or provide the holding power through magnetic fields.

One of the most common types of clutch/brakes used in industry is one that uses friction plates and is activated electrically (see Figure 12–3). Oftentimes it is mounted on the backside of an electrical motor and serves the dual purpose of a

FIGURE 12–3

Cutaway of an electrical shaft-mounted clutch/brake.

Courtesy of Boston Gear.

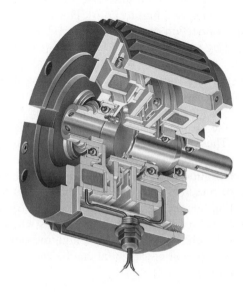

clutch and a brake. They are sometimes referred to as "shaft mounted clutch/brake modules." Primarily they are used as stopping devices. These units can operate on AC or DC. A friction disc electrically activated clutch/brake uses friction to slow and stop the load and an electromagnet to provide the force to apply pressure to the friction surfaces. One part of the unit is stationary and has friction plates attached to its face. Within this member is a coil buried in an insulating material. When voltage is applied to the coil, the pressure plate, which is connected to the other member of the unit, is drawn to the friction plate and the load is slowed or stopped.

Magnetic particle clutches have a rotor with blades operating within a hollow housing. The sealed housing has an electromagnetic coil contained within its perimeter. The housing is filled with steel powder. When the electromagnet is energized, the metal particles are drawn into the shape of the magnetic lines of flux causing a bond between the input rotor and output housing. The locked metal powder prevents the rotor from turning in the housing. Energizing and de-energizing the field controls clutching and braking of the system. High cycle rates and low generation of heat are two advantages of this type of unit.

Eddy current clutch/brakes rely on a noncontacting rotor made from a nonferrous material centered between magnetic discs in the housing. There is no direct mechanical connection between the input and output shafts. The rotor rotates due to the electrically induced magnetic field. Torque and slip can easily be controlled by adjusting the alignment of the magnetic polarity of the discs that are on either side of the rotor. They are primarily applied as an overload clutch and maintain constant speed between the input and output shaft.

Pneumatic and Hydraulic Friction Clutch/Brakes

Pneumatic controlled units use air pressure to actuate the friction discs of the clutch/brake (see Figure 12–4). Dry and clean plant air is delivered by tube to a rotary connection on the mechanism. The volume of air is controlled by a pressure-regulating valve and depends on the size of the unit, the frequency of activation, and the volume of the supply line. The air pressure expands a diaphragm or piston apparatus within the unit. The expanding diaphragm or piston apparatus applies pressure to the mating friction surfaces. The pressure plate engages the face of friction discs, which slow or prevent rotation. Springs are used to disengage the clutch when the air pressure is released. The air is then vented through a quick exhaust valve. These units may be shaft mounted or flange mounted and incorporate the use of a V-belt sheave or flexible shaft coupling.

Hydraulically controlled units actuate cylinders or pressurized bladders similarly to pneumatic types. Hydraulic pistons or cylinders can be used to move an actuating linkage or apply direct force to friction plate surfaces. Springs are often used to return or disengage the plates when the hydraulic pressure is released. Hydraulic units are used where high pressure is required for high-power applications. The rate of engagement can be smoothly and quickly controlled in hydraulic clutch/brakes.

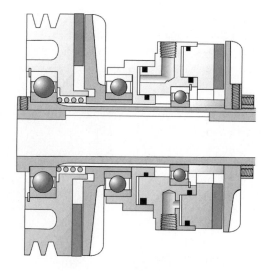

FIGURE 12–4
Pneumatic
shaft-mounted
clutch/brake
incorporating
a V-belt sheave.
Courtesy of Nexen
Group Incorporated.

FIGURE 12–5
Torque limiting clutch
device for use with
roller chain and
sprockets.
Courtesy of Rockwell
Automation.

Torque Limiting Clutches

Torque limiting clutches are used as overload protection devices in mechanical power transmission systems. One of the most common types uses a driving sprocket mounted on the clutch assembly (see Figure 12–5). The plate type sprocket is sandwiched between two friction plates. Pressure is applied to the friction discs and sprocket by a spring plate and adjusting nuts. The amount of torque allowed through the unit is determined by the spring plate pressure and by tightening or loosening the adjustable nuts. When a jam-up occurs in the system for any reason, the sprocket will slip between the friction discs. This type of clutch is designed primarily to prevent damage or breaks to other components within the system. It is not designed for repetitive starts and stops.

Fluid Clutch

Fluid clutches have four main purposes: soft start, delayed engagement, reduced load starting, and overload protection. They are sometimes referred to as "fluid torque converters." Wet fluid clutches have a sealed hollow housing filled with oil

FIGURE 12–6
Fluid clutch/coupling.
Courtesy of Rockwell Automation.

(see Figure 12–6). A rotating impeller with vanes transfers energy by centrifugal force to the viscous oil fluid. The torque is then transferred from the fluid to a runner with vanes that are attached to the output shaft. The amount of fluid contained within the housing can be varied to allow more or less slip in the mechanism. Heat is generated by continued slip, so fluid clutches are not recommended for high-cycle-rate applications. They are more suited for use as a soft start device on highly loaded apparatuses.

Another form of the fluid clutch is the dry type. Dry fluid clutches are similar to oil fluid clutches in their operation (see Figure 12–7). They are sometimes referred to as "dry fluid couplings." The housing of the unit is connected to the prime mover shaft, and a rotor with vanes is connected to the load. The housing has internal flutes and is filled partially with a charge of steel shot instead of oil fluid. When the motor

FIGURE 12–7
Dry fluid clutch/
coupling.
Courtesy of Rockwell
Automation.

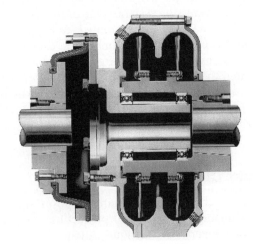

is started, centrifugal force throws the steel shot to the periphery of the housing. It then forms a solid connection between the rotor blades and housing flutes. Slippage occurs for a short period of time during start-up. The amount of steel shot determines the transmission of torque and the time required to accelerate the load. Excessive loads will cause continued slip, overheating of the unit, and deterioration of the shot. The shot charge must be examined periodically and replaced if needed.

Overrunning Clutches

An overrunning clutch is described by a variety of terms. Some of the common terms used to describe this family of mechanical clutches are: cam clutch, freewheeling clutch, backstop, sprag clutch, roller ramp clutch, one-way clutch, or single-direction clutch. With all of these devices, the primary function is to allow rotation in a single direction only to prevent reverse rotation. Overrunning clutches are made in different types of configurations. Pawl and ratchet, ramp, wrap spring, and sprag are the main types in use (see Figures 12–8 and 12–9).

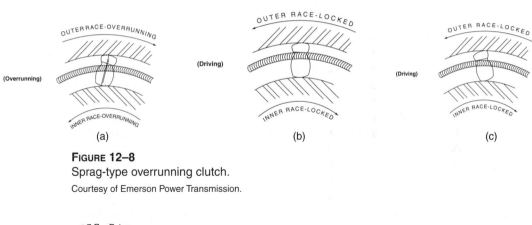

FIGURE 12–8
Sprag-type overrunning clutch.
Courtesy of Emerson Power Transmission.

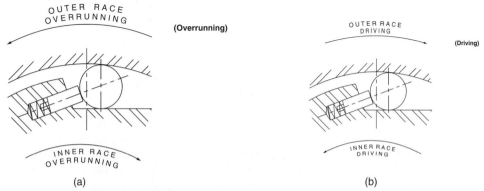

FIGURE 12–9
Roller-and-spring ramp-type overrunning clutch.
Courtesy of Emerson Power Transmission.

Pawl and Ratchet Clutch The pawl and ratchet is the most primitive form of a one-way clutch. It uses a hook-shaped pawl that is spring-loaded or operated by the force of gravity to slide over the teeth of a turning ratchet. The special shape of the ratchet teeth will allow the pawl to catch on the backside of the teeth if it attempts to reverse direction.

Ramp Clutch The ramp clutch uses spring-loaded rollers or balls that rotate on hardened steel ramps. The rolling elements are positioned on the inclined ramp that is part of the inner ring, which is mounted on a turning shaft. The outer ring has a smooth cylindrical surface and can be mounted into a machined housing. A spring-loaded plunger applies pressure to the rolling element to assure continuous contact between the rings and rollers. When either ring is rotated in a direction that forces them into the narrowing space between rings, engagement occurs. The rotational torque produces a wedging action to prevent reverse rotation.

Sprag or "Cam" Clutches Sprag or "cam" clutches incorporate specially shaped locking cams positioned between steel rings that allow only single direction rotation. The cams are uneven in height from one side to the other. They are evenly spaced between rings that have a hardened bearing surface, and held in place by a garter spring and retainer. The design of the cam is such that if it is pivoted in a particular direction, free rotation is permitted while it lightly slides over the race surfaces. If the direction of rotation changes for either ring, the cam is forced to lean toward its high side and wedges between the races, stopping rotation. Either race can be the driving or overrunning member, but typically the inner ring mounted on a shaft is rotating and the outer ring is locked in the housing. Lubrication is essential to lower the degree of friction and heat that is created by the sliding action. Typically this type is used inside a gearbox as a backstop device.

Wrap Spring Clutch The wrap spring clutch is a common light duty mechanism that is simple in its construction. It consists of three parts: input hub, output hub, and spring. The hubs, which have a slightly larger outside diameter than the spring's inside diameter, are inserted into the ends of the hollow spring. Based on the direction of the spring wrap, rotation is allowed in one direction as the spring slides over the hubs. When reverse torque is applied to either hub, the spring tightens on the hub surface and prevents reversed turning.

Overrunning Clutch Applications

Overrunning clutches are used in a variety of applications and on many different types of machines. There are three basic application groupings for overrunning clutches: overrunning, indexing, and backstopping (see Figure 12–10).

If two independent prime movers are connected to a single-driven machine and one of the prime movers is coupled with an overrunning clutch, this would be considered an overrunning application. Oftentimes a large rotary kiln or mills will have a primary and secondary motor connected to them. The primary driver is used during normal operation with the secondary drive connected through an overrunning

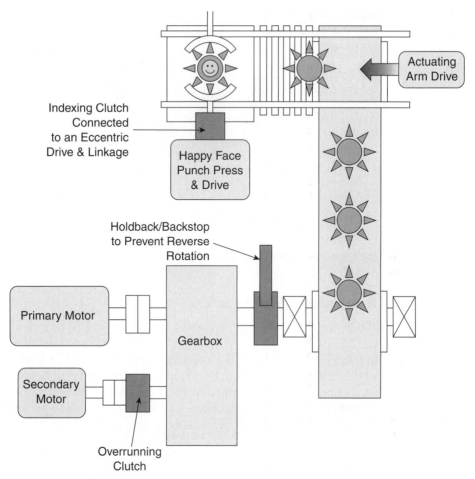

Indexing Clutch
Connected
to an Eccentric
Drive & Linkage

Happy Face
Punch Press
& Drive

Actuating
Arm Drive

Holdback/Backstop
to Prevent Reverse
Rotation

Primary Motor

Gearbox

Secondary
Motor

Overrunning
Clutch

FIGURE 12–10
Wigit conveyor/punch press system showing overrunning, holdback, and indexing clutch applications.

mechanism. The secondary, or "inching drive," is used to start or slowly rotate the equipment. The secondary driver will have a high ratio, slow operating speed. When the primary driver that operates at a higher-base speed is started, at some point it will overrun the other connected shaft. There are shaft-mounted overrunning couplings and foot-mounted units that are coupled between the machines.

On indexing applications, the clutch is used to hold or position an object so it can be altered in some way. Typically an eccentric drive and linkage of some sort will be involved to transform linear motion into rotary motion and vice versa. Oftentimes in the manufacturing of parts, where the process of punching, stamping, and pressing occurs, it is necessary to hold or position the component in place. An indexing clutch is used as part of the machine to assist in this purpose.

FIGURE 12–11
Backstop or holdback
device used on bulk
material conveyors
installed to prevent
reverse rotation of
the shaft.

Courtesy of Falk
Corporation.

If the clutch mechanism that allows only single-direction rotation is mounted securely to the rotating shaft or housing, and its purpose is to prevent reverse rotation, it is a holdback or backstop application.

Backstops/Holdbacks

Industrial backstops can be considered a category of their own. A backstop is a mechanical clutch device that allows rotation in a single direction only and is used as a safety mechanism. It is essentially an overrunning clutch with the sole function of preventing load reversal. Backstops for light to moderate loads can be mounted internally within the gearbox on the back end of the input shaft. A holdback is a backstop that is used on an incline conveyor application to prevent reversal and unloading of the belt when the system is stopped. Heavy-duty, belted bulk material handling conveyors will have a holdback mounted on the head pulley shaft to prevent the conveyors from unloading in the wrong direction if the motor should be stopped (see Figure 12–11). Heavy-duty holdback mechanisms will have a hollow shaft and a keyway for mounting purposes. A short piece of iron beam will be connected to the holdback that will engage the supporting conveyor framework if reverse torque is applied to the device.

Clutch, Brake, and Backstop Installation, Maintenance, and Troubleshooting

Clutches, brakes, and backstops have their own unique installation and maintenance concerns. It is recommended that the manufacturer of the particular unit be contacted to obtain the service and installation manual. These bulletins should be read and the recommended procedures followed. General guidelines applicable to most clutch, brake, and holdback devices are:

■ Follow all state, federal, and plant safety regulations. Lockout, tagout, and verify that the unit has been electrically and mechanically neutralized prior to any work conducted on or around the unit.

- Always install/replace guards to all units to protect operators from injury.
- Contact the manufacturer to obtain maintenance bulletins prior to working on the unit. Some manufacturers will not warranty or assume liability if the unit is opened, modified, or tampered with in any way, shape, or form.
- Heat is an important consideration when dealing with clutches and brakes. Always monitor and observe the operating temperatures of the device. High cycle rates might require additional cooling apparatuses.
- Follow all of the manufacturer installation and adjustment guidelines.
- Dirt, moisture, and oils can interfere with the proper operation of the mechanism. Clean all surfaces of the unit. Oil on the surfaces of friction plates will cause slippage.
- Excessive overhung loads can cause bearing and shaft problems. The distances between the clutch, brake, or backstop and the bearing should be kept minimal.
- Frequently inspect all components such as friction plates and discs for wear or damage.
- High torque loads can destroy the unit. Monitor and correctly calculate the amount of torque the unit is being exposed to. Proper initial selection is imperative.
- Shaft and key fits must be tight and correctly sized. Measure all shafts and bores.
- Internal bearings and bearing surfaces must be inspected and in good working order.
- Certain types of backstops and holdbacks require a lubricant level be maintained within the device. Check oil levels on a routine basis.
- Internal and external parts such as springs or linkages must be undamaged.
- Rotating unions on pneumatic and hydraulic units should be functional. Tubing must be kept short and flexible.
- All internal plates should be concentric and parallel.
- Use caution when releasing any spring tensions.
- Proper alignment is an important consideration. Vibration associated with misalignment will cause clutch/brake failure over time.

Questions

1. What is a clutch?
2. What is a brake?
3. What is a backstop?
4. Define four types of clutches and brakes.
5. What is the simplest and oldest type of clutch?
6. Name the two types of friction clutches.
7. Name two types of holdbacks.
8. What are the different means to activate or engage a clutch?
9. Name three types of electrical clutch/brakes.
10. What are the two basic types of fluid clutches?
11. List some of the important considerations for clutch/brake selection.
12. List six considerations for installation and maintenance of clutches and brakes.

Shaft Couplings and Alignment

Rotating shaft couplings are used to connect two different machines that have shafts in line with one another. They serve the purpose of forming a connection and transferring power and torque. Different types of flexible shaft couplings, why certain types are better suited for particular industrial applications, and how they are selected is explained. Installation and maintenance will be covered in detail, along with the importance of proper alignment and assembly of rotating shaft couplings.

Objectives

Upon completion of this chapter, the student will be able to:
✔ Explain the purpose and uses of flexible shaft couplings.
✔ Describe the differences between a metallic and nonmetallic shaft coupling.
✔ List the various types of flexible shaft couplings.
✔ Describe the advantages and disadvantages of various types of couplings.
✔ Have a basic understanding of flexible coupling selection procedures.
✔ Understand and list the various maintenance and installation procedures that pertain to flexible couplings.
✔ Describe reasons why precision alignment of couplings is important.
✔ Describe the various alignment procedures used with couplings.
✔ Demonstrate the ability to properly align a coupling without special tools.

INTRODUCTION

Shaft couplings have been in use in one form or another for thousands of years. The fathers of the modern flexible coupling are Jerome Cardan and Robert Hooke. These gentlemen invented and utilized an early form of a joint with yokes, cross, and bearings. In 1886, F. Roots thinned down a rigid sleeve which allowed it to flex; it became one of the first flexible couplings. It was not until the early-to mid-twentieth century that a wide variety and types of flexible couplings began to appear in the industrial market. The ever-increasing demand for a flexible connection between high-speed shafts that transmit large amounts of torque and power drove the development of the improved versions that are common today.

A shaft coupling is essentially a sleeve connecting two rotating shafts, transmitting the torque from driver to driven machines. It is a very efficient form of power transmission and capable of handling exceptional amounts of torque and horsepower at various speeds.

The three principle types of shaft couplings are rigid, flexible, and universal joints. The rigid shaft coupling allows no flexibility between shafts and its use is limited. The flexible shaft coupling allows some flexing of the connected shafts and comes in a wide variety of forms and types. Universal joints are used in connecting shafts that have extreme misalignment.

RIGID SHAFT COUPLINGS

Rigid shaft couplings are used when a fixed union is required to connect two rotating shafts. Rigid couplings transfer power directly from driver to driven shafts. The design does not compensate for any misalignment of the shafts. Any misalignment is transferred into bending moments and shearing loads on the shafts and bearings in the machinery. Nearly perfect shaft alignment and bearing support must be maintained to prevent failure of the connected components. Shock loads and vibrations are directly transferred due to its rigid design. Traditionally, overhead line shafts employ the use of these couplings to connect sections of long shaft lengths where the shaft handles the bending and inevitable misalignment. However, they should be used only in extremely slow moving, almost perfectly aligned applications. The advantages are that they require little maintenance and no lubrication. There are three common types of rigid couplings: sleeve, ribbed, and flanged.

Rigid Sleeve Couplings

A rigid sleeve coupling, as the name implies, is a steel sleeve machined from a single bar, with a hole drilled in it (see Figure 13–1). The unit usually has a straight bore of equal diameters on both sides. A continuous keyway, running the length of the inside diameter, and setscrews are commonly used to hold it to the shaft. It also

FIGURE 13–1
A one piece rigid sleeve coupling.
Courtesy of Lovejoy Power Transmission Products.

FIGURE 13–2
Ribbed rigid coupling split horizontally.

Courtesy of Rockwell Automation.

FIGURE 13–3
Flanged rigid coupling split vertically with tapered type bushings.

Courtesy of Rockwell Automation.

can be provided with a spline hole or tapered bore. It is generally used on slow-speed, lightly loaded applications. The rating of the coupling is the same as that of the shaft it is mounted on.

Ribbed Rigid Couplings

Ribbed couplings can also be referred to as rigid split-sleeve compression couplings (see Figure 13–2). The two horizontally split halves are bolted together to join shafts of the same diameter. A single continuous key runs through the bore to help transmit the power. They are available in shaft diameters up to 6″. High loads can be transmitted at low speeds under near-perfect alignment conditions.

Flanged Rigid Couplings

Flanged rigid coupling halves are split vertically and bolted together (see Figure 13–3). The same or different diameter shafts can be connected. They are available as bored and keyed or taper bushing versions. They are the most common type of rigid coupling and are sometimes combined with a flexible half on floating shaft arrangements. Installation is easy and maintenance minimal. They also require precise alignment to prevent shaft and bearing failures.

FLEXIBLE SHAFT COUPLINGS

Flexible shaft couplings are designed to flex under the various loads acting on them. The basic construction of most flexible couplings consists of shaft-mounted hubs (flanges) joined by a flexible element of some type and material or through the engagement of mating parts. Typically, one hub is mounted to the prime mover and the other on the driven machinery. Flexible couplings will accommodate varying degrees of shaft misalignment depending on their construction and element composition. Torsional and shock loads in the system can be absorbed and vibration dampened by flexible couplings. The two basic types are metallic and nonmetallic. Both versions have common and some unique features that act as advantages or disadvantages depending on the particulars of the application.

A form of flexible coupling is the spacer coupling. A spacer coupling has a spacer of different standardized lengths added into the unit to bridge the gap between shaft ends. This in effect makes the coupling have a greater length without increasing the outside diameter. The pump industry requires greater spacing between the pump and motor for servicing components. The "drop-out" spacer facilitates the removal of packing, bearings, and other serviceable components without disturbing the connected machinery bases.

The primary purpose of a flexible shaft coupling is to transmit torque/power between two rotating shafts and to compensate for minor amounts of misalignment while dampening shock and vibration between the connected mechanisms.

The basic purposes of a flexible shaft coupling are:

- Connect two rotating shafts from different pieces of equipment.
- Transmit power and torque.
- Compensate for misalignment.
- Provide torsional dampening.
- Act as a break point between machines.
- Dampen vibration.
- Allow for axial movement of the shafts.
- Minimize backlash.
- Provide electrical insulation.

When a means is required to connect the shafts of a driver (prime mover) and the driven equipment, a coupling is an efficient direct connection. Torque and power are transferred directly without additional force transformers such as V-belts and chain. Little energy is lost from the prime mover to the driven machine when it is direct-coupled.

Wherever two shafts are to be co-linear and connected, alignment becomes an issue. The equipment bases will inevitably move from heat, vibration, and errant forces that cause the centers of rotation to be out of line. Flexible couplings will compensate for minor amounts of this misalignment. Improper installation can also require a union that will accommodate this misalignment. It is imperative to understand that a particular type of coupling may continue to function under a misaligned condition, but the life of the connected equipment and the coupling will be shortened (see the alignment section of this chapter).

Shock and intermittent transient loads can damage connected equipment unless dampened. Sudden and unexpected load reversals need to be compensated by the coupling. If momentary peak overloads cause some part of the system to break, it is less expensive and an easier repair to replace the coupling, rather than shafts and bearings. Vibration can be detrimental to the drive components and must be limited through proper balance, installation, and coupling selection. Thrust loading due to axial movement of the shaft from thermal growth can be handled by the right coupling.

The construction, materials, and type of coupling will determine if the unit is torsionally stiff or soft. Torsional stiffness is a resistance to twisting action between the driving and driven halves of the coupling. A torsionally soft coupling has low resistance to twist or angular displacement about the axis of rotation. An antibacklash coupling used on precision positioning and indexing drives will be flexible to allow for shaft misalignment but have very little, if any, looseness between the mating parts. Backlash is the free movement of parts within the coupling, and is different from torsional softness.

Although most couplings are not designed to be used as an electrical insulator, a nonmetallic version will not allow current to travel between the two shafts. Couplings should never be counted on to provide electrical insulation.

Nonmetallic Flexible Couplings

Nonmetallic flexible shaft couplings, also called "elastomeric" couplings, incorporate a joining member that is not metal. The design of the coupling uses a flexible nonmetallic insert or element within the coupling. Power and torque are transmitted through this element that joins what is typically metallic hubs. The hubs are mounted to each connected shaft. The element or insert is made from a rubber, plastic, or synthetic compound that is resilient to shock, vibration, and overloads. Material composition of the element plays an important role in how well the coupling functions under various loads and operating environments.

Nonmetallic couplings can further be grouped into two types: compression and shear. In some cases both types of loads are acting on the unit at the same time.

Compression types are characterized by a design in which the driving and driven hubs rotate in the same plane and the element is unattached to the hub while being compressed under load. They are generally stiffer than shear types and can tolerate overloads better. A jaw-type coupling with a "spider" insert is an example of a compression coupling. A shear-type elastomeric coupling is characterized by a design in which all parts of the driving and driven hubs rotate in different planes. The element is usually attached to both hubs in some manner. This design generally handles more misalignment and is torsionally softer than a compression type. Examples include toothed sleeve and donut elastomeric couplings. Combination shear and compression loaded designs will use a radially removed, wrapped, flexible element with no metal-to-metal contact if the center element fails.

The advantages of elastomeric nonmetallic flexible shaft couplings are:

■ Torsionally soft. The torsional stiffness of a coupling is a mechanical property of the materials, modulus of elasticity, the shape of the coupling, and the twisting effort (torque) required to deflect a coupling in a circular direction. Coupling insert materials—like various forms of rubber—are torsionally soft compared to urethane. A balance between resilience and load-carrying capacity must be considered based on the nature of the application.

- No lubrication required. Because of a lack of metal-to-metal contacting parts under normal operation, a lubricating film is not required.

- Good vibration dampening and shock absorbing qualities. The softness of the material used in the flexible element cushions vibration resulting from misalignment, unbalance, or backlash. The characteristics of the elastomeric element allow it to absorb torsional and dynamic energy created from operating loads.

- Low bearing loads. The greater the torsional softness of the flexible element, the lower the reactionary loads produced in the connected machine bearings due to misalignment.

- Easy installation and low maintenance. A typical elastomeric coupling has fewer component parts than a metallic coupling and requires minimal installation effort.

- High misalignment capacities. The flexible nature of the insert allows high angular and parallel misalignment. This capacity is for the coupling and not the limits of the operating equipment. The greater the misalignment of the shafts, the lower the life of all components.

- Inexpensive. The initial purchase price of most elastomeric or nonmetallic couplings is low.

The limitations of the elastomeric nonmetallic-type couplings are:

- Environmentally sensitive. Chemicals in the operating environment can adversely react with the flexible element and cause it to deteriorate. Extreme temperatures such as those above 200°F can destroy the element unless it is of a special compound formulated to continuously operate in higher temperatures.

- Torsionally soft. Being too soft in an application that requires zero backlash or positive displacement can be a liability.

- Oversize requirements. Generally a larger outside diameter elastomeric coupling must be selected to accommodate the same torque and power requirements handled by a metallic type.

- Balance. They tend to be less accurate in both their axial and radial run-out tolerances, as well as difficult to perform balancing operations on.

- Less overload capacity than a comparably sized metallic coupling.

NONMETALLIC COUPLING TYPES

Jaw Couplings

Jaw types are one of the most common forms of nonmetallic couplings in use (see Figure 13–4). An asterisk-shaped insert referred to as a "spider" separates two hubs with jaws (stubby protrusions) that loosely interlock. The insert cushions the loads being transferred between the hubs and prevents them from having meshing contact. The hubs are usually straight bored to fit various shaft sizes. A standard

FIGURE 13–4
Standard flexible elastomeric jaw coupling consisting of two hubs and a center member known as a spider.

Courtesy of Lovejoy Power Transmission Products.

keyway with a setscrew helps to hold it in place. Installation is easy and the cost of purchase is low. Replacement of the insert can be done without moving any of the connected equipment when the setscrew is loosened and the hubs are slid back onto the shafts.

The inserts or spiders are available in NBR (Nitrile Butadiene Rubber) as standard material. The temperature range for standard NBR spiders is −40°F (−40°C) to 212°F (100°C). Other materials such as urethane, Hytrell®, bronze, and nylon are available. The application environment and loads will dictate the selection of the spider material best suited to attain maximum life. Manufacturers should be consulted for the particular features and benefits of the different materials. A wrap-around version of a flexible elastomeric insert is sometimes referred to as a "snap-wrap" (see Figure 13–5).

Alignment limits for angular are 1/2 to 1° and from 0.010″ to 0.015″ parallel/offset, depending on the materials used in its construction and the size of the unit.

A curved jaw version, which is widely used in Europe and Asia, is similar to the straight jaw type except the jaws are concave. A corresponding crowned insert must be used to properly fit in place between the hubs.

A relatively newer form of the jaw coupling is a shear type with the jaws rotating in separate planes (see Figure 13–6). There is no metal-to-metal contact if the center insert should fail. A wide wrap-around urethane insert separates the jaws and pushes the hubs apart. A ring encases the element that is made of a high-impact plastic or steel. They are an improved version of traditional jaw couplings.

FIGURE 13–5
Standard jaw coupling flexible elastomeric inserts also known as cushions.

Courtesy of Lovejoy Power Transmission Products.

FIGURE 13–6
Shear elastomeric
wrap-around element
jaw coupling. Notice
the element cover used
to keep it in place.
Courtesy of Falk Corporation.

Flexible Sleeve Couplings

Sleeve couplings are a shear-type coupling that is simple in design. It is composed of two hubs and a sleeve (see Figure 13–7). Power and torque are transmitted through the twisting of the elastomeric sleeve (element). One of the chief assets of this coupling is that significant degrees of misalignment can be tolerated. The center of the coupling is open, so close positioning of the shaft ends is possible. When standard rubber compound sleeves are used as a center member, it is considered a torsionally soft coupling. On the negative side, it does not respond well to torsional vibrations such as those created by reciprocating engines, pumps, and compressors. It also can fail prematurely if it is operated continuously at 25% of its rated torque capacity. This is due to the rubbing action between element and hub teeth created by light loads.

The hubs are available as bored and keyed, or tapered bushing mounting types. The hubs are also referred to as flanges. A spacer version, which extends the hub length, allows the coupling to be mounted on shafts that have a greater than normal distance between ends.

FIGURE 13–7
Flexible sleeve coupling.
Courtesy of T. B. Woods
Incorporated.

The sleeve is sometimes referred to as a donut. The sleeve is a hollow spool with teeth molded on the outside diameter of each end that mesh with teeth on the inside diameter of the hubs. Standard sleeve material is EPDM rubber, but can be obtained in Hytrell and neoprene. EPDM rubber has a temperature rating of $-30°F$ ($-34°C$) to $275°F$ ($135°C$). The torsional flexibility is $15°$ with an angular misalignment capability of $1°$ and a parallel limit from $0.010''$ to $0.062''$. Neoprene sleeves offer better resistance to oil and chemicals than EPDM rubber but are limited to a maximum temperature of $200°F$ ($93°C$). Hytrell is a polyester elastomer designed for high torque and excellent resistance to chemicals. The torque capacity of a Hytrell sleeve is four times that of EPDM or neoprene. It has limited torsional flexibility of $7°$ with an angular misalignment capability of $1/4°$ and a parallel limit from $0.010''$ to $0.035''$. These sleeves are available in three forms: one-piece solid, one-piece split, and two-piece split (see Figure 13–8).

Tire Couplings

Rubber tire couplings are named such because of the connecting elastomeric element resemblance to auto tires (see Figure 13–9). This coupling is often called a donut coupling. The coupling consists of two flanged hubs and the center element.

FIGURE 13–8
Two-piece split flexible
sleeve coupling.

Courtesy of T. B. Woods
Incorporated.

FIGURE 13–9
Standard tire-type coupling
using tapered type bushings
to mount to the shafts.

Courtesy of Rockwell Automation.

Like sleeve couplings, power and torque are transmitted through the twisting of the elastomeric tire. It was initially designed for applications with high peak transient torque and shock loads. A design variation of this form of coupling has an inverted tire shape.

The hubs are available as bored and keyed, or with tapered bushings. The tire is attached to the hubs by clamping plates and fasteners. Another version has the clamping plates molded into the urethane element.

The tire is made from natural rubber or neoprene and is constructed with layers or cords of a synthetic material. Because of centrifugal and axial forces acting on the tire as it revolves, speed capacities should be checked during selection. The compound of natural rubber is suited for operation in ambient temperatures from $-45°F$ to $180°F$ and neoprene handling of $-40°F$ to $210°F$.

Tire couplings generally will accommodate significant amounts of both angular misalignment (up to 4°) and parallel/offset (up to 1/8″).

Donut Couplings

The true "donut" flexible nonmetallic coupling has two shaft-mounted hubs joined to a donut-shaped elastomeric element by cap screws (see Figure 13–10). The element between the hubs transmits the torque and handles misalignment. Metal inserts are bonded into the element to provide a fastening point between the hubs and the center member. The donuts can be square, round, or have a variety of other shapes. They can be solid or "wrap-around" in form. Generally the angular and parallel/offset limits are high on donut-type couplings. These couplings can handle cyclic loads and dampen vibration that passes through the coupling. The major limitation is the need for large diameters to accommodate shaft sizes as compared to other couplings of similar torque capabilities. Material composition of the tire is typically rubber compounds and urethanes.

FIGURE 13–10
Donut flexible coupling.

Courtesy of Lovejoy Power
Transmission Products.

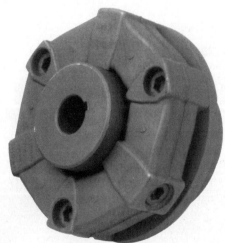

FIGURE 13–11
Pin and bushing
flexible coupling.
Courtesy of Emerson
Power Transmission.

Pin and Bushing Couplings

Pin and bushing couplings have rubber compound bushings mounted within a metal case (see Figure 13–11). These bushings are hollow to allow pins or fasteners to fit through and into the drilled holes in the hubs. The coupling has high capacities for misalignment. The bushings can be made from rubber compounds, urethane, or neoprene.

Pin and Disc Couplings

Pin- and disc-type elastomeric couplings have two hubs and a single solid flexible element (see Figure 13–12). The element is a hollow ring with numerous equally spaced holes in it through which the hub pins pass. The hub pins alternately penetrate the disc to assemble the coupling. The capacities for misalignment are similar to other elastomeric-type couplings. The ring can be made from rubber compounds or urethane.

Elastomeric Sleeve Gear Couplings

Elastomeric sleeve gear couplings have gear teeth on the outside diameter of the hubs and a continuous nylon material sleeve with corresponding internal teeth (see Figure 13–13). The hubs fit into a hollow sleeve made from a low coefficient of friction material like nylon compound. They are primarily used on light-duty industrial applications such as on the input shaft of a fractional horsepower motor-reducer.

FIGURE 13–12
Pin and disc coupling.
Courtesy of Rockwell Automation.

FIGURE 13–13
Elastomeric (polymer)
gear sleeve coupling with
a continuous sleeve joining
the hubs.

Courtesy of T. B. Woods
Incorporated.

METALLIC FLEXIBLE COUPLINGS

Metallic flexible shaft couplings are entirely made from metal alloys of various kinds. Flexibility is achieved through loose fitting parts sliding and rolling against one another. On grid and chain couplings, the joining member is metal. Gear coupling hubs mesh and flex within flanges that are bolted together or slide into a continuous metal sleeve. Lubrication is required in gear, grid, and chain couplings to prevent seizing and minimize wear. In some cases, the mechanical flexing occurs by using a set of metallic discs. Another version of a metallic flexible coupling is a one-piece heavy spring. Power and torque are transmitted through metal elements or from gear-tooth engagement on metallic couplings. The hubs are mounted to each connected shaft by straight bore or taper bushings. Metallic flexible shaft couplings generally have lower misalignment capabilities than elastomeric couplings and vibration dampening is minimal. Metallic couplings are found in applications that require large amounts of power and torque at reasonably high speeds.

The advantages of metallic-type couplings are:

- Torsionally stiff. Load transfer is efficient and not lost on the flexing of a soft element.

- High temperature capacity. Without an elastomeric element to deteriorate from high ambient temperatures, the coupling can run in extreme temperature.

- Good chemical resistance. Standard elastomeric coupling elements are susceptible to attack from oils and process chemicals, resulting in failure.

- High torque capacity. Most metallic couplings are considered "power dense." Power density is the ability to handle high loads in a small package.

- High-speed capacity. Certain types, such as the disc, can handle extremely high speeds especially if they are dynamically balanced. There are no elastomeric elements on metallic couplings to distort and change shape as the speed increases.

- Minimal backlash. Because of the nature of their design, some metallic couplings will not allow any backlash to occur.

- Accurate alignment. A metallic coupling has closely machined surfaces that are held to close tolerances. Run-out is minimal and readings off of these surfaces are more accurate.

- Low cost per unit of torque transmitted.

Limitations of metallic-type couplings are:

- Wear of components. The sliding and rolling action that occurs during operation will eventually lead to metal fatigue.

- Lubrication required. The mating metal parts require a lubricant film to prevent seizing and galling.

- Quantity of parts. A typical metallic coupling will have more component parts and fasteners than a nonmetallic coupling. Assembly is more involved.

- Alignment limitations. The alignment capabilities for metallic couplings are generally less in both angularity and parallel/offset than nonmetallic types.

- Vibration dampening. The ability to dampen vibration is limited because of the metal-to-metal contact within the coupling.

- Electrically conductive. Unless specially modified, they are positive conductors of electricity from shaft to shaft.

METALLIC COUPLING TYPES

Grid Couplings

Grid couplings have two flanged hubs with slots cut axially into the perimeter of the flanges, and a continuous S-shaped grid of steel wrapped around and into the slots of the hubs (see Figure 13–14). Torque and power are transmitted through the grid spring, which is free to rock, pivot, and float within the hub teeth. The grid absorbs impact loads by spreading them over an increment of time and area. The grid design coupling excels in applications where an all-metal type coupling is required and vibration-dampening ability is desired. The serpentine grid allows axial float and moderate amounts of angular and parallel/offset misalignment. The recommended angular installation maximum is $1/16°$. The parallel/offset installation maximum varies with the size of the coupling but averages from 0.006″ to 0.012″. Lubrication is required within the coupling due to the metal-to-metal contact.

FIGURE 13–14
Metallic grid coupling with
a cover.
Courtesy of Falk Corporation.

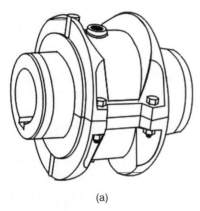

(a)

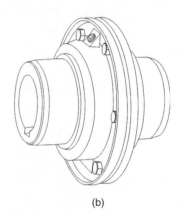

(b)

FIGURE 13–15
Metallic grid coupling covers: (a) split horizontally and (b) split vertically.
Courtesy of Falk Corporation.

The sealed coupling cover serves a dual purpose: It assists in retaining the lubricant and helps to hold the grid in place. There are two designs of covers: the horizontal split and the vertical split (see Figure 13–15). The horizontal cover design is made from an aluminum alloy and is ideal for low to moderate running speeds. It is the most common cover design currently in use because of its ease of installation. The vertical split steel cover is designed to operate at higher speeds than the horizontal type. Vertical covers must be installed prior to the hubs during assembly and cannot be removed without pulling the hubs from the shaft. Both cover designs require their own unique seals and gaskets to form a barrier to contaminants and retain the lubricant. Grade 5 or 8 cap screws hold the halves together. The covers are drilled and threaded to accept standard pressure lubrication fittings.

Standard hubs are straight bored and keyed with a setscrew over the keyway. Tapered bushing shaft hubs are available from some companies. Tapered hub bores with standardized AC and DC mill motor specifications can also be obtained.

Different versions of the grid-style coupling are available. The spacer grid coupling incorporates a spacer hub bolted to the shaft hub (see Figure 13–16). The shaft hub is bored and keyed for the driver or driven shaft. The grid wraps around

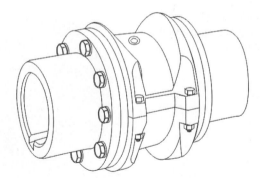

FIGURE 13–16
Metallic grid
spacer coupling.
Courtesy of Falk
Corporation.

the teeth of the spacer hub. Floating shaft types accommodate large spans between shaft ends (see Figure 13–16).

Gear Couplings

Gear couplings have mating parts with meshing gear teeth. Gear couplings are "power dense" and will handle very high amounts of torque and horsepower. The term power dense is used to describe power transmission components that have a high power capacity-to-size ratio. They are an excellent choice for heavy industrial applications where the transfer of significant amounts of power and torque take place.

Torque and power are transmitted from the driving hub's teeth to the sleeve teeth and onto the driven hub. Shock and vibration tend to be directly transferred through the unit because of the lack of a dampening element. Lubrication in gear couplings is mandatory to lessen the wear caused by the metal-to-metal contact among the teeth. A certain amount of backlash is designed into the gear teeth by making the gear teeth narrower than the gaps between the teeth. This backlash is in part what gives the coupling the ability to flex. The crowned shape of the hub teeth prevents binding and distributes the load over a broader contact area. The curve in the tooth form contributes to its misalignment capabilities.

The gear teeth provide for misalignment because they are loosely fitted together and crowned. The average static misalignment capability of gear couplings is 1° per mesh, but the recommended installation and operating misalignment is limited to 1/8° per mesh plane. Exact alignment will yield longer equipment life.

Gear couplings are available in essentially two styles: two-piece sleeve and one-piece sleeve. The two-piece style consists of two shaft hubs and two separate sleeves (see Figure 13–17). External gear teeth are cut on the circumference of the hub and mesh with internal teeth on the sleeve. Each hub mates with its own sleeve and together are bolted to the other coupling half by through holes in the sleeve (flanges). Two-piece sleeve couplings are available in exposed or shrouded bolt design. The standard configuration is the "full flex" or "double engagement" coupling with two hubs and two sleeves yielding two flex planes. This style of gear coupling is also available as a spacer coupling, consisting of two flexible hub and sleeve assemblies connected by a tubular spacer of varying lengths (see Figure 13–18).

FIGURE 13–17
Two-piece sleeve gear
coupling.
Courtesy of Falk Corporation.

FIGURE 13–18
Spacer gear coupling.
Courtesy of Falk Corporation.

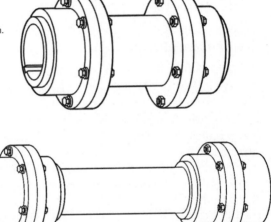

FIGURE 13–19
Floating shaft gear coupling.
Courtesy of Falk Corporation.

A "flex rigid" or "single engagement" configuration has a rigid flange bolted to a flexible gear coupling half. They are sometimes used in pairs mounted on opposite ends of a floating shaft arrangement (see Figure 13–19). Floating shaft configurations are used on applications like overhead cranes, and paper machine and steel mill roll drives.

The hub teeth sliding within the coupling sleeve teeth handle axial movement of shafts due to thrust loading or thermal growth. The distance between the ends of the driving and driven shaft can vary with the application and connected equipment. Most gear couplings have the advantage of being set up with a variable gap between hubs. The hub may be engaged in the flange in a "normal arrangement" or with one or both hubs "reversed."

The two halves of a flanged sleeve gear coupling are bolted together with grade 8 hardware specially made for that purpose. It is not recommended that standard

FIGURE 13–20
One-piece or continuous
sleeve gear coupling with
two hubs joined by one
sleeve.
Courtesy of Falk Corporation.

bolts be used in their place. The best coupling design should use the bolts to ensure clamping force to provide face friction, and not allow the bolts to transmit the load in shear.

The other style gear coupling consists of a continuous sleeve that holds both of the hubs (see Figure 13–20). A common cylindrical sleeve with internal gear teeth acts as the female part receiving two male hubs from each end. Because there are no fasteners on the continuous style, it is easier to install than a flanged type. The lack of a flange also makes it smaller in diameter than comparably rated flanged types. This single sleeve style is referred to as a "continuous sleeve" gear coupling. With both styles the gear coupling hubs are available with straight, tapered, or bushed bores, depending on the shaft and mounting requirements.

Certain brand gear couplings are interchangeable with other brands. Ratings of torque and power, along with critical dimensions, should be checked before substituting gear couplings. Flanged sleeve gear couplings will mate half-for-half with other gear couplings if they are made to standard AGMA dimensions and both are of the same shrouded or exposed design. Outside diameters, bolthole patterns, fastener size, and flange thickness dimensions should all be checked before interchanging and assembling different brand halves. Hubs of one brand will not fit into the flange of another brand, due to differences in the size, shape, and number of teeth.

Disc Couplings

Disc couplings use a pack of layered metal alloy discs as the flexible element; the pack is bolted between hubs. The disc coupling can consist of two flanged hubs separated by a single disc pack; or two hubs, one or more disc sets, and a center spacer (see Figure 13–21). The disc pack is connected to the hubs or spacer by alternate bolts. Layered discs are clamped together using disc attachment bolts, nuts, bushings, and washers. Tight clamping is necessary and critical to ensure that forces are transferred via friction from the hub flange to the disc pack and from the disc pack to the flange to prevent the bolts from being loaded in shear. The bushings and washers serve to equalize and distribute the load over a broader area, preventing concentration of those forces.

FIGURE 13–21
Disc pack coupling. This
disc coupling uses two
disc packs separated by
a center spacer.
Courtesy of Falk Corporation.

Torque is transmitted through pure tension in the flex elements. It has limited end float capacity and the limits are available from manufacturers. The disc coupling is backlash free and torsionally stiff. They are an ideal choice for precise synchronization and on positioning drives such as machine tools and printing applications. The speed ratings are very high compared to other comparably torque rated and sized couplings. They are machined to be very concentric and with close tolerances, resulting in excellent balance. Disc couplings are well suited for high-speed applications like turbines, compressors, generators, and test stands.

Disc couplings have very few environmental limitations. They are not limited to the operating temperature ranges of greases because lubrication is not required. Chemicals, dirt, and moisture will not be detrimental to the life of the flexible element the way it would be to elastomeric types.

Flexibility is achieved by the bending of individual discs in the pack over the crowned bushing and washer faces. A single disc pack will accommodate only angular misalignment and axial movement. A two-disc pack coupling has generous misalignment capacity for both angular and parallel misalignment, as well as axial displacement. Actual misalignment capacities vary depending on the manufacturer, size, and type.

Spring Couplings

Spring couplings use a tension-fitted spring attached to hubs or flanges on both ends (see Figure 13–22). Multilayered springs allow torque to be transmitted through the springs in either direction. Generally these are small couplings that handle limited amounts of torque, but speeds as high as 30,000 rpm with smaller units.

FIGURE 13–22
Spring coupling.
Courtesy of Lovejoy Power
Transmission Products.

FIGURE 13–23
Beam couplings.

Courtesy of Lovejoy Power
Transmission Products.

Beam Couplings

Cutting a helix in a hollow bar, which forms a curved beam spring (see Figure 13–23), makes a beam coupling. It is sometimes referred to as a "constant velocity" coupling. The coupling formed from a single bar of aluminum or stainless steel can be cut with various coil thickness and helix starts. The beam coupling has a moderate to high degree of torsional stiffness, which makes it appropriate for motion control applications. Drives that are frequently starting and stopping or reversing direction will benefit from the use of beam couplings. Hubs are bored and setscrewed or use a split clamping hub for shaft mounting. Beam couplings will operate with severe misalignment (3°, 4°, or 5°) with some variations functioning as U-joints. Parallel alignment limits range from 0.004″ to 0.035″ depending on the size and configuration. They can be run up to speeds of 10,000 rpm. Beam couplings will allow axial loading up to 0.010″ but care must be taken not to install the coupling compressed or expanded.

Chain Couplings

Chain couplings consist of two hardened tooth sprockets fastened together by a chain of some type (see Figure 13–24). The chain can be roller, silent, or a special plastic material. The most common type uses a special doublewide roller chain wrapped around the two hubs and joined together with a connecting link. Hubs are available bored to size or with tapered bushings. A cover is required to hold in the necessary lubrication. Chain couplings have limited misalignment capacity and low

FIGURE 13–24
Chain coupling shown
with cover.

Courtesy of Rockwell Automation.

shock and vibration dampening capability, as well as being torsionally stiff. They are still commonly used on older slow moving unsophisticated applications, though from a practical standpoint very few new applications specify chain couplings.

MISCELLANEOUS COUPLINGS

Fluid Couplings

Fluid couplings are mechanical two-piece apparatuses consisting of an impeller and a runner contained within a casing (see Figure 13–25). The casing is filled to a pre-determined amount of oil or special fluid. The impeller is connected to the driving shaft and the runner is connected to the driven shaft. There is no mechanical connection between the two.

Both the impeller and the runner have internal blades, or flat vanes. When the impeller is revolved, the oil is thrown to the periphery of the casing away from the coupling center. As the prime mover begins to accelerate, the impeller centrifugally pumps the fluid to the stationary runner. The fluid impinges on the blades of the runner, producing a torque proportional to the weight and rate of the fluid. A series of baffles and chambers is designed into the unit to control the fluid flow, producing a gradual increase in torque and allowing the prime mover to accelerate rapidly to its running speed. When the transmitted torque reaches the amount of resisting torque, the runner starts to rotate and accelerates the driven machine. The time required to reach full speed is dependent on the following: the inertia of the driven load, the resistive torque, and the amount of torque being transmitted by the coupling. Once the oil is propelled to the outside circumference of the casing and full running speed is achieved, there will be some slip. This slip is usually less than 4 percent. If the load torque should increase because of an overload, the amount of slip will increase, and cause the runner to drop in speed.

Slip allows fluid couplings to provide overload protection. It is, in effect, a "soft start" device and a "slip clutch." In the event of a prolonged overload situation, the continuous slip will create high oil temperatures. The core of a fusible oil plug in the casing will melt and release all of the oil. This will disconnect the power to the output shaft. A proximity cutout switch or thermal trip plug and limit switch are recommended. Various mechanical connections such as sheaves and couplings are

FIGURE 13–25
Fluid coupling.
Courtesy of Falk Corporation.

FIGURE 13–26
Dry fluid coupling.

Courtesy of Dodge-Rockwell
Automation.

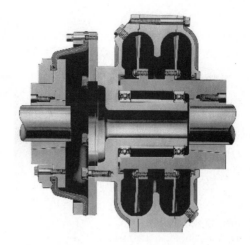

available to provide the most applicable mounting arrangement for the application. When properly applied and selected, a fluid coupling will provide reasonably smooth acceleration of high-inertia loads.

Dry Fluid Couplings

Dry fluid couplings are similar to oil fluid couplings in their operation (see Figure 13–26). They are sometimes referred to as dry fluid clutches. The housing of the unit is connected to the prime mover shaft and a rotor with vanes is connected to the load. The housing has internal flutes and is filled partially with a charge of steel shot instead of oil fluid. When the motor is started, centrifugal force throws the steel shot to the periphery of the housing. It then forms a solid connection between the rotor blades and housing flutes. Slippage occurs for a short period of time during start-up. The amount of steel shot determines the transmission of torque and the time required to accelerate the load. Excessive loads will cause continued slip and overheating of the unit and deterioration of the shot. The shot charge must be examined periodically and replaced if needed.

UNIVERSAL JOINTS

Universal joints were actually one of the first types of flexible couplings invented. The principle of the universal joint was conceived by Leonardo da Vinci and put into practice by Robert Hooke in the late 1600s. The primary purpose of a universal joint is to connect two shafts, transmit power/torque, and allow for extreme misalignment. The applications where universal joints are used are innumerable.

The basic design of a U-joint consists of two shaped blocks (hubs or forks) bored and keyed to accommodate mounting on the driver and driven shafts. A cross-shaped centerpiece of the cross-and-bearing or block-and-pin type joins these two hubs. This centerpiece, referred to as a yoke, joins the two hubs and allows for

(a)   (b)

FIGURE 13–27
(a) Single universal joint, and (b) double universal joint.
Courtesy of Lovejoy Power Transmission Products.

FIGURE 13–28
Universal joint installation
considerations.
Courtesy of Lovejoy Power
Transmission Products.

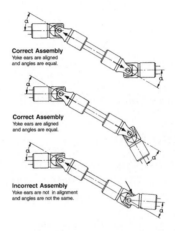

misalignment of the shafts. The yoke contains bearings or acts as the load-bearing
surface and requires lubrication to function properly. There are two basic types: the
single and the double universal joints (see Figure 13–27). A single type U-joint will
accommodate angular misalignment but not parallel/offset misalignment. A double
U-joint has a center section connected on each end with separate yokes. This allows
for accommodation of both angular and parallel/offset misalignment. The operating angle limitations are up to 35° on single joints and 70° on double joints. It is imperative when installing double universal joints that the yoke ears are aligned and
the angles equal (see Figure 13–28).

FLEXIBLE COUPLING SELECTION

Flexible coupling selection is primarily determined by the ratings of the coupling
along with the application/operating requirements and constraints. The published
horsepower and torque ratings of couplings are based on the endurance and yield
strength of their components. The application/operating requirements and constraints are: type of connected equipment, mounting dimensions, space limitations,

environment, installation, maintenance, load, torque, power, speed, alignment, cost, and numerous other factors. Proper selection should be based on all or some of these specifications and issues. The critical factors must be determined and prioritized to make a properly engineered selection.

Most flexible shaft couplings are multipurpose and will satisfactorily function at acceptable levels in a variety of locations. The perfect coupling for all environments and applications does not exist. Each type of coupling has attributes and performance features that will make it a better choice under certain operating circumstances. The critical requirements for a particular application must be met by the coupling to prevent failure. The coupling ultimately selected should be a practical compromise of primary and secondary attributes. Most manufacturers of flexible couplings can provide the design/selection engineer with the proper manual or software that eliminates any guesswork. The goal in the selection process is to choose the right size and type of coupling to provide a long, trouble-free life.

As a preliminary step to selection of a coupling, the following factors need to be evaluated and information gathered as part of the process:

- Power and torque. The amount of torque and horsepower to be transmitted by the coupling should be calculated and checked against the manufacturer's published ratings.
- Speed. The operating speed of the equipment's shafts should be checked against the maximum speed rating of the coupling.
- Prime mover. The type of prime mover and electrical or internal combustion engine will determine the amount of torsional pulses being generated.
- Installation constraints. The mounting dimensions of the driver and driven equipment, along with any space limitations, should be considered.
- Type of service. The type of service and prime mover must be known to select a "service factor."
- Alignment. The amount of misalignment that can be reasonably expected to occur during operation must be compared with the coupling capacity for angular and parallel misalignment.
- Shaft dimensions. The diameter, length, and shape of the shaft—plus keyway dimensions—should be measured and cross-checked against their capacities listed in the manufacturer's catalog. The method of mounting, straight bored or tapered bushing, should be decided on.
- Axial dimensions. The distance between the shaft ends of the connected machines must be measured and considered. The lateral float of the shafts created by thermal growth or mechanical loads must be compared to the capabilities of the coupling.
- Backlash. How much, if any, backlash is acceptable in the system.
- Vibration and shock. The levels of vibration in the system and peak shock loads transmitted should be known and compared with the dampening ability of various couplings.

- Installation and maintenance. The ease of installation is important if time is a constraint and numerous couplings need to be installed or replaced. The level of required maintenance, such as lubrication intervals, and parts replacement can be a determining factor.

- Operating environment. Take into account all environmental factors such as temperature, atmospheric conditions, and exposure to harsh chemicals, moisture, and abrasives. Compare that to the sensitivity of the coupling to operate under these adverse conditions.

- Torsional stiffness/softness. The degree of twisting action that will occur during starts/stops and operation must be compared to what is acceptable.

- Reactionary loads. Consideration must be given to the amount of misalignment in the parallel/offset plane that will be exerted on the shafts and transferred to the supporting bearings.

- Driven load characteristics. The type of machine the coupling is connected to will affect its performance.

- Accuracy and balance. Certain types of couplings are manufactured to closer machined tolerances, which is an asset in precision alignment. High operating speeds require a well-balanced design to limit vibration.

- Expense. The purchase price should be balanced against the overall installation and maintenance costs measured over the life of the coupling.

The method used to select the coupling once all of the operating and application considerations have been met, regardless of whether it is a new installation or replacement, can be done by one or more of the following procedures:

- Exact replacement. If the coupling did not prematurely fail and yielded a satisfactory life, an identical coupling should be selected and installed.

- Quick selection tables. Coupling manufacturers can provide service factor charts that are to be used in conjunction with ratings tables. These tables list the horsepower and torque ratings for the coupling operating at a variety of speeds.

- Formula selection. The transmitted horsepower, which has had a service factor applied to it, is multiplied by one hundred and divided by the revolutions per minute of the connected shafts. The calculated number is compared to the published ratings tables that are available from the coupling manufacturer. The ratings of the selected coupling must equal or exceed the requirements of the application.

$$\text{HP/100 RPM} = \frac{\text{Transmitted HP} \times 100 \times \text{Service Factor}}{\text{RPM}}$$

- Torque. Calculate the torque of the application and choose a coupling that has a torque rating that meets or exceeds the calculated torque.

COUPLING INSTALLATION AND MAINTENANCE

To attain maximum life from a flexible coupling, proper handling, installation, and maintenance of all parts is required. Most manufacturers can provide information on the installation and maintenance procedures relevant to their couplings. These bulletins should be available and read by maintenance mechanics involved in handling flexible couplings. Assembly, hub mounting, alignment and required follow-up maintenance, properly done, will ensure trouble-free operation. It is recommended that a regular inspection and maintenance schedule be established for the various direct-coupled drives in the plant.

Removal and Installation of Shaft Coupling Hubs

Flexible coupling hubs must be properly secured to the shaft to be effective. Shafts can be cylindrical, tapered, spline, and a variety of other shapes they must match. The connection between hub and shaft must be able to transmit torque and loads without coming loose, slipping, vibrating, or contributing to any unbalance in the system. The hub to shaft interface is extremely important.

There are three basic types of hub to shaft interface: clearance fit, interference fit, or taper bushing mount. Clearance fits, also known as loose or slip fits, have a bore size in the hub that is larger than the shaft diameter. Interference fits, also known as press, shrink, or tight fit, have a slightly smaller bore diameter than the shaft it is to be mounted on. Taper bushings rely on the wedging and compression forces of the bushing to shaft fit (see V-belt chapter). The majority of coupling hubs have clearance fits on smaller shaft sizes and interference fits on larger shaft sizes. The application requirements and the preferences of the design engineer dictate the fit or interface type.

Prior to mounting or dismounting coupling hubs to the shaft, the following practical steps are required to ensure safety and effectiveness:

- Lockout, tag-out, and verify that the machine is electrically and mechanically neutralized. Follow all plant safety procedures.
- Inspect and remove the coupling guard. All exposed rotating couplings must be shielded with OSHA-approved guards. The guard must be reinstalled prior to start-up.
- Clean all parts and the work area to avoid contamination and eliminate hazards.
- Accurately measure the hub bore and shaft diameters to ensure proper fits.
- Check the fit of the key in both keyways. The key should fit snugly against the sides of the keyway.
- Inspect for, and remove any nicks, burrs, or rises, on the shaft, keyways, and hub that will interfere with the mounting.
- Gather all of the required coupling parts and tools needed for complete assembly.

Clearance fit hub installation and removal is a relatively simple process. The hub is bored to a diameter greater than the shaft per standard specifications (see

Shafting chapter). The looseness is equal to approximately 0.0005″ per inch of shaft diameter. Overboring the hub can result in excessive looseness, vibration, and fatigue at the corner of the keyway or setscrew hole. A clearance fit hub uses a key and a setscrew tightened over the keyway to hold the hub in place.

Interference fit hubs require heat or force to mount and remove the hub from the shaft. They also require a hub bore diameter that is smaller than the shaft diameter, and tightness equal to approximately 0.0005″ per inch of shaft diameter. The correct specifications can be determined by viewing shaft fit tables in the chapter on shafting.

To remove the hub from a shaft, apply even heat from a rosebud torch to approximately 300°F to 500°F while applying pressure. Heat sensitive crayons can be used to indicate the hub temperature. Avoid placing the flame directly on the teeth or on any one spot for an extended period of time. Take care that the heat does not burn out any equipment seals. A mechanical jaw puller should be attached and used to apply removal pressure (see Figure 13–29). Never strike the hub with a hammer. A hydraulic press may be used by properly blocking the hub and pressing on the shaft to push it out. Care must be taken to not pull metal that would gall the shaft during removal.

Hub installation is best done with heat; however, a cold press may be done with a hydraulic press if the shaft diameter is small and the amount of interference is light. A hydraulic press may also be used in combination with heat to assist in the mounting. Heat sources may be ovens, oil baths, or torches. The hub should have heat applied evenly to raise it to a temperature of 275°F to 300°F. The temperature differential for shrink-fit installations can be calculated at 160°F per 0.001″ of shrink per inch of shaft diameter, plus 50°F to 75°F. When using an oven, allow at least one hour for each inch of wall thickness.

If a rosebud-tipped torch is used to heat the hub bore to expand it for installation, direct the flame into the bore (see Figure 13–30). Avoid overheating a specific spot—to ensure that the hub expands evenly—by keeping the torch in motion. Monitor the temperature by marking several locations on the hub (see Figure 13–31) with heat sensitive crayons, also referred to a "temp-sticks." The painted marks from the crayon will melt when the temperature reaches the rating of the stick. Care must be taken not to overheat the hub because the "temp-stick" indicates that the temperature is *at least* that which is indicated on the crayon. Mount the hub on the shaft quickly to avoid heat loss.

FIGURE 13–29
Coupling hub removal
with a jaw-type puller.
Courtesy of Falk.

FIGURE 13–30
Heating a coupling hub for installation using a rosebud torch.
Courtesy of Falk Corporation.

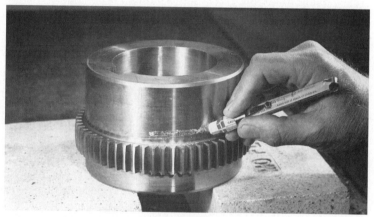

FIGURE 13–31
Coupling hub marked with a temperature crayon.
Courtesy of Falk Corporation.

If seals and covers are required, mount them prior to the hubs to make sure they do not come into contact with the hot hubs. Install the key in the shaft prior to hub installation. Position the hub keyslot to line up with the shaft keyslot. Slide the hub onto the shaft until the face is even with the shaft end. If it is necessary to overhang the hub, the shaft engagement must equal or exceed the shaft diameter. Light taps from a soft hammer may be required to drive the hub onto the shaft. Avoid excessive pounding that may damage the hub, shaft, and connected equipment bearings.

The additional component parts of the flexible coupling may be installed after the hubs are mounted. Center members must be installed without being "wound-up" or loaded in any way. Never forcibly position or attach the flexible element by any means other than called for in the coupling specifications. (No duct tape or spot welding is allowed.) Gear sleeves must slide freely over the mating teeth on the hub. All fasteners must be replaced with the correct type and tightened to the specified torque.

A thorough inspection and replacement of all worn parts are required to bring the coupling back to a "like-new" condition. Worn and damaged couplings will fail and potentially cause additional damage to connected equipment.

METALLIC COUPLING LUBRICATION

Grid, gear, and chain metallic flexible couplings all require lubrication to function without premature failure. The correct type and adequate amounts are essential for satisfactory operation. Lubricate couplings upon installation and at least once every 6 months. Time and centrifugal forces work to separate the oil from the thickener causing degradation of the lubricant. The only effective way to relubricate a coupling where the grease has separated is to disassemble the coupling and thoroughly clean the components. After cleaning, the coupling must be repacked with lubricant.

When installing a new or reinstalling a rebuilt metallic coupling, handpack all surfaces to ensure penetration of lubricant to all metal-to-metal contact points. After assembling the cover or sleeves, attach a low-pressure grease gun to the fittings and pump grease into the unit. Remove the drain plug and continue to pump until fresh lubricant is seen exiting the plug hole (see Figure 13–32). Couplings that operate in adverse environments, high heat, moisture, or if subjected to shock loads may require more frequent lubrication. Extreme pressure (EP) greases are often recommended. For normal service and speeds use NLGI #1 grease or NLGI #2.

Significant advances have been made in developing special greases with the right characteristics required for coupling applications. These "long term greases" are designed to function for extended periods of time in couplings without losing their lubricating properties.

FIGURE 13–32
Lubricating a metallic flexible shaft coupling through the cover with a grease gun.
Courtesy of Falk Corporation.

SHAFT COUPLING BALANCE/UNBALANCE

The topic of balance is important to all rotating equipment, especially when it comes to flexible shaft couplings. Balancing rotating equipment is a complex field best left to the trained experts who use equipment designed for that purpose. The rotating object (rotor) is usually a sum of various shapes and materials with a variety of different forces acting on it that can make determination of the level of unbalance difficult. But practical knowledge of the topic is useful in assembly and maintenance of rotating machinery.

Unbalance, along with misalignment, is one of the most significant sources of vibration in plant machinery. An unbalanced machine or component causes vibration and stress on all connected equipment within the system. Because a flexible coupling is connecting two machines usually operating at a relatively high speed, any unbalance in the component will adversely affect both machines. If the application is critical to production—or vibrations can cause errors in the manufacturing process—a balanced coupling is required. Balanced rotating machinery increases bearing life, reduces power consumption, helps yield precisely manufactured products, and can prevent catastrophic failures.

Causes of Unbalance

To understand coupling balance, an understanding of unbalance is necessary. Machines have a variety of rotating parts within them, such as impellers, couplings, and rotors. Every rotating component has a center of gravity and some degree of unbalance. The center of gravity is that point in a body around which its mass or weight is evenly distributed (balanced) and through which the force of gravity acts. Ideally the center point would lie on the axis of the shaft, which is the geometric center of the shaft. If the center of gravity of a rotating body (coupling) does not coincide with the axis of rotation of the shaft that it is mounted on, the part is in a state of unbalance.

The unbalance creates centrifugal forces at the rotating speed of the equipment that results in vibration at that frequency. Isolating and measuring the frequency can be useful in investigating coupling balance. The higher the speed, the greater the centrifugal force. Centrifugal force is an apparent force tending to pull an object outward as it rotates around a center. Centrifugal force increases proportionally to the square of the speed increase: For example, if the speed doubles, the centrifugal force quadruples. Unbalance in the coupling will cause increased vibrations in the system as speed goes up. The faster the operating revolutions, the more concerned one needs to be about the issue of balance.

The potential for unbalance results from displacement of mass with respect to the center of rotation. The displacement of mass can occur by nonconcentric shapes (runout), nonuniform density of the material (voids), nonsymmetric geometry (keyways), or by machining tolerance stack-up. Another cause for unbalance in elastomeric couplings is that the element changes shape under speed and load. The higher the concentricity of the object, the closer the coupling will be to a balanced condition. Cylindrical surfaces that are concentric with the center of the coupling,

---

Recommended Run-Out Guidelines

Machine Speed (RPM)	Maximum allowance Total Indicated Run-out (T.I.R)
0–1800	4 mils (0.004″)
1800–3600	2 mils (0.002″)
3600 and up	less than 2 mils

FIGURE 13–33
Run-out guideline.
Courtesy of Northern Michigan University.

along with faces that are perpendicular to the shaft axis, will contribute to good balance. The run-out tolerances applied at the factory will determine what amount of nonmounted, inherent unbalance is in the coupling. From a practical standpoint, the more round and true—or less radial and axial—run-out a coupling has, the more balanced it will be. If the material or mass of the coupling is shifted away from the geometric center of the coupling due to density differences or voids, the center of gravity is not its geometric center. When the center of gravity of the coupling is not the same as the center of rotation, unbalance exists. Bores that are concentric and in line are also necessary to provide a balanced coupling. Loose fits between the hub bore and shaft may cause a mass shift. The ultimate balance of the coupling cannot be determined until it is mounted on its operating shaft and subjected to load. Figure 13–33 is a general run-out guideline.

Quality couplings are designed and produced to be within ISO, DIN, and AGMA balance specifications, and these organizations provide the guidelines for balancing. For most applications and speeds below 1800 rpm, the standard coupling supplied by a manufacturer is sufficiently in balance by design and manufacture. If additional dynamic balancing is required for high-speed or critical applications, the coupling must be specially ordered. A balance machine is sometimes used to fine-tune a coupling to further reduce the unbalance. This is done by adding or removing mass equal and opposite to the unbalance.

Balance/Unbalance Types

Static, dynamic, single plane, and two plane are all terms used to describe types of balance. Static balance consists of placing the part with its shaft on knife-edges so that the heavy side rotates to the bottom. By trial and error, weight is added or removed to the point where the piece no longer rotates to the "low side." This crude method seeks to compensate for unbalance as a single point in a single plane of rotation. Static balance should not be confused with single-plane balance. Dynamic balancing involves spinning the coupling at its operating speed on a turntable, or between bearing supports on a special balancing machine. Instrumentation is used to measure the amount of unbalance and determine corrective actions.

A better way to think of balance is in terms of single-plane and two-plane balance. Single-plane balance is measured dynamically in a single plane of rotation. Single-plane balancing is best represented by thinking of a rotating disk with a hole located off the geometric center, which would cause it to be out of balance because the hole would shift the center of gravity off the center of rotation.

Two-plane balancing is where balance corrections are made at two locations or planes on the coupling axis. The length of the part is the primary factor as to whether single-plane or double-plane balancing should be performed. Generally the longer a component is to its diameter, the greater the need for two-plane balancing. Some engineers recommend single-plane balancing for parts with lengths less than 1-1/2 times the diameter. It is best to check for unbalance in two planes when the rotor length is 5″ or more. Most modern balancing equipment is capable of separating the effects of unbalance in one plane from that of another and achieving a truly dynamically balanced coupling in multiple planes.

Flexible Coupling Practical Field Balancing

The methods and practices used in the assembly of the rotating coupling in the field, as well as the condition of the components, will have an effect on the amount of dynamic unbalance to the unit during operation. Most maintenance mechanics do not have access to, nor the training and time needed to precisely dynamically balance a coupling in the field. But there are practical inspection, correction, and assembly steps that can be implemented to yield a reasonably well-balanced coupling. Usually none of these seemingly insignificant issues will be the difference between a very rough running machine and a very smooth running one, but the combined effect can be significant. The following issues can be addressed and procedures implemented for a practical approach.

- Check the radial run-out of both the driving and driven shafts and coupling hubs. Run-out problems usually fall into one of three categories: off-center hub bore, bent shaft, or hub skew bored (see Figure 13–34a). When checking the run-out, a magnetic base dial indicator must be mounted stationary and the shaft/hub rotated 360° (see Figure 13–34b). A strap wrench rather than a pipe wrench may be used to rotate the equipment to prevent the damage of metal surfaces. The total indicator run-out must fall within the coupling alignment specifications. As a general guideline the following specifications may be used: 0–1800 rpm = 0.004″ maximum, 1800–3600 rpm = 0.002″ maximum, 3600 and higher = 0.001″.

- Factory balanced couplings are marked with match marks and must be assembled with the mating match marks aligned. Component parts of assembly balanced couplings must not be replaced without rebalancing the component assembly. If parts are substituted, the degree of balance will be compromised.

- The fasteners, their mating holes, and the assembly process have an affect on the overall balance of the coupling. All fasteners and holes/threads should be inspected for damage and roundness. If a screw or screw hole is deformed or

FIGURE 13–34a
Off-center, bent shaft,
and skewed bore
coupling hubs.

Courtesy of Northern
Michigan University.

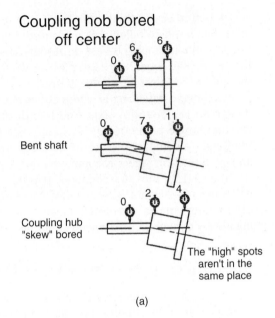

Coupling hob bored
off center

Bent shaft

Coupling hub
"skew" bored

The "high" spots
aren't in the
same place

(a)

Checking shaft and/or coupling hub "runout"

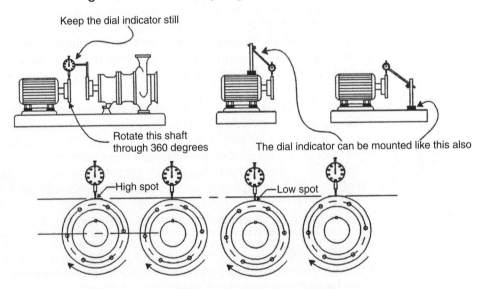

Keep the dial indicator still

Rotate this shaft
through 360 degrees

The dial indicator can be mounted like this also

High spot

Low spot

"Runout" problems usually fall into one of these categories

(b)

FIGURE 13–34b
Run-out check of coupling hub.

Courtesy of Northern Michigan University.

damaged, it must be replaced or corrected. Bolts should be tight-fitting to prevent them from moving about during operation. The coupling bolts, washers, and nuts should all be of uniform size and shape. Proper torque tightening values should be used with a cross-tightening sequence pattern.

- The keyway and the key used to assist in fastening the rotor (hub) to the shaft contributes to the total assembly's dynamic balance. The differences in mass of both the driving and driven keys—based on length, width, and height—must be compensated for. There are two options available to the maintenance technician. The first is to measure the length, height, and width of the keyway within the coupling hub, and the straight portion of the shaft keyway out of the hub. Cut and grind a step key that will fill the void in the hub and the straight portion of keyway outside of the hub. The second option is to measure the entire length of the straight portion of the key along the shaft. Measure the keyway length within the hub. Add the two lengths together and divide by two. This calculated number is the length of the key to be used. This method is best applied when both the driving and driven shafts have the same key shape design. In some cases the differing shapes and sizes of connected shaft keyways can make the process of balancing—from the keying perspective—more complicated. All keys must be sized to properly fit in the keyway. The keys of the two shafts should be positioned 180° apart.

- Hub to shaft fits should preferably be line-to-line or interference fit. This will prevent any wobbling of the hub on the shaft while rotating. If a clearance fit hub is used, make sure it is not overbored beyond the recommended tolerances. The bore of the hub must be centered and not out-of-round or skewed. Setscrews used to hold the unit in place must be tightened carefully; overtightening may cause distortion of the hub shape.

- All foreign material should be removed from any rotating object. Dirt and compacted materials filling any voids will cause unbalance and vibration.

FLEXIBLE COUPLING ALIGNMENT

Flexible coupling alignment is actually rotating shaft alignment. Perfect alignment is a condition in which the centerlines of two rotating shafts from closely coupled machines are co-linear. Shaft misalignment is the deviation of the shaft centerline position from a co-linear axis. Every shaft has a rotating center, which is sometimes referred to as the "shaft axis." The actual alignment of the shaft axis is almost always done within a specified tolerance. A tolerance is a permitted variation. Misalignment occurs when the shafts are not aligned within that specific tolerance. The tolerance limits are dictated by the operating parameters of the equipment and the flexible coupling joining the shafts. It is important to note that the operating position of the shafts and equipment will not necessarily be the same as the installation position. Various factors like thrust loads, temperature, piping strains, and overhung loads will affect the relative position of both shafts when running. Compensation for these movements during the set-up process needs to be taken into account when aligning the shafts.

Alignment is a process in which multiple factors need to be considered and addressed. The preliminary steps and the procedures taken by the technician will be determining factors in the operating alignment condition. The alignment process must be viewed from a three-dimensional perspective. The shafts must be co-linear in the parallel/offset vertical and horizontal planes. They must also have no angularity in the horizontal and vertical planes. The distance between the shafts, known as the gap, must be set to a specified amount. The reality of multiple circumstances and variables almost always causes a misalignment.

The goal of the maintenance technician is to align the shafts/couplings to the best possible condition within the economics and constraints of the machine. The coupling should not be used as the basis for permissible misalignment because it will tolerate many times more than that of the equipment. The system alignment must be based on the minimum requirements of the driving and driven equipment. The connecting coupling will allow varying amounts of misalignment depending on the type, but ultimately the system life will be extended if the rotating centers of the equipment are aligned.

Misalignment Types

There are three basic forms of misalignment: parallel (offset), angular, and a combination of both (see Figure 13–35). Parallel or offset misalignment is a condition where the two shafts are parallel but not in the same axis. The parallel misalignment may be in the vertical or horizontal plane. Angular misalignment is a condition where one shaft is at an angle to the other. Angular misalignment can exist in the vertical and horizontal plane. Shaft misalignment is rarely purely angular or parallel, but is usually a combination of both types. If a shaft of one machine is higher (parallel vertical plane) and pointing down (angular vertical plane) toward the other, it is in a combined misalignment condition.

From a couplings perspective, there are "planes of flexibility." The planes of flexibility within a coupling are actually pivot points. A full-flex coupling, such as a gear coupling, has two pivot points with one attached to each of the connected shafts. The actual pivot point can be between loosely separated coupling parts, or in the bending section of a flexible disc element. The flex plane of an elastomeric coupling is within the elastomeric element itself. Spacer or floating shaft couplings will have widely spaced flex planes that allow maximum misalignment within the coupling.

Another consideration is the gap between the faces of the shaft or coupling hub faces. This is sometimes referred to as the axial gap. Almost all couplings specify a gap setting that must be set within tolerance. As a guideline, the gap should be, set within 10% of the recommendation. The gap allows for thermal growth or axial movement of the two shafts to not push against each other and damage bearings and internal machine components. Another reason to ensure the correct gap is to allow proper assembly of connecting center members.

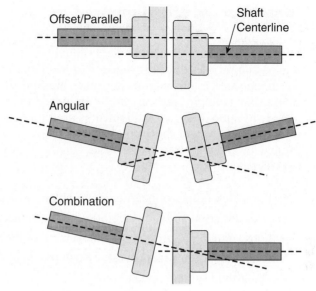

FIGURE 13–35
Types of shaft misalignment.

Causes of Misalignment

The causes of misalignment are numerous and usually combine to adversely affect the alignment condition during installation and operation. Primarily they can be grouped into the following: equipment setup and installation, mounting bases, thermal growth, damaged components, and applied loads. The proper setup and installation of the rotating equipment is important to achieve good alignment. The machines should be level with respect to one another. The distance between the shaft ends should fall within the specified gap for the coupling in use. The machine base must not be positioned so as to be "bolt bound" and prevent adjustment of the machines. The base plates of both pieces of connected equipment are the underlying foundation or support. The plates must be rigid and strong enough to support the weight of the machines and prevent flexing under operating loads. There can be no defects, rises, cracks, or loose fastening points to make alignment difficult to maintain. The eventual alignment will be only as good as the bases the rotating machines are mounted on. When running under load, machine temperatures change because of friction, hot moving materials, or extreme ambient conditions. This thermal growth will cause the equipment to change shape and position relative to the other machine. Damaged and bent shafts or bearings with excess clearance will make precision alignment nearly impossible. Piping strains from attached pipes or conduits also can pull a machine out of alignment. Finally, the actual installation of the coupling and amount of adjustment done or not done by the mechanic will cause misalignment. All of these issues need to be considered when determining the root cause of the misalignment before corrective action may be taken.

Detection of Misalignment

Misalignment is not easy to determine on machines that are running. Detecting misalignment of rotating equipment should take a multifaceted approach. The technician should use all of his senses and available tools in determining whether a coupling/shaft is misaligned. A single method of detection is insufficient and unreliable; multiple steps should be employed to gather all the evidence. Sophisticated vibration detection equipment is useful for detection purposes. The evidence of misalignment will show up as symptoms of the cause. The cause can be determined by paying attention to some of the following conditions that point to misalignment:

- Premature failure of bearings, seals, shafts, or couplings.
- Excessive axial and radial vibration of the connected machines and components.
- Loose mounting fasteners and movement of the base during operation.
- High temperature readings in the equipment or on the coupling.
- Oil or fluid leakage around seals.

The Importance of Precision Alignment

The reason for precision alignment is primarily economic. Precision alignment takes time and effort to accomplish. Production and operating concerns sometimes drive the need for a "quick-fix" that allows the machine to be up and running in a short time. But ultimately a "duct tape and bailing wire" mentality will *increase* costs over time. Precision for the sake of precision is nice, but hardly justifiable in today's competitive marketplace. Precision alignment produces overall cost savings resulting from less downtime. The following reasons justify precision alignment:

- Optimal machine operating performance. Machines will function at their peak capacity and highest level of accuracy when aligned. More product and better product is possible with accurately aligned shafts.
- Misalignment causes a waste of energy. Electrical consumption is increased due to intensified loads on the motor. Precision alignment can potentially reduce energy loss by 5% to 10%.
- Increased bearing life. Offset of the shaft centerlines, relative to one another, will cause reactionary loads on the bearings. These loads can cause an exponential decrease in the life of the bearings.
- Reduced vibration. Vibration is detrimental to the life of the machine and the connecting components.
- Minimized shaft bending and shearing. Misalignment generates overhung loads on shafts, which can cause fatigue or catastrophic failure.
- Minimized coupling wear. Mating coupling parts that are binding or subjected to uneven cyclic loads will wear out or fail from fatigue prematurely.

- Better seal performance. Seals, gaskets, and packing are less likely to leak or allow contaminants into the machine if they are running true.
- Safety. Safety is not compromised by hurried installation and maintenance practices.
- Increased production and less downtime equals more profit.

Alignment Tolerances

Alignment tolerances are usually based on the recommendations of the manufacturer, the speed of the machine, and the operating requirements. Most plant engineers have determined a tolerance limit suited for the specific application and require that the maintenance mechanic meet those expectations.

Offset alignment tolerances are usually measured in thousandths per inch (mils). An offset is a deviation of a position from a known reference line. In shaft alignment, offset pertains to the difference of one shaft centerline to the other shaft centerline. Essentially it is the measurement of how much higher or lower, left or right, one shaft is to the other. Flexible coupling manufacturers will list the recommended limits in tables. As a general guideline, Figure 13–36 lists tolerances that may be used for parallel/offset alignment limits.

There are several methods used to define angular misalignment. Angular misalignment tolerances can be measured as the difference between the gaps on coupling hub faces 180° apart. The manufacturer publishes the allowable difference

	Short Couplings						Spacer Shafts	
	Excellent			Acceptable			Excellent	Acceptable
	Offset	Angularity		Offset	Angularity		Offset Per Inch	
RPM	(mills)	(mills/in.)	(mills/10 in.)	(mills)	(mills/in.)	(mills/10 in.)	(mills/in. of spacer length)	
600	5.0	1.0	10.0	9.0	1.5	15.0	1.8	3.0
900	3.0	0.7	7.0	6.0	1.0	10.0	1.2	2.0
1200	2.5	0.5	5.0	4.0	0.8	8.0	0.9	1.5
1800	2.0	0.3	3.0	3.0	0.5	5.0	0.6	1.0
3600	1.0	0.2	2.0	1.5	0.3	3.0	0.3	0.5
7200	0.5	0.1	1.0	1.0	0.2	2.0	0.15	0.25

These suggested tolerances are the maximum allowable deviations from desired values (targets), whether such values are zero or nonzero. These recommended tolerances should be used in the absence of in-house specifications or tighter tolerances from the machinery manufacturer.

FIGURE 13–36
General alignment tolerances.
Courtesy of Northern Michigan University.

between the gaps for various couplings. The measurements are usually taken on the shafts or hubs at the twelve o'clock and six o'clock positions for vertical angularity, and the three o'clock and nine o'clock positions for horizontal angularity. Angular misalignment can also be defined as the slope relationship of the two shaft centers. The slope can be quantified by first determining the offset difference in any two measuring planes. These measuring planes are points of measurement, such as the center of the mounting pads. Once the offset differences at various points along the shaft axis are determined, slope in mils per inch is known.

Some manufacturers list the suggested tolerances in degrees, minutes, and seconds. The angularity in degrees may be calculated by using the following formula:

$$\text{Angularity in degrees} = \frac{57.3(A - B)}{D}$$

Where: A = large gap

B = small gap

D = diameter across the gap

Remember that the flexible coupling alignment tolerances are not the same as the connected machines' tolerances. A flexible coupling may allow gross amounts of misalignment and still function, but the equipment ideally requires near perfect alignment.

Preliminary Alignment Steps

Alignment of rotating shafts and couplings involves some preliminary procedures that must be completed regardless of which alignment method is used. Oftentimes the ultimate accuracy of the alignment will be based on the initial setup procedures and the amount of effort given to the details. The following procedures are general guidelines that are applicable to most alignment work.

- Always lockout, tagout, and verify that the equipment is neutralized. Mechanically block any loads or flow of product that may cause a movement of shafts during the alignment process. Follow all plant safety procedures.

- Gather and organize the tools that are required to complete the job. Besides the standard maintenance mechanic tools, some specialized items may be required. Dial indicators are used to determine run-out, soft foot, and the relative position of the shafts. Taper gauges that measure in thousandths of an inch can be used for measuring the gaps between the faces of the coupling hubs.

- Keep accurate records. Consult any prior documentation pertinent to the installation and alignment of the equipment. Keep accurate records of the process, detailing the steps and specifications. Verify the alignment specifications for the equipment and the allowable tolerances of the flexible coupling. Graphing of the two machine shaft centerlines, relative to one another, and the mounting pad positions can be useful.

- Take multiple measurements at consistent reading points. Always take multiple measurements to confirm consistency in the numbers. Mark the location of the reading point to be exact in the location for repeatability. All measurements must be repeatable to within one thousandth of an inch.

- Inspect all mounting surfaces. A rigid and solid foundation is essential to maintain smooth running and aligned equipment. Flat steel mounting pads should be grouted into a concrete base. This will allow the unit to be more accurately aligned and positioned. A steel foundation should be designed and engineered to provide sufficient rigidity, preventing induced loads and mass of the machines that could cause flexing and misalignment of the rotating elements. Machined steel fabricated bases that support the driver and driven equipment can provide compactness, proper alignment, and ease of installation. The feet of motors, gearboxes, pumps, etc., must be checked and found free of paint, rust, debris build-up, and defects. All of these imperfections must be filed or machined off to provide a level and clean mounting surface. Any equipment with cracked mounting pads should not be used.

- Proper bolting is imperative. All damaged and worn fasteners should be discarded and replaced with new bolts with the correct specifications. Bolts should be tightened in sequence and to the proper torque specifications for their size and grade. The anchoring characteristics must be proper in order to ensure a good alignment and prevent the machines from coming loose during operation. A drilled and tapped base plate should have a minimum thickness of 1-1/2 times the root diameter of the anchoring fastener. If the base plate is less than 1-1/2 times the bolt diameter, hardened steel washers and nuts are required. The length of the bolt relative to the depth of the hole is important. The unthreaded portion of the bolt must not be too long nor should the overall length of the bolt exceed the depth of the hole. Fasteners must not be "bolt bound" (see Figure 13–37). A bolt-bound fastener will prevent the horizontal movement of a machine, due to the contacting of the anchor bolts to the sides of the machine anchor holes. The equipment should have slightly oversized through holes and be centered to allow slight movement in all directions. The threaded holes should not be stripped or wallowed out to cause the bolt to be cocked (see Figure 13–37). If the mounting hole of the machine is too large, the bolt head will force the washer into the hole, causing a "dowel" effect (see Figure 13–37). Thicker washers will help correct this problem of cupped washers.

- Jacking or pusher bolts aid in machine movement. Jacking bolts should be installed to ease horizontal movements that will be measured in thousandths of an inch. A jacking bolt is a bolt that is installed in a threaded hole of a plate that is attached to the side of the mounting base. Typically a rectangular piece of 1/2″ steel is welded onto the base at all four corners in line with the feet of the installed equipment. These bolts are designed for movement of the equipment only and not for securing purposes. If jacking plates/bolts cannot be installed, a pry bar or light taps from a soft mallet can be used to adjust the position of the machine. Never strike the equipment with a steel hammer to move it.

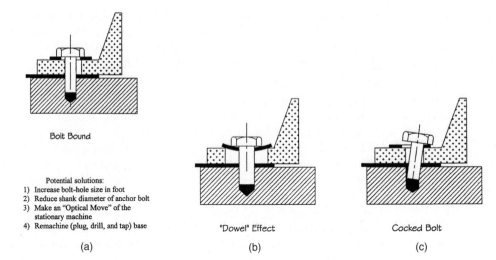

Potential solutions:
1) Increase bolt-hole size in foot
2) Reduce shank diameter of anchor bolt
3) Make an "Optical Move" of the
   stationary machine
4) Remachine (plug, drill, and tap) base

Bolt Bound

"Dowel" Effect

Cocked Bolt

(a)   (b)   (c)

FIGURE 13–37
(a) Bolt bound, (b) dowel effect, and (c) cocked bolt.
Courtesy of Northern Michigan University.

- Eliminate any piping strains. Improperly installed conduit and pipe connections can adversely affect machine alignment. Machines should be aligned initially unattached from any piping. Plumbing and conduit must be aligned with the connections. Machines should be supported by structurally sound overhead or floor supports. Do not use winches to bring any connections together for fastening. If a forced connection is made to properly aligned equipment, the strain will pull the machine out of alignment. Flexible connections can be used when the piping unions are not capable of being aligned. These flexible joints will also dampen vibrations that are created by uneven flowing product.

- Replace worn bearings. If the shaft bearings have excessive looseness, they should be replaced. When an alignment is done on a machine with plane bearings, the shaft should be positioned in its mechanical center. On electric motors with plane bearings, the rotor will hunt for its magnetic center when operating. This position is the optimum setup location for the coupling, as long as it is not thrusting against the bearings. The magnetic center can be found by running the motor with the coupling center member disconnected, and scribing a line on the shaft. This will locate the operating position of the shaft.

- Inspect and measure the shaft and coupling hubs. All shafts, hub bores, and keys should be inspected, measured, and cleaned. Follow proper hub to shaft mounting procedures. A run-out check of the shaft and coupling hubs should be performed. Set up a magnetic base dial indicator to check the TIR (total indicator run-out) to be sure it is in within the recommended tolerance limits.

- Calculate the thermal growth of equipment if necessary. Certain hot running applications, such as boiler feed pumps and kiln drives, require the calculation and consideration of machine movement from thermal expansion. Heat rise variances across the machine and relative to the other connected machine will cause an uneven change in height across the plane of centers. A temperature change between cold startup and hot operating conditions can influence the alignment.

For linear expansion of steel shafting and cast iron base equipment, the following formula may be used to calculate the amount of growth.

Expansion in inches = 0.0000063* × L × T

> Where: L = Length in inches
>
> T = Temperature increase in degrees Fahrenheit
>
> * = Constant for steel and cast iron

For vertical rise of equipment, determine or estimate the average rise in temperature of each piece of equipment from its shutdown condition to its operating temperature. Measure the height of the shaft above the base plate for each piece of equipment. Allow for a change in this height of approximately 0.007″ for each foot of height and each 100°F temperature change. The height will increase with a rise in temperature and decrease for a drop in temperature. Most of the position changes will occur in the first 10 to 15% of the cooling and heating cycle. Make allowances for this change in elevation when aligning the equipment cold so that the shafts will be aligned at operating temperatures. This is called the cold offset allowance. If a temperature variance from front to back occurs across a single piece of equipment, an average temperature and the differences across the machine need to be determined. The average can be calculated from readings taken at four different vertical positions in line with the pedestal pads. The offset that may occur from the front to the back of the machine can be determined by taking the series of temperature readings in line with all mounting pads. Because the pedestals may not heat evenly, the difference in change should be determined. Corrective shimming will allow for uneven growth and properly position the shaft axis to be co-linear when operating. The following formula for thermal change can be used to approximate rise:

$$\Delta C = 0.007'' \times \frac{H}{12''} \times \frac{\Delta T}{100°F}$$

> Where: ΔC = Change in height
>
> H = Measured height base to center line of shaft in inches
>
> ΔT = Change in temperature from cold start to hot operating in degrees Fahrenheit

■ Use quality shim stock. Stainless steel precut horseshoe-shaped shims are the easiest and most efficient shims to use (see Figure 13–38). The shims are available in an assortment of sizes, such as 2″ × 2″, 3″ × 3″, etc. Typically a variety of thicknesses, which are clearly marked on the face of the shim, will come in a kit. Old shims that are bent, damaged, or have curled edges should be discarded and replaced with new ones. Never use washers or thin sheet steel because the thicknesses can vary greatly and they are not always flat. Brass shim

FIGURE 13–38
Stainless steel
horseshoe shim kit.

Courtesy of Precision Brand
Products.

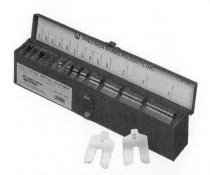

FIGURE 13–39
Elastomer shim
used to correct
angular soft foot.

Courtesy of Precision
Brand Products.

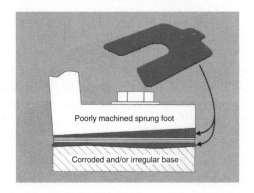

Poorly machined sprung foot

Corroded and/or irregular base

stock is an acceptable alternative, but it must be cut and deburred. Most standard plastic color-coded shim stock is not designed to bear up under heavy loads and tends to slide out during movements due to its slippery qualities. Plastic sheet must be resistant to oils and solvents that may be present around the bases or it may degrade. Special plastic material precut shims are now available to correct angular soft foot problems (see Figure 13–39).

■ Measure and correct soft foot conditions. Soft foot is a condition where one or more of the mounting pads of the equipment do not make complete contact with the base plate surfaces (see Figure 13–40). This condition can be caused by machine mounting pad or base plate irregularities. Ideally the base will be machined flat and true and not be the cause of the soft foot problem. Soft foot is called by several names and takes various forms, such as angular, induced, springing, or parallel. All or several of these types can exist on a given piece of equipment. In some cases the underside of the foot or mounting pad of a motor or driven piece of equipment is not parallel and may exhibit a wedge-shaped gap known as angular or sprung soft foot. Other times the mounting pad is parallel to the machined base surface, but an air gap on one of the four pads exists; this is known as short foot or parallel soft foot. External forces like piping strains, improper bracing, or grossly misaligned couplings create induced soft foot. Springing soft foot—or squishy foot—is a condition that is created by burred shims or from loose foreign material between the pads and base. If the problem of soft foot is ignored and the machine is tightened down, the frame or machine housing will twist. This creates internal misalignment and load distortion resulting in

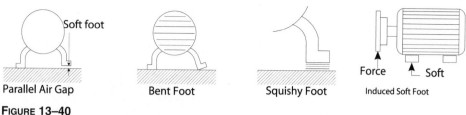

Parallel Air Gap	Bent Foot	Squishy Foot	Induced Soft Foot

FIGURE 13–40
Soft foot.

Courtesy of Northern Michigan University.

premature bearing failure and shaft deflection. It also causes frustration during the actual alignment process when attempting to take accurate measurements. Soft foot must be corrected to achieve accurate alignment of rotating equipment.

The problem of soft foot is corrected by re-machining base plate and mounting pad surfaces or by properly shimming the spaces of noncontact. There are two methods for determining soft foot: the at-each-foot method and the shaft deflection method. Both can be used to verify the existence and degree of soft foot. Both methods have their advantages and disadvantages. The at-each-foot method is a good way to determine angular soft foot and get an accurate map of each foot. The shaft deflection method is quick and is a good way to determine if distortion of the machine housing is occurring due to soft foot. A practical means to correct the soft foot condition is to follow one or both of the procedures detailed below.

At-each-foot method:

1. Center and rough align the equipment on a clean base/pads so as not to be bolt bound or twisted. Make sure all rust, paint, dirt, burrs, etc. are removed from the base plate and machine mounting pads prior to installation.

2. Hand tighten the mounting bolts. Slide a feeler gauge blade (see Figure 13–41) between the base plate and machine mounting pads to determine where contact is being made at each foot. Map the area on a sheet of paper, noting the locations and gap size.

3. Install shims under the foot with the greatest gap first. Proceed to the other feet and do the same. Gaps greater than .002″ should be corrected. The shape of the air gap might be a wedge; a pyramid of shims should be inserted. This is a check for angular soft foot.

FIGURE 13–41
Checking for soft foot
with a feeler gauge.

Courtesy of Northern Michigan
University.

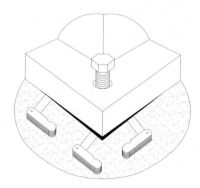

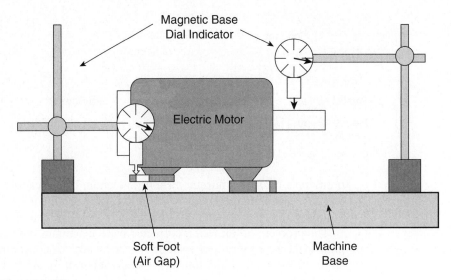

Magnetic Base
Dial Indicator

Electric Motor

Soft Foot
(Air Gap)

Machine
Base

FIGURE 13–42
Dial indicator setup for soft foot check. Using a dial indicator at the foot or on the shaft
to record deflection is acceptable.

4. The soft foot condition can be confirmed with a dial indicator. Remove all shims and foreign material under the mounting pads. Tighten the mounting bolts enough to prevent any unintentional movement, but not excessively. Place a dial indicator over the foot believed to have the worst soft foot condition, with the stem as perpendicular to the mounting plane as possible (see Figure 13–42). Adjust the indicator to zero and watch for movement as the bolt is carefully loosened. Record the results and retighten the bolt.

5. Repeat the procedure at each foot, recording the gaps to determine the correct amount of shims required.

6. Install all the required shims and leave them in place during the rest of the alignment process. Lightly tighten the bolts.

Shaft deflection method:

1. Center and rough align the equipment on a clean base/pads so as not to be bolt bound or twisted. Make sure all rust, paint, dirt, burrs, etc. are removed from the base plate and machine mounting pads prior to installation. Tighten the mounting bolts firmly.

2. Secure a magnetic base dial indicator to the base plate. Adjust the stem to be perpendicular to the shaft axis and take a reading off of the shaft or coupling hub (see Figure 13–42). Zero out the indicator.

3. Slowly loosen the bolt on one of the feet. Observe and record any movement of the dial indicator. The amount of movement shown by the dial indicator is the amount of shims to be placed under that foot.

4. Continue checking and shimming under each foot until movement of the indicator is less than .002″.

■ Dial indicator use. Dial indicators are excellent tools to provide a means to accurately determine the relative position of shafts and coupling hubs. They are also useful in measuring soft foot conditions. But they must be used properly and be in good working order. The two face designs are the balanced and continuous types. The balanced type has positive side numbers and negative side numbers, where as the continuous type starts at zero and reads to ninety-nine. Both types can be employed successfully as long as the total movement (TIR) is read and regardless of whether the needle begins before, after, or at zero. The placement of the magnetic indicator base should be on a rigid nonmovable surface. The tip (probe) of the indicator must be on tight and maintain contact on the surface at all times without bottoming out. Readings need to be taken off of clean, undamaged surfaces. The plunger should be positioned at right angles to the shaft/hub axis for offset (rim) readings, and parallel to the shaft axis for angular (face) readings. The plunger should take measurements at consistent locations, the 12, 3, 6 and 9 o'clock positions. Indicators should be run through their movements at least three times to ensure correct readings. If the indicator assembly is chained or bracketed to the shaft/hub, it must be secured to prevent unintentional movement. The assembly consists of clamp, riser block, spanning rod, indicator rod, and indicating device.

An important consideration when using dial indicator assemblies clamped to the shaft/hubs is "rod sag." The weight of the indicator, the span length, the rigidity of the bar, and the specific hardware arrangement can cause rod sag. The longer the span between the clamp and the opposite shaft/hub, the more the weight of the indicator will cause the rod to sag. Rod sag must be taken into account when reading across the span length of the coupling. The sag, if not accounted for, can throw off the readings. A 3/8″ rod spanning a distance of 10″ can affect readings by .008″ to .012″. The amount of sag can be determined by installing the indicator assembly on a solid shaft or rigid pipe. A separate continuous bar is used because it is not misaligned. It must be assembled and clamped in the same way it will be used during the alignment process, and read at the same distance or span (see Figure 13–43). The tip of the indicator contacts the bar at the 12 o'clock position

FIGURE 13–43
Setup for checking
indicator rod sag.

Courtesy of Northern
Michigan University.

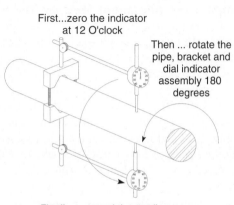

First...zero the indicator
at 12 O'clock

Then ... rotate the
pipe, bracket and
dial indicator
assembly 180
degrees

Finally ... record the reading
on the bottom

and is zeroed out. The bar is rotated 180° to the 6 o'clock position and will show a negative reading, which is the total amount of indicated sag that must be compensated for in the readings. The measured sag can be calculated by mathematical manipulation. It will be added as an absolute value to the offset alignment TIR reading taken at the 6 o'clock position during the actual process. An easier way to compensate for the sag measured at the 6 o'clock position is to dial it in as a positive value on the indicator when it is in the 12 o'clock position.

Alignment Methods

There are at least five methods used to align rotating shaft couplings, some of which are: the straight edge/feeler gauge, feeler gauge/taper gauge, rim-face indicator, reverse indicator, laser systems, and combinations of each. Basically the method is named for the primary tools used to determine and correct the alignment condition. Some methods are quick and easy; others are more involved and time-consuming. Certain methods, such as those incorporating the use of the laser or dial indicators, are more exact.

Many factors will determine the method used to align rotating shafts and couplings. Some of the critical factors are cost, time, the level of the mechanic's training, the tools available, and the characteristics of the application. Cost and time are obvious constraints on any alignment job. The cost of labor, equipment, and downtime must all be weighed to determine the alignment method, as well as the level of precision required. The training of maintenance technicians in proper alignment procedures is imperative. If the technician is unfamiliar with the use of dial indicators or lasers, he/she should not attempt to use them. When a maintenance mechanic has only a straight edge and feeler gauges and does not have access to sophisticated electronic alignment equipment, the tools available will dictate the method. On critical high-speed production equipment precise alignment is necessary to provide continuous, accurate operating conditions. In these cases lasers, dial indicators, and all the time needed to properly align the shafts should be provided.

Straight Edge Alignment Method The simplest and oldest type of alignment procedure is the straight edge method. A straight, smooth, graduated scale—in conjunction with a feeler gauge—is used to align the coupling hubs. The steel straight edge is placed across the edges of the coupling hubs or shafts to determine the parallel/offset differences. A feeler gauge blade is inserted between the hub rim and the straight edge to measure the offset difference. Readings are taken at the 12, 3, 6, and 9 o'clock positions. A feeler gauge is inserted between the faces of the hubs at the same clock positions to determine angularity. The adjustment of shims and movement of jacking bolts are made to correct differences in the readings. This method is crude and does not offer accuracy or repeatability. It is quick and can be used effectively to rough in a coupling/shaft.

Straight Edge/Feeler Gauge/Taper Gauge Alignment Method A taper gauge is added to the mix when a feeler gauge and straight edge are used for angular checks. A taper gauge is a flat, tapered strip of steel with thousandths of an inch and millimeters marked on its side. A straight edge is used to determine parallel offset differences between the rotating hubs/shafts. The gap between the straight edge and the rim of the hub can be measured with a feeler gauge or taper gauge. Using the taper gauge to measure gaps can check angularity in both the vertical and horizontal plane. The taper gauge is inserted into the gap between the hub faces to measure the space and the differences at the various clock positions. This method is not extremely accurate, but the taper gauge is an easy instrument to use and makes the alignment process less time-consuming than more sophisticated methods.

Single Dial Indicator Method All of the above mentioned tools are used along with a single dial indicator arrangement. The single dial indicator assembly is used to take readings off of the rim of the coupling to establish parallel/offset alignment. It has the advantage of being used when only one shaft can be rotated. It is a more accurate method than just using straight edges, taper gauges, and feeler blades. A more detailed explanation of this method is written in the "Simplified Basic Alignment" section of this chapter.

Rim/Face Alignment Method The rim/face alignment method is an alignment method in which the offset and angular gaps are measured using two dial indicators. Typically both dial indicators are mounted on a single rod, and take simultaneous rim (peripheral) and face readings (see Figure 13–44), This method is sometimes referred to as combination rim/face. One of the indicators—measuring at the 12, 3, 6, and 9 o'clock rim positions—determines the parallel/offset differences. The other indicator plunger reads off of the face—at the standard clock positions—to determine angularity. Rim and face readings must be taken with the coupling disconnected and adequate space between the shaft/hub ends to allow for positioning of the indicator plunger. Both shafts should be rotated simultaneously to prevent distorted readings from run-out. By rotating both shafts/hubs together, shaft centerlines are measured. This method allows for a single move to correct angular and parallel/offset differences. However, when one of the pieces of equipment cannot be rotated because of a locked rotor or impeller, this complicates matters. Ensure that neither the driver, nor driven shafts, moves axially when taking readings, because this also will distort the readings.

Graphing and plotting important locations and distances of the shaft centerlines should coincide with the rim/face alignment method. Graphing of the shaft centerlines involves measuring the diameter of face indicator travel, the distance from the rim indicator to the front mounting pad bolt center, and the distance between the mounting pad bolt centers. These dimensions are plotted on graph paper and assigned vertical and horizontal values. The horizontal measurements are in inches and the vertical measurements in thousandths of an inch. Plotting can be done to view the top and side relative positions. With graphing, the technician is determining the slope angle and position of the shaft centerlines to allow for a single move by installing the required amount of shims under each foot. Shifting and shimming of the movable machine is based on the amounts recorded on the indictors. Plotting of alignment measurements produces a hard copy record and helps

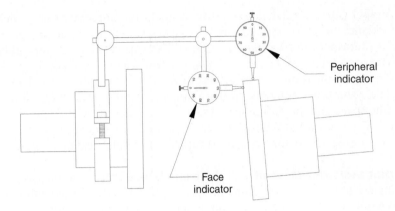

This method can only be used for couplings installed with a spacer, and involves the use of two dial indicators supported on a bracket or two bracket set up across the coupling.

Procedure...

1. Attach the alignment bracket firmly to one shaft and position the indicator(s) on the face and perimeter of the other shaft.

2. Zero the indicators at the 12 o'clock position.

3. Slowly rotate the shaft and bracket arrangement through 90 degree intervals stopping at the 3, 6, and 9 o'clock positions. Record each reading (plus or minus).

4. Return to the 12 o'clock position to see the indicator(s) re-zero.

5. Repeat steps 2 through 4 to verify the first set of readings.

Indicator readings log

Peripheral readings

Face readings

FIGURE 13–44
Rim/face or peripheral dial indicator alignment technique.
Courtesy of Northern Michigan University.

the technician visualize the required moves. It is time-consuming but less expensive than a laser alignment system.

Reverse Dial Indicator Method The reverse dial indicator method uses two dial indicators to take readings off of opposing sides of the coupling hub rims (see Figure 13–45). One dial indicator is mounted on the coupling hub of the driver and takes measurements on the driven machine. The second dial indicator is mounted

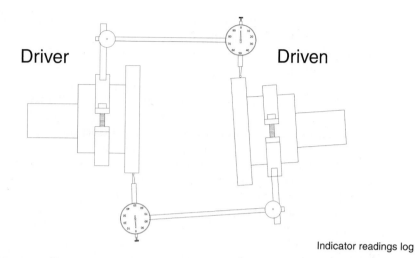

Driver

Driven

Procedure...

1. Attach the alignment bracket(s) firmly to one (both) shaft(s) and position the indicator(s) on the perimeter of the other shaft.

2. Zero the indicator(s) at the 12 o'clock position.

3. Slowly rotate the shaft and bracket arrangement through 90 degree intervals stopping at the 3, 6, and 9 o'clock positions. Record each reading (plus or minus).

4. Return to the 12 o'clock position to see if the indicator(s) re-zero.

5. Repeat steps 2 through 4 to verify the first set of readings.

6. If one bracket was used, mount the bracket onto the other shaft and repeat steps 1 through 5.

Indicator readings log

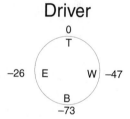

Driver

0
T

−26   E   W −47

B
−73

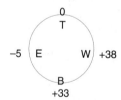

Driven

0
T

−5   E   W +38

B
+33

Figure 13–45
Reverse indicator technique.

Courtesy of Northern Michigan University.

on the coupling hub of the driven machine, positioned 180° from the first indicator, and takes readings on the coupling hub of the driver. Be sure to consider bar sag, also known as road sag. Both shafts should be rotated simultaneously during the operation.

Readings are taken at 90° intervals, starting with zeroed indicators at the 12 and 6 o'clock positions. The readings, along with relevant machine dimensions, can be plotted on graph paper as in the rim/face method. The technician should establish which machine will be considered the movable machine—usually the motor—and which machine will be the stationary machine. A line is drawn on the graph paper that represents the stationary equipment shaft center. Plotting the mounting bolts and indicator positions in the horizontal plane on the graph paper, along with the vertical offset, allows for a single set of shim adjustments on the movable machine for vertical angular and parallel corrections. Raise or lower the movable machine by adding or subtracting shims under the mounting pads to correct the height difference. Horizontal corrections are accomplished by zeroing both indicators in a horizontal position (3 and 9 o'clock). Both shafts are rotated 180° to the other side and readings are recorded. The indicators are then adjusted to read one-half of what they read after the 180° rotation. Push and pull the movable machine with the jacking bolts until both indicators read zero.

The reverse rim indicator method has advantages and disadvantages, as is the case with any alignment method. Space is not required between the hub faces for another indictor to record face readings, and shaft end-play does not affect the readings. Although it is somewhat time-consuming and relies on the accuracy of the technician to correctly measure and graph, it can be efficient and accurate.

Laser System Alignment Sophisticated and computerized laser alignment systems can be used to achieve extremely accurate alignment (see Figure 13–46). The modern systems essentially are extensions of the rim/face graphing method, with all the measuring, plotting, and calculating done by the laser microprocessor. The initial cost of the equipment is high but is quickly recovered from energy, up time, and maintenance savings. Sending/receiving heads are bracketed to both the driver and driven shaft/hubs. Beams projected to receiving transducers are converted to impulses for the calculator. Readings are taken at various positions and entered into the computer, which is connected to the laser heads.

Full rotation is generally not required with most units. All offsets and angles are read and measured accurately without the use of indicators or special tools. Bracket sag, distance, and axial float are all nonissues with lasers. The more sophisticated systems can determine and account for soft foot and thermal growth. Certain span and bolt center measurements need to be entered for calculation of required shims and movement. A real-time display will show the required shims

FIGURE 13–46
Laser alignment system.
Courtesy of SKF Industries.

and movements necessary to bring the equipment into near perfect alignment. The alignment requirements, conditions, and movements can be recorded and saved within the machine for future reference.

Care must be taken when handling the system to avoid damaging it. It must be kept free of contamination from dirt, moisture, heat, chemicals, and sunlight, all of which can adversely affect the performance and accuracy of the unit. In the hands of a trained technician, it is an extremely accurate and efficient method of alignment.

Practical Rotating Shaft/Coupling Alignment

For a practical and simplified approach to alignment, most maintenance technicians can use the following general guidelines with minimal resources and within a limited time frame to accomplish an acceptable alignment. These steps should be used in conjunction with the preliminary alignment guidelines.

1. Lock out, tag out, and verify that the equipment is electrically neutralized. Machines that have a potential for shaft rotation from a back pressure or load surges should be mechanically blocked.

2. Assign one individual as the team leader to document and coordinate the alignment process.

3. Gather and organize the required tools. In addition to the standard mechanics tools, a feeler gauge set, magnetic base dial indicator, taper gauge, and straight edge are needed.

4. Clean and inspect the mounting bases of both connected machines, if possible. The machine bases must be in good shape. It is preferable that both machine bases are checked, but in many instances only one machine is removed from the foundation while a new unit is being installed. Clean away any debris, paint, or old shim stock from the base mounts. Slowly run a straight edge along the machined mounting surfaces of the base pads to find any rises or burrs. Carefully dress these off with a fine tooth file. This process should also be done for the mounting pads of the equipment that is being installed. Do not grind off substantial amounts of metal from the base or pads because it will create a soft foot condition.

5. Inspect all the fasteners and washers for damage and replace them as needed.

6. Refer to any prior alignment records. Keep records of the process and graph the centers of rotation, relative to one another, if time allows. Graphing is helpful to visualize the positions of the shafts and will determine the required shim placement for alignment.

7. Look up the alignment tolerances for the machine and the couplings. Strive for perfect alignment.

8. Check the shafts and coupling hubs for the amount of radial run-out on both the driven and driving machines. Use a dial indicator on a magnetic base for this purpose. Excessive run-out will prevent good alignment and cause a balance problem.

9. Inspect the shafts and coupling hubs for mechanical damage or wear and replace/repair, if necessary.

10. Install the coupling hubs using proper procedures. Be sure that the coupling hub engages the shaft a minimal distance (length through bore) equal to or greater than the shaft diameter it is mounted on. Do not hammer the hubs onto the shaft.

11. Inspect the bearings of the equipment for excessive clearance and end-play.

12. Eliminate any piping strains that will twist the equipment frames or induce soft foot.

13. Center the equipment/bolts in their slots to prevent a bolt-bound condition. Centering the equipment will allow for movement in any direction.

14. Rough-align the equipment with a straightedge, placing it in the approximate running position.

15. Check for and correct any soft foot conditions. With the equipment bolts removed, determine the amount of soft foot under each mounting pad with a feeler gauge blade. Map the amounts and areas under each foot on a piece of paper. Start with a .003″ blade and pass it under the feet to measure the air gaps. Progressively measure with thicker blades to determine the amounts and condition of the soft foot. Verify these conditions with a dial indicator, using the "at-each-foot" or "shaft deflection" method. Correct the worst foot with the required shims and advance around the bases, measuring and correcting as you go. Shaping and pyramiding of shims might be required to fix angular and irregular conditions. Regardless of which method and procedure is used to determine soft foot, find it and correct it.

16. Use undamaged stainless steel precut horseshoe-shaped shims, if at all possible. Place a .020″ to .030″ shim pack under each foot in addition to any amounts placed to correct soft foot or overall height differences between machines.

17. Determine if thermal growth of the equipment during operation will cause movement. Calculate the amounts and differences between the machines and compensate for the differences by shim adjustment.

18. Determine which machine will be the movable machine and which will be the stationary one. This might be predetermined by the conditions and constraints of the equipment and installation.

19. Set the hub gap. If the coupling hub faces are relatively close to one another (1/8″ to 1/2″), place a precut shim equal to the approximate required gap between the coupling hub faces. Larger diameter hubs might require multiple shims. Push the machines together to set the prescribed coupling hub gap. Be careful not to put the machines into a bolt-bound condition. Overhanging of the hubs a small amount to set the gap is acceptable. If the coupling is a spacer type, an inside micrometer or caliper can be used to measure and set the gap approximately equal at the 12, 3, 6, and 9 o'clock positions. Setting the gap with a shim or calipers brings the coupling hubs into rough angular alignment.

20. Rough align the machines using a straightedge across the hub sides. Raising one of the machines by adding additional shims might be necessary. Check the alignment at the 12, 3, 6, and 9 o'clock positions. This process rough-aligns the coupling in the parallel/offset planes.

21. Mark both hubs with a pencil or soapstone at the 4 o'clock positions. Line up the 12 o'clock position marks on both hubs. Always return the hubs to their respective positions if they should move during the process. Measure at consistent locations on the hubs.

22. Lightly tighten the bolts to prevent unintentional movement of the machine during the alignment process. Initially the bolts should be snug but not excessively tightened. Proper and consistent tightening sequences should be followed.

23. At this point, a rim/face or reverse dial indicator alignment method could be employed. The graphing of centers of rotation, indicator positions, and feet locations could also be done. The slope angles would then be determined and visualized. Theoretically, once accurate measurements are taken and correctly plotted on graph paper, a single move and shim placement can be done to properly align the shafts. Space, time, tools, and training limitations, however, often prevent this from happening. Single movement accurate alignment is rarely successful because of the realities of most machine operating and adjustment conditions. The maintenance technician must fall back on a simple, practical, and methodical adjustment process.

24. The easiest methodology corrects misalignment in the following order: vertical angularity, vertical parallel/offset, horizontal angularity, and horizontal offset.

25. Determine the angularity of the hub faces. Use a taper gauge or feeler gauge to accurately measure the gap between the hub faces at the 12, 3, 6, and 9 o'clock positions. Record the measurements on a drawn circle at their appropriate locations. Visualize or graph what the measurements reflect about the vertical and horizontal positions of the shaft/hubs relative to one another. Remember that the target measurement is the recommended coupling gap. Most manufacturers will allow a 10% variance in the gap. Good coupling alignment measurements will be within .004″ of the target and .004″ of each other. Repeat the measuring process.

26. Determine if the shaft/hubs are within an acceptable amount of angularity. Most couplings require that operating angularity be held to less than a 1/16 of a degree. The degrees of angularity can be determined by using the following formula:

$$\text{Angularity in degrees} = \frac{57.3(a-b)}{D}$$

Where: a = large gap

b = small gap

D = hub diameter (linear distance between the clock positions)

27. The shim adjustment of the machine's mounting pads should be done at one mounting pad location at a time. Simultaneous adjustments made by multiple individuals will result in confusion and unintended movement. Loosen the mounting bolts and carefully pry up the base just enough to slide the required amount of shims in or out. Do not apply extreme leverage with the pry bar; the case may crack or a personal injury may occur as a result of a slip. Be careful not to move the machine out of alignment during shim adjustment. Resnug the bolts after each move.

28. Correct the angular misalignment in the vertical plane first. On the movable machine, measure the distance between the center of the mounting bolts and the coupling hub face diameter. Calculate the difference between the top and bottom gaps. Take the distance between the movable machine mounting bolt centers and divide it by the coupling hub face diameter. That ratio is multiplied by the difference between the large gap and the small gap. (The gaps should be measured at the 12 o'clock position and the 6 o'clock position.) The resulting answer is the required amount of shims to be installed or removed under the front or back feet to correct vertical angular misalignment. Whether the front two feet or back two feet of the movable machine require the shim adjustment depends on the slope angle. Shim adjustments over .010″ should be equalized between the front and back feet. Splitting the difference will help prevent dramatic offset/parallel changes. Once the initial shim adjustment is made, recheck the angular measurements.

Use the following formula to determine the required amount of shims needed to correct the angularity differences in the vertical plane.

$$\frac{DBMB}{FD} \times GD = \text{Required Shims}$$

Where: DBMB = Distance between the mounting bolts

FD = Coupling hub face diameter

GD = Difference between the large and small gaps

For example, the correct gap setting between the coupling hub faces is .125″. A gap of .135″ is measured at the 12 o'clock position, and a gap of .115″ is measured at the 6 o'clock position. The distance between the movable machine, which in this case is the motor, is 5.5″. The coupling hub face diameter, which is the linear distance between the clock positions, is 4″.

Thus: $(5.5 \div 4 = 1.375) \times (.135 - .115 = .020) = .0275$ shims.

Therefore .0275″ of shims need to be placed under the back feet or removed from the front feet of the motor. Because .0275″ is considered a significant amount of shim adjustment, it would be advisable to split the difference between the front and back sets of feet. An amount of .013″ of shims could be taken from under both front feet (one foot at a time) and added to the rear feet of the motor. This split is done to avoid changing the rough parallel/offset alignment that was done initially with a straightedge.

29. Determine the parallel/offset alignment differences between the driving and driven equipment shaft hubs in the vertical plane. Place a straightedge across the hub surfaces at the 12 o'clock position and measure the difference in height with feeler gauges or a taper gauge. The movable machine will need to be lowered or raised this amount at all four of its mounting pads. If a dial indicator with a chain bracket is available, mount it securely to one of the hubs. Allow for bracket sag and ensure that good dial indicator positioning and practice are used. If the bracket is mounted on the motor, the indicator point will be reading the position of the driven equipment relative to the motor. Set the indicator to zero at the 12 o'clock position. It is recommended that both hubs be rotated together for a more accurate reading, but this is not always possible. Oftentimes only the motor shaft can be rotated because the pump (driven machine) has a locked shaft/rotor. In that case rotate the motor shaft hub with the dial indicator bracket to sweep the other hub with the stem/plunger. Read the numbers at the four relevant clock positions while it is reading the peripheral of the other shaft. Draw a circle and record the measurements in the circle. Repeat the sweep, recording the numbers to check for repeatability. Verify

FIGURE 13–47a

Three examples of a validity rule check.

Courtesy of Northern Michigan University.

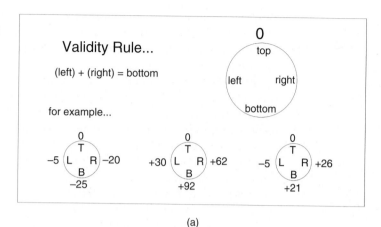

(a)

FIGURE 13–47b

Validity rule problems.

Courtesy of Northern Michigan University.

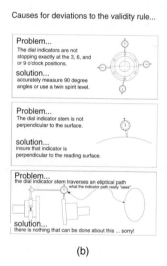

(b)

the "validity rule." The validity rule states that left plus right equals the bottom. The sum of the readings at the 3 and 9 o'clock positions must equal the 6 o'clock position number (see Figure 13–47a). If the sum is not within one thousandth of an inch, the deviation could be caused by one or more of the reasons shown in Figure 13–47b. The problems include: an inexact reading at the four appropriate clock positions, the indicator stem is not perpendicular to the hub surface, the indicator brackets are loose, the indicator is traversing an elliptical path, or extreme angularity exists. Attempt to correct any of these problems to make the validity rule work.

30. Correct the parallel/offset difference in the vertical plane. If a straightedge is used, the height change is equal to the measured space between the straightedge and the hub surface. A dial indicator bracketed to one hub and its plunger reading the outside diameter of the other will give you accurate parallel/offset readings. As the indicator traverses the hub from 12 to 6 o'clock, it will record both of the offset differences between the top of one hub and the bottom of the other (see Figures 13–48a and 13–48b). Because the centers of rotation are being aligned, the recorded amount must be halved. Again, one-half of the recorded amounts at the 6 o'clock position is the required move. If the indicator is bracketed to the motor shaft and rotated to the 6 o'clock position, and the number is positive, the plunger is being forced in. This inward movement of the indicator plunger (positive number) tells us that the driven equipment is lower than the motor and will need to be raised. If this piece of equipment is immovable, the motor will need to be lowered. If the number on the indicator is negative, the plunger is moving outward. The negative number means that the driven equipment is higher than the motor. If the driven machine cannot be moved, the motor must be raised at all four mounting pads one-half of the amount recorded on the indicator face. Caution: Make the shim moves at one mounting foot/pad at a time.

31. After the initial shimming process to correct the angular and parallel/offset in the vertical plane is completed, recheck the position of the equipment. Measure the gap with a taper gauge for an angularity check, and sweep the indicator to verify parallel/offset alignment. The numbers should be repeatable and all within .004″ of the target. The target for angularity is the specified coupling gap, and the target for parallel/offset alignment is zero.

FIGURE 13–48a
Indicator setup side view.

Courtesy of Northern Michigan University.

Indicator Readings

Notice that the dial indicator reads the actual centerline to centerline distance as it traverses from the 12 o'clock to 6 o'clock position on the circumferential readings. This is due to the fact that the indicator "sees" both of these distances as it moves from one side to the other.

Side View

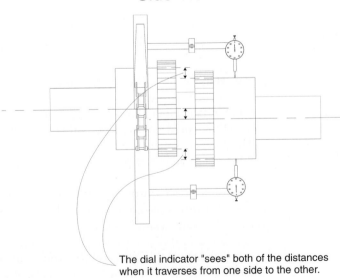

The dial indicator "sees" both of the distances when it traverses from one side to the other.

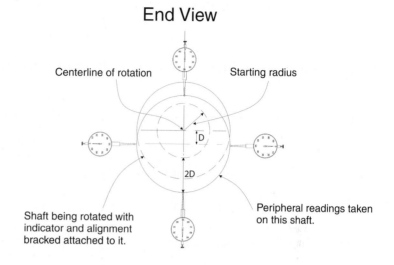

End View

Centerline of rotation

Starting radius

D

2D

Shaft being rotated with
indicator and alignment
bracked attached to it.

Peripheral readings taken
on this shaft.

32. Correct the horizontal plane alignment. Use a straightedge along the sides of the hubs at the 3 and 9 o'clock positions to determine the approximate condition of the alignment. Examine and interpret the horizontal angularity numbers determined with the taper gauge, and the horizontal parallel/offset readings from the indicator sweeps. Position the indicator at 3 o'clock and zero it out. Rotate the indicator on the hub/shaft to the 9 o'clock position and record the amount (rotate both of the shafts simultaneously if at all possible). Adjust the indicator to read one-half of the recorded offset amounts. Use the jacking bolts to twist and turn the movable piece of equipment to equal the gaps at 3 and 9 o'clock. One of the bolts can be left reasonably tight to act as a pivot point. When the indicator reads zero at the 9 o'clock position, the machine is aligned in the horizontal parallel/offset plane. An additional magnetic base indicator, placed with the plunger at a right angle to the shaft axis and parallel to the mounting surface to read off of the movable machine's foot, will help track movement. Make incremental adjustments. Make sure the machine mounting bolts are loose enough to allow for movement of the equipment. Caution: Pushing against a machine that is tightly bolted down with jackscrews will cause the welds on the jacking screw plates to crack. If the bolts are too loose the machine might inadvertently move beyond the required amount. Care must also be taken not to slide off of the shim pack(s) or cause them to pyramid when moving horizontally. Continue to adjust the movable machine until the gap at 3 and 9 o'clock is within .004″ of the target, and the horizontal parallel/offset is within .004″ of zero.

33. Recheck the vertical and horizontal numbers and adjust if necessary.

34. Tighten down the machine mounting bolts to the required specifications. This will cause only a slight change in position if care is taken during the process not to inadvertently twist the machine.

35. Assemble or install the remaining coupling parts carefully. All fasteners should be tightened to their proper torque amounts. Metallic couplings that require lubrication should be filled with the appropriate coupling grease. During the assembly process of grids, sleeves, flexible elements, etc., do not use extreme force that might cause damage to the equipment and move it out of alignment.

Troubleshooting Quick Guide: *Metallic Type*

Problem	Probable Cause	Corrective Action
Hub Breakage	Installation damage	Properly mount hubs with correct bore size. Do not use excessive force.
	Overload/shock load	Verify coupling ratings, increase size. Eliminate shock loads or jam-ups.
	Improper shaft-to-hub fits	Make sure shaft-to-hub fits are correct.
Excessive Vibration & Noise	Misalignment	Align coupling to close tolerances.
	Loose fasteners	Properly tighten all coupling fasteners to correct torque specifications.
	Poor base/foundation conditions of driver & driven equipment	Inspect base & foundation; level & account for soft foot conditions. Repair damaged footings.
Excessive Wear of Meshing Parts	Inadequate lubrication	Properly lubricate all grid, gear, & chain couplings at regular intervals. Check gaskets & seals.
	Entry of contaminants	Clean, lubricate, & replace seals.
	Misalignment	Align coupling to close tolerances.
Grid Fracture	Excessive loads and/or shock	Verify coupling ratings, increase size. Eliminate shock loads or jam-ups.
	Extreme misalignment	Align coupling to close tolerances.
Lubricant Failure	Extreme temperatures	Lubricate with appropriate lubricant for the operating environment.
	Entry of contaminants	Clean, lubricate, & replace seals.
	Inadequate supply	Lubricate on regular basis, pull purge plugs during lubrication process.

Troubleshooting Quick Guide: *Nonmetallic/Elastomeric Type*

Problem	Probable Cause	Corrective Action
Hub Breakage	Installation damage	Properly mount hubs with correct bore size. Do not use excessive force.
	Overload/shock load	Verify coupling ratings, increase size. Eliminate shock loads or jam-ups.
	Improper shaft-to-hub fits	Make sure shaft-to-hub fits are correct.

Troubleshooting Quick Guide: *Nonmetallic/Elastomeric Type*

Problem	Probable Cause	Corrective Action
Excessive Vibration & Noise	Misalignment	Align coupling to close tolerances.
	Loose fasteners	Properly tighten all coupling fasteners to correct torque specifications.
	Poor base/foundation conditions of driver & driven equipment	Inspect base & foundation; level & account for soft foot conditions. Repair damaged footings.
Excessive Wear of Meshing Parts	Misalignment	Align coupling to close tolerances.
	Entry of contaminants	Clean & shield coupling.
Element Failure	Misalignment	Align coupling to close tolerances.
	Overload/shock load	Verify coupling ratings, increase size. Eliminate shock loads or jam-ups.
	Improper installation	Install per specifications.
	Operating environment too severe for element material	Check material specifications & limits.

Student Exercise

Select the high- and low-speed flexible shaft couplings required for the following application (see Figure 13–49). Fill out the data sheet. Use a flexible shaft coupling manufacturers' catalog or software to assist in the process ("FALK"). An electric motor is connected to a gearbox that is driving a paper mill press roll. The drive runs 24 hours per day, 7 days a week. Calculate the speed and the torque for each shaft. The horsepower of the NEMA "B" type motor is 30. The motor speed is 1750 rpm. The gearbox ratio is 47:1. Fill out the required information.

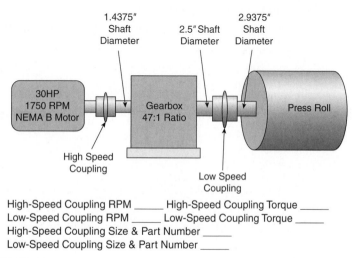

High-Speed Coupling RPM _____ High-Speed Coupling Torque _____
Low-Speed Coupling RPM _____ Low-Speed Coupling Torque _____
High-Speed Coupling Size & Part Number _____
Low-Speed Coupling Size & Part Number _____

FIGURE 13–49

Questions

1. What is a shaft coupling?
2. What is the difference between a flexible and rigid shaft coupling?
3. List the primary purposes of the flexible shaft coupling.
4. What are the two basic categories of flexible shaft couplings?
5. List four types of both categories.
6. What are the advantages and disadvantages of using an elastomeric flexible coupling?
7. What are the advantages and disadvantages of using a metallic-type flexible coupling?
8. What is the difference between a coupling loaded in shear and one that is loaded in compression?
9. What criteria and methods are used in the selection of flexible couplings?
10. What can cause unbalance in a coupling hub? Give several reasons.
11. Describe the difference between clearance and interference fit shaft hubs.
12. What is a universal joint?
13. List seven important installation and maintenance steps for flexible couplings.
14. Why is it important to align flexible couplings?
15. What causes misalignment?
16. What are the different types of misalignment? Describe them.
17. What is soft foot and how is it corrected?
18. List four safety procedures that should be done before working on a coupling.
19. What is shaft and machine thermal growth?
20. Name four alignment methods.
21. Why would one alignment method be chosen over another?

CHAPTER 14

Gear Drives

This chapter covers open and enclosed gear drives. Gearing has been used to transfer power, increase torque, and change speed in machinery for hundreds of years. Gears come in a wide variety of sizes, shapes, and forms. Gear terminology and the different types and forms of open gears used in machinery will be addressed in some detail. Methods used to align ring gears and the subject of gear failure will be examined. Types of enclosed gear drives and their various configurations will be made understandable for the student.

Objectives

Upon completion of this chapter, the student will be able to:
✔ Understand the basic principles of gear function and operation.
✔ Understand and explain common gear terminology.
✔ Identify the various types of gears.
✔ Describe the various methods used to align open ring gears.
✔ Identify the different types of enclosed gear drives.
✔ Have a basic understanding of gear reducer selection procedures.
✔ Understand and demonstrate the basics of proper gear reducer repair and maintenance.

INTRODUCTION

Gears are toothed machine parts, such as a wheel or cylinder, that mesh with another toothed part to transmit power and to change speed or direction. Gears have been used as a form of power transmission since primitive times. Aristotle recorded the use of meshing wooden pegged wheels in use as early as 300 BC. The Romans used gears to drive grinding mills and this practice continued throughout the Middle Ages. With the advent of the steam engine as a prime mover, iron gears began to replace wooden systems. Today's gears are precision machined or molded parts made from a wide variety of metals, plastics, and compounds.

Gears provide a non slip, positive means to transmit power. Changes in speed, shaft rotation, and direction are accomplished with gear drives. They can be used on multiple shafts whose centerlines are parallel or at right angles to each other. Gears

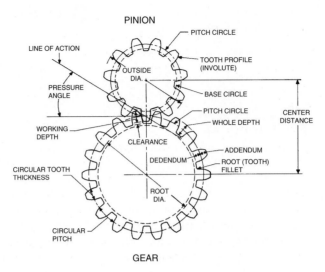

FIGURE 14–1
Pinion and gear set, with terminology.
Courtesy of Boston Gear.

"mesh" by having the faces of several teeth come into contact with each other by a combined sliding and rolling motion. Lubrication is required to minimize wear and seizing that occurs during this action. Gears are a relatively efficient means to transmit high loads at various speeds in a compact area.

When two gears with differing numbers of teeth are used in a drive, the gear with the lesser number of teeth is called the "pinion" and the other is referred to as the "gear." The difference between the gears' size or number of teeth is called the "ratio" (see Figure 14–1). For example, a 10-tooth pinion mating with a 20-tooth gear has a ratio of 2:1. If they both had 10 teeth, the ratio would be 1:1. A ratio difference changes the speed and torque being transmitted by the drive. The direction of shaft rotation is opposite when two gears are meshing. Typically gears are used to decrease the speed and multiply the torque of the system.

Compound gear trains are two gears mounted on a common shaft meshing with other gears. Compound gear trains are typically used in enclosed gear drives ("speed reducers") to minimize space requirements, achieve high ratios and multiply the output torque.

GEAR CONSTRUCTION

The material that the gear is made from depends on the operating conditions, the type of gear, the application, and the cost constraints. Ferrous materials such as steels and cast and alloy irons are widely used. Nonferrous materials such as brass, composite fibers, or plastics are used for special purposes where friction and weight are important considerations.

The shape and size of the gear tooth is critical to how it handles the various loads that are acting on it. Tensile stresses on the contact side of the tooth and compressive stresses on the opposite side cycle continuously on the gear as it functions. The metal composition, size, and shape all are key in this process. Gear geometry is of a complex nature, and gear teeth can be made in a wide variety of shapes and profiles. These geometric shapes are referred to as tooth forms. Because of the variety of gear shapes and systems, identification of a gear for replacement can be difficult without a part number. A "gear gauge" (see Figure 14–2) can be useful in determining the correct gear. Uniform shape or "form" of the tooth is important to transmit power efficiently and smoothly from one gear to another.

The producers of gears use special cutters and hobs on computerized machine tools to generate the selected tooth profile. The involute tooth profile or form is one of the most common. An involute shape is a curve that is traced by a point on a taut line unwinding from a base circle (see Figure 14–3). An involute curve drawn from a large base circle will be less curved than one drawn from a smaller base circle. Thus small gears with an involute profile will have a significant curve as compared to large gears, which will be less curved. A gear rack will have a straight

FIGURE 14–2
Gear gauge used to size gears.
Courtesy of Emerson Power Transmission.

FIGURE 14–3
The commonly used curve that gear teeth match, known as an "involute tooth curve."
Courtesy of Boston Gear.

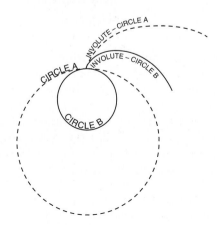

20 Teeth   48 Teeth   Rack

FIGURE 14–4
Involute tooth-shaped gear patterns.
Courtesy of Boston Gear.

profile because it is essentially an infinitely large gear (see Figure 14–4). Two mating gears with a different number of teeth—but the same tooth size—that are made with an involute shape will mesh without binding.

GEAR TERMINOLOGY

Various terms are used in describing the size, shape, and application of gears. Some of the terms are specific to the type of gear (spur, helical, worm, etc.); other terms are generic to all gears. The American Gear Manufacturers Association (AGMA) is composed of member companies that produce gears, enclosed gearing, and an assortment of gear-related products. They establish standards and define various technical terms that are applicable to the design and application of gearing. The following are a few of the significant terms that are common in the gear industry.

Diametral Pitch System The "diametral" pitch (DP) of a gear is the number of teeth in the gear for each inch of pitch diameter. The diametral pitch determines the gear tooth size. For example, a number 12 pitch gear with 12 teeth has a 1″ pitch diameter. Therefore, the higher the diametral pitch number, the smaller the size of the teeth. Common diametral pitch sizes range from 3 DP to 48 DP (see Figure 14–5). Mating gears must be of the same pitch size.

$$\text{Diametral Pitch} = \frac{\text{Number of Teeth}}{\text{Pitch Diameter}}$$

Pitch Diameter and Pitch Circle The pitch diameter is the diameter of the pitch circle (see Figure 14–1). With parallel shaft gears, it can be determined by the ratio of the number of teeth divided by the diametral pitch. The pitch circle is an imaginary circle containing the pitch point. The pitch point is the optimum location where force is applied during the meshing of gear teeth.

Another term used in describing gear sizes is circular pitch. The circular pitch is the distance from a point on a gear tooth to a corresponding point on the next tooth, measured along the arc of the pitch circle.

$$\text{Circular Pitch} = \frac{3.1416}{\text{Diametral Pitch}}$$

FIGURE 14–5

Spur gear pitch size chart showing the relative size of the teeth (not to scale).

Courtesy of Boston Gear.

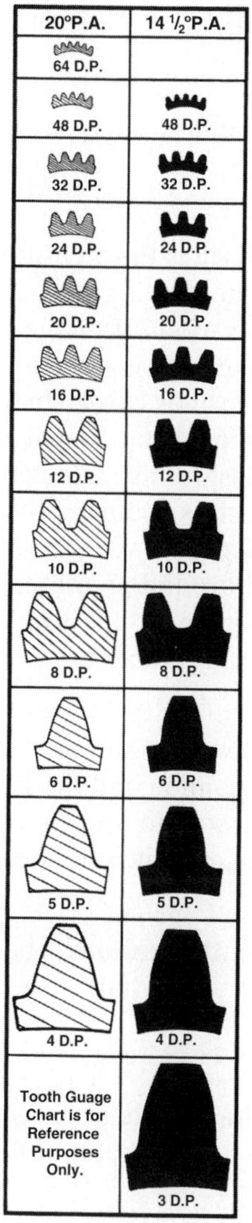

20°P.A.	14 ½°P.A.
64 D.P.	
48 D.P.	48 D.P.
32 D.P.	32 D.P.
24 D.P.	24 D.P.
20 D.P.	20 D.P.
16 D.P.	16 D.P.
12 D.P.	12 D.P.
10 D.P.	10 D.P.
8 D.P.	8 D.P.
6 D.P.	6 D.P.
5 D.P.	5 D.P.
4 D.P.	4 D.P.
Tooth Guage Chart is for Reference Purposes Only.	3 D.P.

Pressure Angle Pressure angle (PA) is generally related to the shape of the gear tooth (see Figure 14–1). It is the angle at which pressure, or power, from the tooth of one gear is passed on to the tooth of another. Standard gears have either a 14-1/2° or 20° pressure angle. All mating gears must have the same pressure angle to prevent binding. The 20° PA is currently a more common pressure angle. It gives the gear tooth a wider base and allows it to carry a higher load capacity. The 14-1/2° PA is an older pressure angle, but it is still in use.

Center Distance On parallel shaft gearing, the center distance is the distance between nonintersecting axes of the mating gears. On right angle nonintersecting shafts, it is the distance from the centerline of one shaft to the centerline of the other. Standard center distances for spur gears can be defined by the following formula:

$$\text{Standard Center Distance} = \frac{\text{Pinion PD} + \text{Gear PD}}{2}$$

Backlash Backlash is the measurable space at the pitch circle between two meshing gears on the nondriving side of the face (see Figure 14–6). The face is the axial length of the gear tooth. If the gear is locked in place, and the pinion moved back and forth, the "play between the gears" is the backlash. All gears require a specified amount of backlash, or "play," based on the diametral pitch and center distances. A change in center distance will cause an increase or decrease in backlash. The amount of backlash is also affected by gear tooth wear or from thermal growth of the gear. Backlash is required to prevent the gears from binding and abnormal wearing, and to allow space for lubrication. Proper backlash aids in preventing noise, heat build-up, and premature failure of the gears. Backlash also is an indication of precise alignment if it is equal from side to side on the gear face. On gearing with nonfixed or adjustable center distances, backlash may be adjusted.

Clearance The working, or root, clearance is the radial distance between the top of a tooth of the pinion and the bottom of a tooth of the gear. A given amount of clearance is necessary to prevent the teeth from bottoming out during operation. It can be used as an alignment indicator when measured side to side.

Addendum and Dedendum The addendum is the height of the tooth above the pitch circle. The dedendum is the depth of the tooth below the pitch circle (see Figure 14–1).

Modular Gearing Modular gearing ("module system") is the system used for sizing metric gears. It is the ratio of the pitch diameter in millimeters to the number of teeth in the gear.

FIGURE 14–6
Gear backlash and gear root clearance between gear teeth.

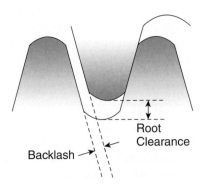

Root
Clearance

Backlash

GEAR TYPES

Spur Gears Spur gears (see Figure 14–7) are one of the first gear types produced and are still in use. The teeth of a spur gear are cut parallel to the shaft axis that they are mounted on. Spur gears are relatively easy to produce because of their straight tooth design. The teeth are usually the involute tooth form. Spur gears may have teeth that are external, internal, or on a rack. External gears have teeth that are cut on the outside perimeter of the gear. Internal gears have teeth cut on the inside circumference of the gear. They are used in combination to form a planetary gear arrangement (Figure 14–8). A rack is a gear with teeth cut on a straight-line bar (Figure 14–9). Rack and pinion arrangements convert rotary motion from a driving pinion to linear motion on the rack. Because of the significant amount of friction generated by the sliding action of spur gears meshing, they are used in low to moderate speed applications. The faster spur gears are driven, the more potential there is for noise and vibration. Simplicity in design and manufacturing make them an economical type of gear.

Helical Gears Helical gears (Figure 14–10) are an improved variation of a spur gear. The teeth are cut on a helix angle to the shaft axis. Helical gears generally have

FIGURE 14–7
Spur gears.
Courtesy of Boston Gear.

FIGURE 14–8
Internal spur gears.
Courtesy of Boston Gear.

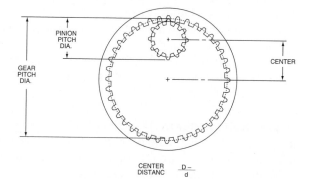

NOTE: The difference in tooth numbers between Gear and Pinion should not be less than 12.

FIGURE 14–9
Spur gear rack.
Courtesy of Boston Gear.

FIGURE 14–10
Helical gears.
Courtesy of Boston Gear.

greater load carrying capacity than spur gears of comparable size because the tooth of the gear is cut at an angle instead of a straight line. This creates a longer tooth—with multiple teeth in mesh—and overlapping of pitches. The result is a quieter, smoother, and more efficient gear that can be run at higher speeds than spur gears. Helical gears are available in either right- or left- "hand" angles. The teeth of a left-hand gear lean up and to the left when the gear is placed on a flat horizontal surface. The teeth of a right-hand gear lean up and to the right when it is placed on a flat horizontal surface. Mating gears operating on parallel shafts require the same helix angle, but opposite hands (Figure 14–11a). Mating gears operating on shafts at right angles require the same helix angle, but the same hand (Figure 14–11b). The disadvantage to helical gearing is the thrust loading that is created as a result of their design. The larger the helix angle, the greater the axial or thrust loads. The correct application and installation of supporting bearings can compensate for this side loading. Helical gears are used extensively in enclosed gear drives.

Double Helical Gears Double helical gears have a set of right- and left-hand helix teeth cut on the same gear and separated by a space (see Figure 14–11c). The helix angle and pitch are the same for both sets of teeth. The axial thrust problem associated with single helical gears is cancelled out by the equal but opposing forces generated by the opposite-hand teeth. Gear drives using double helical gearing are capable of transmitting large loads at significant speeds.

FIGURE 14–11a
Left-hand and right-hand
helical gears.

Courtesy of Boston Gear.

FIGURE 14–11b
Two left-hand helical gears.

Courtesy of Boston Gear.

FIGURE 14–11c
Double helical gear tooth
pattern.

Herringbone or Chevron Gears Herringbone gears are an older type of double helical gearing (see Figure 14–11d). The significant difference is that the right and left helix teeth join in the center to form an apex. Typically they are cut with a high helix angle that allows them to handle extreme loads. The drawback is they are more difficult and expensive to manufacture, as well as being sensitive to misalignment.

FIGURE 14–11d
Herringbone or chevron
helical gear pattern.

The gear must be allowed to float back and forth on the bearings to prevent it from binding in the center. Herringbone gears are rarely encountered in newer machines, but at one time were used in enclosed gear drives with sleeve bearings.

Bevel Gears   Bevel gears (see Figure 14–12) are cut on cones and usually operate on shafts that intersect at 90°. The teeth may be straight or spiral. There are several types of bevel gears: straight, miter, spiral, zerol, and hypoid.

Straight bevel gears are cut on conical blanks with tooth forms that resemble spur gears. If an inexpensive gear is needed to transmit power between right angle shafts, straight bevel gears are a good choice. However, there are speed limitations to straight bevel gears because they tend to be noisy and rough running at high speeds. Ratios are generally 4:1 or less.

Miter bevel gears are mating bevel gears with the same number of teeth (1:1 ratio). There is no change in speed between the driver and driven shafts.

Spiral bevel gears also are cut on cones but the cutter travels in an arc of a circle to produce curved teeth (Figure 14–13). Helical gears are to spur gears what spiral bevel gears are to straight bevel gears. The spiral curvature of the teeth provides a gradual, smooth meshing action. Multiple engagement of teeth allows greater amounts of power to be transferred at increased speeds as compared to straight bevel gears. Spiral bevel gears are used extensively in right-angle speed reducers.

Zerol bevel gears incorporate the design merits of both straight and spiral bevel gears. Although the teeth have a curved design, they have a zero degree spiral angle. They are smooth, quiet running gears that produce no axial thrust.

FIGURE 14–12
Bevel gears.

Courtesy of Emerson Power
Transmission.

FIGURE 14–13
Spiral bevel gears.
Courtesy of Boston Gear.

Hypoid bevel gears have pitch surfaces that are hyperboloids of revolution. The teeth are nonsymmetrical spirals. Hypoid bevel gears are similar to spiral bevel gears, except they operate on shafts that are not intersecting. The pinion shaft is on a different plane than the gear shaft, which allows for a supporting bearing on both ends. The contact action is a combination of rolling and sliding friction. The hypoid design allows stronger and smoother operation than standard spiral bevel gears. They are common in automotive differential drives.

Worm Gears Worm gears (see Figure 14–14a) are a combination of gears that consist of a worm and a worm wheel mounted on two right angle, nonintersecting shafts. Worm gear sets come in a wide variety of pitches and pressure angles. The worm is a cylinder with threads cut on the outside. The worm wheel (gear) has teeth that are milled at an angle that matches the angle of the screw thread on the worm and resembles a helical gear. Generally the worm gear is made of a brass alloy. The worm is always mounted on the input shaft and the worm gear is mounted on the output shaft.

A typical worm is made from steel and then ground or polished. The worm may have one or more threads or "starts" spiraling continuously along its length (see Figure 14–14a and 14b). One, two, three, or four start worms are available. The start or lead can be a right- or a left-hand type. To determine whether it is a right or left hand, place the worm in a vertical position; if the thread spirals up and to the right

FIGURE 14–14a
Worm and worm gears.
Courtesy of Emerson Power
Transmission.

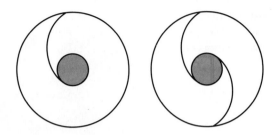

FIGURE 14–14b
Worm gear starts, single and double, viewed from the end.

it is right-handed, and if it spirals up and to the left it is left-handed. The worm "lead" is the distance the thread travels laterally in one revolution of the worm. The "worm lead angle" is the angle formed between the worm thread and a plane perpendicular to the centerline of the worm.

Worm gears are considered one of the smoothest and quietest forms of gearing. This is because contact between the worm and worm gear is purely sliding action. However, this results in energy being converted to heat and a loss of efficiency. The advantages of worm gear sets are many and include high ratios in a compact area, resistance to shock loads, quiet running, and a tendency to be self-locking when back-driven. The ratio of the worm gear set can be determined by dividing the number of teeth on the gear by the number of starts on the worm. For example, a 60-tooth worm gear running with a two-start worm will have a ratio of thirty to one (30:1). At ratios of thirty-one and greater, the drive is considered "self-locking" and cannot be back-driven without extreme force. Under no circumstances should a worm gear drive be used to hold a load or prevent the system from reversing.

OPEN GEAR ALIGNMENT

Open gear alignment is a complex and involved process that should be performed by trained, knowledgeable specialists. Proper procedures pertaining to installation and alignment of large, open gearing is available from manufacturers, and all pertinent manuals should be consulted. This text will cover only the basic alignment procedures and is not meant to replace manufacturer bulletins.

Large ring gears and pinions are used to drive drying kilns and grinding mills. The ring gears are typically made in sections that are bolted to a flange that encompasses the rotating mill. The mill, or kiln, is supported by trunnion bearings or rollers. The pinion connected to a driving mechanism meshes with the gear to turn the entire kiln or mill at a predetermined speed. These gears and pinions are expensive and their successful operation is generally crucial to the continued production of the plant product. That is why care must be taken to install and align the gears correctly to prevent costly failure and downtime.

It is assumed that the proper procedures for the installation of the gear segments to the shell have been followed and the gear teeth are in reasonably good condition. Prior to the installation of the gear segments, the mounting flange should be checked for both axial and radial run-out (TIR). Acceptable limit tolerances are available from the manufacturer upon request. With the mounting flange data, it is possible to position the gear segments selectively to duplicate the "as-manufactured" run-out of the gear. Because the gear is made in sections, it is required that the gear be assembled in exactly the same manner as when the teeth were cut. Careful observance of all match marks during the assembly process is crucial to alignment. It is recommended that the shell temperature be uniform prior to mounting gear segments. All mounting surfaces should be clean and smooth. Numbered stations around the circumference of the gear are helpful in the installation and alignment process. Measurements of the gap between the mounting flange outside diameter and the counter bore of the gear at each station should indicate the amount of needed adjustment. Jackscrews are used for radial adjustment of the gear during erection. These jackscrews can cause distortion of the gear rim and usually must be backed off and pinned or removed prior to operation. On "hinged spring" mounts, all chairs, supports, springs, and other mounting apparatuses must be installed and oriented correctly to ensure that the gear is supported and under tension during operation. All clearance and alignment bolts, sleeves, keys, dowels, and studs must be properly fitted and torqued as needed, regardless of which type of mount is used. This portion of the procedure is intricate and it is imperative that trained professionals do the work.

The methods for gear alignment are numerous and involve various approaches. The type and condition of the gears and available equipment, as well as application variables and personal experience dictate the method best suited for the application. It is important that the proper alignment procedure be selected to match the type of application. For the sake of discussion, methods can be grouped into static or dynamic, and further grouped into backlash, root clearance, contact pattern, and thermal. Stop-action photo and strobe lights can also be used. It is my opinion that several of these methods should be used in conjunction with one another.

Assuming that the ring gear has been installed properly, pinion adjustment is the main focus whether aligning new pinions to new gears, new pinions to old gears, or realigning old pinions to old gears. Prior to the actual alignment process, regardless of which method is chosen, some preliminary steps must be done.

1. Lockout, tagout, and verify that the unit is electrically shut off. Mechanically neutralize any connected loads to prevent shaft movement. Any connected "inching" drives used for small and slow moves must have the necessary safety interlocks and devices attached and working. Follow all plant safety practices applicable to the project. Do not attempt to bypass or shortcut safety procedures!

2. Study the available documentation. Service manuals will list the procedures that are to be followed. Keep accurate records of what was done for future reference. Organize and gather the appropriate tools needed.

3. Clean the pinion and ring gear sections as required. Inspect the teeth for any burrs and remove them without damaging the teeth further.

4. Check the radial run-out of the gear. Dial indicators should be placed in several locations to read off of machined surfaces (see Figure 14–15). There are two

FIGURE 14–15
Locations for
checking radial
run-out.

Courtesy of Falk
Corporation.

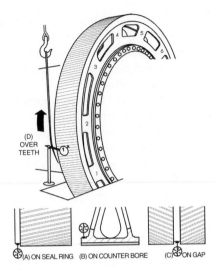

considerations when evaluating the radial runout: The first is the total installed run-out of the gear and the second is the run-out from station to station. The manufacturer establishes these limits based on gear diameter, number of stations, and mounting type. Consult the manufacturer for published tolerances so that the installed values are within the allowable limits.

5. After reviewing the readings, make adjustments to the ring gear by adjusting jackscrews to minimize radial run-out.

6. Check the axial run-out of the gear rim face. Dial indicators should be placed in several locations to read off of machined surfaces (see Figure 14–16). There are two considerations when evaluating the axial run-out: The first is the total installed run-out of the gear and the second is the run-out from station to station. The manufacturer establishes these rim face limits based on gear diameter, number of stations, and mounting type. Consult the manufacturer for published tolerances so that the installed values are within the allowable limits.

FIGURE 14–16
Locations for checking
axial run-out.

Courtesy of Falk Corporation.

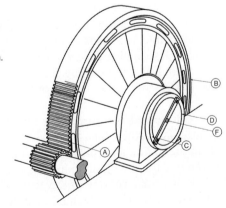

7. After reviewing the readings, make adjustments to the ring gear by shimming between the gear and mounting flange to minimize the axial run-out of the rim face.

8. When the radial and axial run-out values are within acceptable limits, torque all mounting bolts to the manufacturer's specifications.

9. Determine the gear tooth pitch size.

10. If necessary, consider thermal growth and calculate the thermal growth factor based on gear diameter and the temperature rise from ambient. The actual operating temperature of the ring gear varies with the type of enclosure, equipment, lubrication, and product being processed. Manufacturers can provide data and charts to assist in this calculation.

$$\text{Thermal Backlash Factor (inches)} = \text{Center Distance (inches)} \times (\text{Temperature Rise } (^\circ F) / 150{,}000)$$

11. Locate the "high tooth area." The high tooth area is the section of gear teeth with the greatest amount of radial run-out. It should be marked and rolled into mesh with the pinion.

12. Check the condition of the driving pinion bearings and replace them if needed. Set up expansion and nonexpansion bearings and allow the required amounts of float, especially if the gears are the herringbone type. The floating bearing should be centered in its seat. Check for and correct any soft foot conditions under the bearings. Install an approximate .030″ pack of stainless steel shims under the pillow blocks to allow for adjustment. Scribed lines on the foundation and bearing housings can be useful for preliminary positioning of the pinion.

13. Bring the pinion into approximate parallelism of the gear axis by preliminary shimming, shifting, and leveling of the bearing pedestals. Make sure the pinion is centered to the gear so as not to protrude on either side. Set the pinion to the gear using the reference scribe lines so they do not overlap (see Figure 14–17).

14. Fix the gear to prevent rotation and apply a light load to the pinion in the direction of rotation to achieve contact against the mating gear faces.

15. Set up dial indicators on magnetic bases, reading at appropriate locations to measure and track any movements of the pinion.

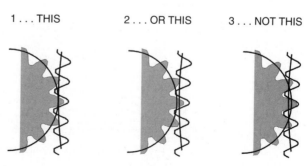

FIGURE 14–17
Gear scribe line check. Do not overlap lines.
Courtesy of Falk Corporation.

16. Determine which method will be used based on the type of mounting: trunnion-mounted shells (grinding mills), low temperature roller-mounted shells, or high temperature roller-mounted shells (kilns).

Backlash Alignment Method

Backlash is the allowed clearance, measured at the pitch line, between the gear tooth and the pinion tooth as they mesh. The backlash method is generally recommended for setting new pinions with new gears. The total required amount of backlash is specified by the manufacturer and is based on the mounting method, pitch size, thermal factor, and wear allowance. The design and maintenance engineers usually predetermine the backlash target number from a combination of calculations and their experience.

The following are general guidelines:

1. If the application is a trunnion-mounted shell or a low temperature roller-mounted shell, the required backlash varies with the pitch, center distance, and temperature. Determine temperatures and calculate the thermal backlash factor from the tables (see Figure 14–18). Determine the diametral pitch backlash factor from the tables provided by the manufacturer (see Figure 14–19). These two numbers added together would be the required amount of backlash.

2. If the assembly is for a high temperature roller-mounted shell, consider a wear allowance factor for shell riding ring and roller wear. As these wear, the teeth of the ring gear and pinion are forced into a tighter mesh. This value must be selected to meet the unique installation requirements of each gear set to ensure that a

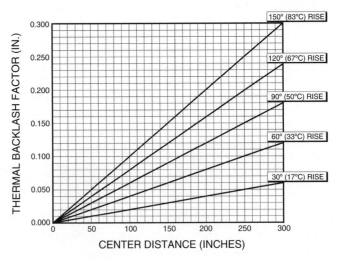

FIGURE 14–18
Thermal backlash factor table based on center distance and temperature rise.
Courtesy of Falk Corporation.

— Diametral Pitch Backlash Factor

Diametral Pitch (See Gear Drawing)	Diameter Pitch Backlash Factor	
	(in)	(mm)
5/8	.055	1,40
3/4	.050	1,27
7/8	.045	1,14
1	.045	1,14
1-1/4	.040	1,02
1-1/2	.040	1,02
1-3/4	.035	0,90
2	.035	0,76

FIGURE 14–19
Diametral pitch backlash factor table.
Courtesy of Falk Corporation.

proper safety margin is maintained. Total these two factors to establish the target backlash requirement.

3. Torque the pinion to the gear in the direction of rotation that it will be operating in, at the station with the highest positive value of radial run-out. Preferably multistation readings should be taken and charted. Under certain restrictive conditions where the mill cannot be rolled, a single set of measurements will have to suffice. This is less than ideal and could lead to improper backlash settings and meshing problems.

4. Take contact and backlash measurements at the pitch line on an undamaged and clean tooth using a feeler gauge blade (see Figure 14–20). Some guidebooks

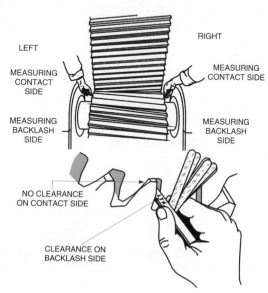

FIGURE 14–20
Taking contact and backlash readings at the pitch line.
Courtesy of Falk Corporation.

recommend using a dial indicator to measure backlash, but that method is unreliable and should be done only to confirm the amounts measured with a feeler gauge blade. Another method is to use a lead wire or sinker, and measure the thickness after it has been crushed. The backlash measurements can be done on the same tooth, or within the same plane on a different tooth. Use consistent measuring points and depths, and mark the locations. Repeat all measurements to establish an average. Record your contact and backlash readings for both sides. The setting of contact and backlash must be done simultaneously.

5. Determine the allowable difference between contact left and contact right. Keep in mind that the term "contact" is somewhat of a misnomer in that you are measuring a gap or space between the mating face of the gear tooth. The ideal "contact" would be zero. The allowable difference is determined by the following formula:

$$A = (F \times R)/D$$

Where A = Allowable difference between contact left and contact right

F = Face width of the gear

R = Allowable gear rim face run-out from manufacturer's tables

D = Gear outside diameter

6. Interpret the readings. The total of contact left plus backlash left should equal the total of contact right and backlash right. A general tolerance guideline is between 0.004″ and 0.010″ difference, but ideally the numbers should be the same. If the totals are not equal, the side with the lowest total is in a position that is closer in center distance than the side with the higher total. Figure 14–21 illustrates gears misaligned in the plane of centers. If the contact left and contact right numbers are not equal, one end of the pinion is at a different elevation than the other. Figure 14–22 illustrates gears misaligned at right angles to the plane of centers. The readings should be as close to zero as possible and certainly within the calculated allowable difference.

7. Set up dial indicators on the bearings and shafts to measure and track movements in all directions. Movements on one side can show up as a change on the other side, depending on the tightness of the hold-down bolts.

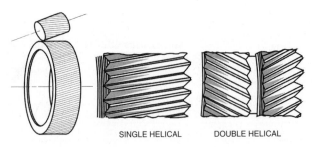

SINGLE HELICAL   DOUBLE HELICAL

FIGURE 14–21
Gears misaligned in the plane of centers.
Courtesy of Falk Corporation.

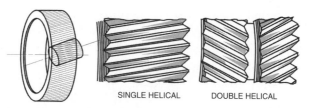

SINGLE HELICAL   DOUBLE HELICAL

FIGURE 14–22
Gears misaligned at right angles to the plane of centers.
Courtesy of Falk Corporation.

8. Adjust the bearing pedestals by shifting and shimming until the difference between total backlash left and right is at the target and within the specifications and tolerances provided by the manufacturer and engineers. Attempt to have zero contact (gap) on the driving side of the tooth faces. Because of varying amounts of radial run-out from station to station on the gear, move the pinion to the optimum position to correct the contact and backlash for the entire assembly. A secondary set of readings and adjustment are advisable.

9. Take into account the change in shape and deflection of the shell when the mill is "charged" and under load. This can be a difficult and complex problem. Qualified engineers can determine this factor based on loads and their experience.

10. Check the tooth mesh for uniform contact to assure accurate gear alignment. Apply a thin, smooth layer of a contact marking medium to five or six clean pinion teeth. Cover the entire tooth profile across the face width and length. Roll the pinion back and forth through the mesh several times. Examine the contact pattern on the gear teeth. The pattern of contact should be across 80% of the gear face width and 50% of the tooth profile height. This process should be repeated at a minimum of three more equally spaced positions on the ring gear. It may also be used as a preliminary indication of present alignment prior to beginning the alignment process. For bi-directional rotating mills, data must be collected for both directions.

11. All connected drive components must be realigned to compensate for any changes in position of the pinion assembly.

Root Clearance Method

Root clearance is defined as the distance between the tips of the pinion teeth and the roots of the ring gear teeth. The root distance should be measured at the center of the gear tooth fillet radius. The contact will still need to be measured at the pitch line between the gear tooth and the pinion tooth as they mesh. The root clearance method is generally recommended for setting new pinions with old gears (see Figure 14–23). The required amount of root clearance is specified by the manufacturer and is based on the condition of the teeth faces, mounting method, pitch size, thermal factor, and wear allowance. The design and maintenance engineers usually predetermine the root clearance target number from a combination of calculations and their experience.

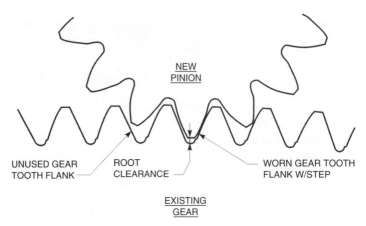

FIGURE 14–23
Root clearance for new pinion with used gear.
Courtesy of Falk Corporation.

The following are general guidelines:

1. If a new pinion is going to mate with a worn gear tooth face, the root clearance on the existing set must be taken before the old pinion is removed. This value is the root required for the new installation. This will ensure that the new pinion will not engage any steps that have been worn into the face of the used gear teeth. As the gear wears while in service, "lands" or steps form on the base of the gear teeth. The new pinion teeth must not be allowed to contact these steps, or premature failure will result.

2. If a new pinion is to be run on unused tooth faces of an old gear, on trunnion mounted shells or on low temperature roller-mounted shells, two factors must be considered. The thermal backlash factor, taken from the graph in Figure 14–18, allows for normal thermal expansion of the gear set during operation. The tooth form root clearance factor based on the tooth size and form must also be considered (see Figure 14–24). The root clearance requirement can be calculated by adding the two together.

3. On pinions mating with unused gear tooth faces on high temperature roller-mounted shells, the tooth form root clearance factor must be added to the wear allowance factor.

4. Torque the pinion to the gear in the direction of rotation that it will be operating in, at the station with the highest positive value of radial run-out. Preferably multistation readings should be taken and charted. Under certain restrictive conditions where the mill cannot be rolled, a single set of measurements will have to suffice. This is less than ideal and could lead to improper clearance and meshing problems.

5. Take contact measurements at the pitch line on an undamaged clean tooth using a feeler gauge blade. These measurements can be done on the same tooth, or within the same plane on a different tooth. Use consistent measuring points and depths, marking the locations. Repeat all measurements to establish an average.

Tooth Form	Approximate Tooth Height		Tooth Form Root Clearance Factor (Excluding Thermal Allowance)	
	(in)	(mm)	(in)	(mm)
1 DP, Mill	1.925	48,9	.222	6,64
1-1/4 DP, Mill	1.675	42,5	.204	5,18
1-1/2 DP, Mill	1.395	35,4	.173	4,39
3/4 DP, FHD	3.025	76,8	.356	9,04
7/8 DP, FHD	2.585	65,7	.311	7,90
7/8 DP, FHD-LA	2.710	68,8	.196	4,98
1 DP, FHD	2.270	54,7	.272	6,91
1-1/4 DP, FHD	1.820	42,2	.220	5,59
1-1/2 DP, FHD	1.520	38,6	.186	4,72
1-1/4 DP, UFD	1.590	40,4	.204	5,18
1-1/2 DP, UFD	1.330	33,8	.172	4,37
1-3/4 DP, UFD	1.140	29,0	.150	3,81
2 DP, UFD	1.000	25,4	.129	3,28

FIGURE 14–24
Tooth form root clearance factor table.
Courtesy of Falk Corporation.

Record your contact readings for both sides. The setting of contact and root clearance must be done simultaneously.

6. Determine the allowable difference between contact left and contact right. Keep in mind that the term "contact" is somewhat of a misnomer in that you are measuring a gap or space between the mating face of the gear tooth. The ideal "contact" would be zero. The allowable difference is determined by the following formula:

$$A = (F \times R)/D$$

Where A = Allowable difference between contact left and contact right

F = Face width of the gear

R = Allowable gear rim face run-out from manufacturer's tables

D = Gear outside diameter

7. Interpret the readings. The clearance left should equal the clearance right. A general tolerance guideline is between 0.004″ and 0.010″ difference, but ideally the numbers should be the same. If the root clearance left measurements are consistently above or below the root clearance right, the side with the lowest total is in a position that is closer in center distance than the side with the higher total. Figure 14–21 illustrates gears misaligned in the plane of centers. If the contact left and contact right numbers are not equal, one end of the pinion is at a different elevation than the other. Figure 14–22 illustrates gears misaligned at right angles to the plane of centers. The readings should be as close to zero as possible and certainly within the calculated allowable difference.

8. Set up dial indicators on the bearings and shafts to measure and track movements in all directions. Movements on one side can show up as a change on the other side, depending on the tightness of the hold-down bolts.

9. Adjust the bearing pedestals by shifting and shimming until the difference between root clearance left and right is at the target and within the specifications and tolerances provided by the manufacturer and engineers. Attempt to have zero contact (gap) on the driving side of the tooth faces. Because of varying amounts of radial run-out from station to station on the gear, move the pinion to the optimum position to correct the contact and root clearance for the entire assembly. A secondary set of readings and adjustment are advisable.

10. Take into account the change in shape and deflection of the shell when the mill is "charged" and under load. This can be a difficult and complex problem that is best addressed by qualified plant and manufacturer engineers.

11. Check the tooth mesh for uniform contact to assure accurate gear alignment. Apply a thin, smooth layer of a contact marking medium to five or six clean pinion teeth. Cover the entire tooth profile across the face width and length. Roll the pinion back and forth through the mesh several times. Examine the contact pattern on the gear teeth. The pattern of contact should be across 80% of the gear face width and 50% of the tooth profile height. This process should be repeated at a minimum of three more equally spaced positions on the ring gear. It may also be used as a preliminary indication of present alignment prior to beginning the alignment process. For bi-directional rotating mills, data must be collected for both directions.

12. All connected drive components must be realigned to compensate for any changes in position of the pinion assembly.

Dynamic Alignment Methods

The most accurate measurement of good alignment and proper tooth contact should be done under normal operating conditions. This is because variables such as temperature, loads, deflections, and bearing/shaft movement come into play while it is running. The initial alignment of the pinion to gear—regardless of which method is chosen—is important, but the dynamic check of contact is the most accurate reflection of how the load is being distributed across the gear teeth. If the load is distributed unevenly, the gear will fail prematurely.

There are at least two dynamic methods: contact patterns using dye and temperature readings of the pinion.

Check the contact pattern of the teeth by painting them with layout dye. Clean the teeth of the gear and pinion. Paint three teeth at six equally spaced positions around the gear with a layout dye of some type. Install all of the gear guards and safety enclosures. Lubricate the teeth of the gear with the correct type of lubricant. Operate the equipment for approximately 8 hours at a minimum of 50% of full load. After the gear has been stopped and the prime mover locked out, remove the lubricant from the teeth but do not remove the layout dye from the teeth. If 95% of the layout dye has been worn off the face width of the gear teeth at each station, the proper alignment between the pinion and gear has been achieved. Excessive axial run-out on the gear will cause the contact pattern to vary from one side of the gear face to the other in a sinusoidal pattern. If a balanced contact of 95% of the gear face is not evident, realignment will be necessary. Evaluate the contact patterns to determine the required movement of the pinion. Misalignment in the plane of centers and misalignment at right angles to the plane of centers must be corrected before the equipment is operated at normal operating speeds and loads.

Temperature readings across the face of the pinion are a good indicator of the contact pattern of the gear and pinion teeth. A pyrometer or infrared thermometer of some type may be used to determine the temperature and thus the amount of contact. Both instruments should be calibrated and accurate, and capable of measuring up to 400°F (200°C). Misaligned gears produce nonuniform load distribution across the face of the gearing, which results in higher operating temperatures at the point of highest load. Equal temperatures across the face of the pinion, end to end, indicate a uniformly distributed load. To determine accurate temperatures using a contact pyrometer, the gear must run for least 24 hours and readings must be taken immediately before any significant cooling can occur. The temperature readings are taken by a contact pyrometer at the pitch line on the loaded flank of the pinion tooth at the pitch line. Readings should be taken through the lubricant at approximately five evenly spaced points. Measuring point number one should be on the nondrive end of the pinion. When using an infrared radiation thermometer, position it approximately 3 feet (one meter) from the mesh of the gear set. Aim the device at five positions across the face while the equipment is operating. The instrument must read the measurement while perpendicular to the axis of the pinion and aimed at the load flank along the pitch line (see Figure 14–25). Another method requires that the readings be taken at the shaft axis line. The differences between these two methods relate to concerns regarding "emissivity." Emissivity is a property of a material that describes its ability to radiate energy by comparing it to a different black-colored surface at the same temperature. Caked-on layers of dark-colored open gear lube on the pinion teeth can cause problems in obtaining accurate readings. Regardless of which method is used, graph the gradient showing both the temperature and location. If the gradient is less than 15°F (8°C), the gear set is aligned. Figure 14–26 shows a solid black line representing a temperature gradient that reflects optimum gear/pinion alignment. For bi-directional rotation applications, operating data must be collected for both directions of rotation. This data can be sent to the manufacturers of the gear set for alignment recommendations.

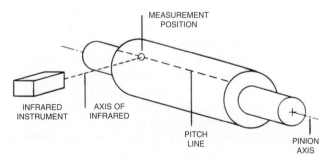

FIGURE 14–25
Measuring positions for temperature readings on a pinion.
Courtesy of Falk Corporation.

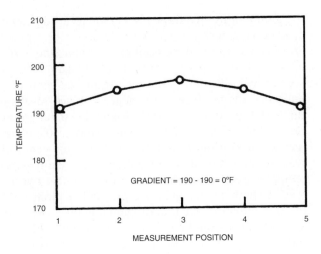

FIGURE 14–26
Acceptable pinion temperature gradient.
Courtesy of Falk Corporation.

GEAR WEAR AND FAILURE ANALYSIS

A gear's function is to transmit torque and rotary motion between the prime mover and driven equipment. Inherent in the use of gears for power transmission are certain levels of noise, temperature, and vibration. When these operating characteristics exceed normal limits, the gear train should be examined for causes and effects related to the problem. Periodic inspection of the gear teeth for visible signs of abnormal wear fatigue or breakage will assist the technician in the diagnostic process and in forming a remedial plan.

Premature failure can usually be traced to one or more of the following: load, environment, installation, and maintenance. Continuous overloads, momentary shock loads, and vibratory loads on the system must be analyzed to see what effect they are having on the system. Excessive loads transferred by the gear teeth result in both subsurface shearing forces and tooth surface distress, resulting in failure. Detrimental operating environmental conditions affecting the performance of the drive—such as extreme ambient temperatures, abrasive contaminants, or corrosives—will quickly destroy the lubricant's ability to cool and provide a film. Abrasive particles, moisture, and chemicals react with the metal rolling elements to corrode and abrade their surfaces. A gear drive that has not been properly installed, aligned, and maintained will have a shortened life. If a load is nonuniformly transmitted across the gear teeth because of misalignment, tooth fracture and surface distress will result. Condition monitoring tools such as oil analysis, sound, temperature, and vibration readings should be included in the diagnostic package. Examination of contact patterns on the surface of the gear teeth, along with other drive monitoring data, will often reveal what role any of the above-mentioned failure causes played in the failure.

Gear Distress and Failure Modes

There are four basic categories for gear failure: surface fatigue (pitting), wear, plastic flow, and breakage.

Surface Fatigue Surface fatigue failure is caused by repeated surface and subsurface stresses beyond the endurance limit of the material. It is characterized by the removal of metal and formation of pits on the gear tooth contact area. The degree of pitting can depend on the hardness of the gear and the number of stress cycles. These cavities or pits may be one of three types: initial (corrective), destructive, or normal. Initial pitting is caused by localized areas of high stress due to loading on the gear tooth high spots (see Figure 14–27). Corrective pitting develops in a short time along a narrow band at the pitch line. As the high contact spots are progressively removed, the pitting stops; thus the term "corrective" applies. Some initial pitting of through-hardened gears is considered normal. If the gearing is to run in a highly loaded application, it is recommended that a reduced load run-in period take place.

Destructive pitting usually starts below the pitch line in the dedendum area of the tooth. Destructive pitting can appear to begin as initial pitting, but as time elapses the severity increases. The cavities are generally larger in diameter and more in quantity (see Figure 14–28). Eventually a bending fatigue crack can originate from one of the larger pits and cause a tooth to break off. System overloads or the tooth hardness not being within the specified hardness are usually the cause of this type of pitting.

Spalling is the term used to describe extensive deep pits over a broad area of the tooth (see Figure 14–29). It is characterized by large areas of material breaking off the contact face. Spalling can be caused by high contact stress, subsurface defects (inclusions), or improper heat treatment. Case crushing is associated with highly

FIGURE 14–27
Corrective pitting.
Courtesy of Falk Corporation.

FIGURE 14–28
Destructive pitting.
Courtesy of Falk Corporation.

FIGURE 14–29
Gear spalling.

Courtesy of Falk
Corporation.

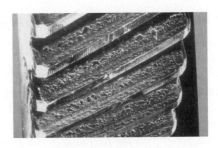

loaded case-hardened gears. Longitudinal cracks on the tooth surface appear which eventually break away (see Figure 14–30). They originate in the case core and make their way to the surface. The cause relates to the subsurface shearing forces and the strength of the material.

Wear Wear is a general term used to describe a somewhat uniform loss of material from the contacting surface of the gear tooth. There are varying degrees of wear. Polishing or light wear is a very slow, normal process of the wearing away of any surface asperities (see Figure 14–31). The cause is usually due to metal-to-metal contact from thin oil films. Moderate or normal wear is a more advanced wear over a long period of time. The degree of wear depends on the lube film, surface hardness, and the amount of contaminants in the oil. Excessive or destructive wear is surface destruction that has changed the profile of the gear tooth shape (see Figure 14–32). The teeth become pointed and have steps in the dedendum portion of the tooth.

FIGURE 14–30
Case crushing/
cracked gear.

Courtesy of Falk
Corporation.

FIGURE 14–31
Light wear.

Courtesy of Falk
Corporation.

FIGURE 14–32
Destructive
wear.

Courtesy of Falk
Corporation.

FIGURE 14–33
Abrasive wear.

Courtesy of Falk Corporation.

Major wear types are abrasive, corrosive, and scoring. Abrasive wear is from hard particle contaminants rolling and sliding across the gear tooth surface under pressure (see Figure 14–33). The source of the particles may be dirt, metallic debris from the gear and bearing systems, or foreign contaminants infiltrating the lubricant. Corrosive wear is deterioration of the surface due to chemical or moisture action. It is caused by active agents in the lubricant, such as acids or water, reacting with the gear steel. Scoring wear is from welding and tearing of gear surface metal (see Figure 14–34). High temperatures, pressures, and marginal oil films produce a galling and seizing effect. This failure mode tends to spiral out of control as temperatures rise and lubricant film deteriorates further. Using the proper viscosity for the application and filtering the lubricating fluid will help control wear.

Plastic Flow Plastic flow is the cold working of tooth surfaces caused by high contact stresses and the rolling and sliding action of the mesh (see Figure 14–35). It is a

FIGURE 14–34
Scoring wear.
Courtesy of Falk
Corporation.

FIGURE 14–35
Plastic flow.
Courtesy of Falk
Corporation.

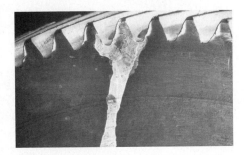

surface deformation associated with softer gear materials or heavily loaded gears. The tips or ends of the gear teeth will have a finned appearance or be rounded off.

Cold flow, rippling, and ridging are all variations of plastic flow. Cold flow takes place when the surface or subsurface material has been pushed over the tips or ends of the gear teeth, producing fills (see Figure 14–35). It is sometimes called rolling. Rippling is a wave-like formation at right angles to the direction of sliding or motion (see Figure 14–36). Ridging is the formation of deep ridges by either wear or plastic flow of material (see Figure 14–37). Contact stresses, loading cycles, material hardness, speeds, and lubricant film thickness all interact to form these various failure modes.

Breakage Breakage is the ultimate failure mode where a tooth breaks off. Fatigue fracture is the most common mode (see Figure 14–38). When repeated stress cycles exceed the endurance strength of the material at critical locations, fatigue cracks will develop. Typically there is a point of origin: The faces of these fractures usually have a series of "beach marks" caused by a progressive advance of the crack with each load cycle. These beach marks are a series of somewhat concentric lines that start at the point of origin and are of increasing diameter, similar to those found on the

FIGURE 14–36
Rippling.
Courtesy of Falk Corporation.

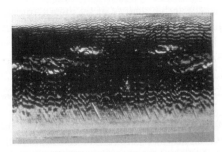

FIGURE 14–37
Ridging.
Courtesy of Falk Corporation.

FIGURE 14–38
Tooth fracture/breakage
attributed to shock load.
Courtesy of Falk Corporation.

surface of a clam shell. Stress risers—such as those created by inclusions, manufacturing surface defects, and heat-treat cracks—will create or aggravate this condition. Impact fractures from single, extreme overloads will have a coarse, fibrous, and torn appearance. Pitting associated with fracture originates from areas that have become severely pitted. The damaged surface area acts as a stress riser and crack origin. Tooth chipping is a fracture mode where the top of the tooth breaks away. Tooth flaws or large foreign objects passing through mesh can cause tooth fractures that do not originate in the tooth root. Severe misalignment or incorrect backlash and clearance settings can also produce chipped teeth.

Several of the above discussed failure modes and types typically combine to result in loss of satisfactory gear life. The overall load, application variables, and gearing installation/maintenance must be considered by experienced experts to analyze the root cause of a failure.

ENCLOSED GEAR DRIVES

Enclosed gear drives, known also as gearboxes or speed reducers, are gears that are mounted on shafts supported by bearings, and enclosed and sealed in a case. They can be spur, internal, bevel, worm, or helical gears. The bearings supporting the shafts and gears are usually radial antifriction types. Enclosed gear drives provide speed reduction and torque multiplication in a compact space within a variety of different shaft arrangements. The fact that they are self-contained, efficient, and safe—and an economical means of power transmission—makes them a popular choice for many applications.

Enclosed Gear Terminology

Certain terms are common to both open and enclosed gearing. The following are additional terms pertaining exclusively to enclosed gearing.

Shaft Configuration The number of shafts and their locations is referred to as the shaft configuration or shaft assembly. Enclosed gear drives (gearbox/speed reducer) can have input shafts that are inline, right angle, or parallel in a vertical or horizontal plane. Multiple input and output shafts are available for right angle and parallel drives. Gearboxes might have a single input shaft and two output shafts or vice versa. Any gearbox that has the input shaft at 90° to the output shaft is called a "right-angle" gearbox. Manufacturers use an alphanumeric coding system to designate the shaft configuration for their units.

Hand of Drive On a gear reducer, the location of the output shaft relative to its input shaft is called the hand of drive. When facing the input shaft, the relative location of the output shaft determines the hand of the drive.

Mounting Method The mounting method of the gearbox is determined by how it is connected to the prime mover and the driven equipment. The two most basic methods are foot mounted (base) (see Figure 14–39) and shaft mounted (see Figure 14–40).

FIGURE 14–40
Shaft-mounted gearbox showing twin tapered bushings.
Courtesy of Rockwell Automation.

FIGURE 14–39
Foot- or base-mounted helical gearbox.
Courtesy of Rockwell Automation.

FIGURE 14–41
Gearbox with top-mounted motor connected by a V-belt drive.
Courtesy of Rockwell Automation.

FIGURE 14–42
Gearbox with scoop-mount motor bracket.
Courtesy of Rockwell Automation.

A gearbox housing with feet (mounting pads) that are bolted to the floor or a base is considered a foot-mounted unit. It may be connected to the prime mover or driven equipment by a chain or belt drive. It can also be direct-coupled to the motor and driven equipment with flexible couplings. The bracket or base the motor is mounted to is referred to as the "motor mount." Some motor mounts are attached to the top of the reducer to allow a belt drive connection (see Figure 14–41). Other motor mounts are attached directly to the reducer supporting the motor that is direct-coupled to the unit. This type is referred to as a "scoop mount" (see Figure 14–42).

A gear motor is a speed reducer with an integrally connected drive motor. The two are bolted together and considered a unit. The motor shaft can fit into a hollow input shaft on the gearbox, or it may be closed-coupled in an enclosing bracket. A variation of the gear motor is the "C-faced" speed reducer (see Figure 14–43). Standard National Electric Manufacturers Association (NEMA) motors with a C-face bolt pattern are attached to C-faced gearboxes with the same bolting dimensions. This

FIGURE 14–43
A gearbox and connected motor—a "Gear motor."
Courtesy of Rockwell Automation.

eliminates the need for couplings, saves space, and makes interchangeability between products easy.

Hollow output shaft gearboxes are mounted directly onto and supported by the driven shaft and bearings. A tie-rod or "torque-arm" attached to the unit and connected to a stationary rigid surface holds the reducer in place and prevents it from spinning. Hollow output shafts, used for direct-mounting onto a driven shaft, eliminate the need for additional supporting structure. These gearboxes are referred to as "shaft mounts."

Standard gearboxes are shipped from the factory to be mounted in the horizontal plane. If the unit is to be tilted to the front or side more than 10°, the location of breathers and plugs might need to be changed. This also applies to units with input and output shafts in the vertical plane. Lubrication levels will need to be adjusted to ensure all bearings are receiving a flow of oil. Special seals might be needed to keep the unit from leaking. Any gearboxes that are to be mounted in a vertical and off-angle plane should be ordered to meet that need.

Efficiency   The power delivered out of the gearbox is always less than that put in because some of the power is converted to heat from the friction of gears and bearings. This loss is measurable and is determined by the type and ratio of the gearing. The efficiency of the drive must be taken into consideration when selecting the proper gearbox for a specific application. Efficiency in percent is expressed as the power out divided by the power in, multiplied by 100.

$$\% \text{ Efficiency} = \frac{\text{Power output}}{\text{Power input}} \times 100$$

Mechanical Rating   The maximum power or torque a speed reducer can transmit, based on the strength of its parts, is its mechanical rating. The limits depend on the mechanical capacity of the gears, shafts, bearings, and case. Reducers are made with a safety margin of 200 to 300% of their mechanical rating to prevent failure during start-up and brief overloads.

Thermal Rating   The thermal rating of a gearbox is the maximum power or torque that can be transmitted continuously through the drive, based on its ability to dissipate the heat generated internally. If the thermal rating of a gearbox is lower than the required continuous output torque and power, a means of cooling must be provided or a larger unit should be selected. The heat generated within the unit is absorbed by the lubricant and transferred to the housing. The surface area of the gearbox case (housing) acts to radiate the internally generated heat to the surrounding atmosphere. Cooling mechanisms such as fans, heat exchangers, and coolant systems will provide cooling to the gear lubricant and case thereby raising the thermal capacity of the unit.

Horsepower and Torque Capacity   The horsepower capacity of a gearbox is its ability to transmit the horsepower of the prime mover. The input horsepower is the amount of power applied to the input shaft of a gearbox from the prime mover. It represents the maximum amount of power that can be safely transmitted by the gearbox. The input horsepower is one of the criteria used in selecting a reducer from published ratings tables. The output horsepower is the power available at the output shaft of a reducer. The output horsepower rating of a reducer is lower than the input horsepower because of efficiency losses. These figures are available from selection tables provided by drive manufacturers. The horsepower/torque ratings tables are product and type specific and are based on the unit operating at a given set of conditions. The torque rating of a gearbox is its capacity to transmit torque at a set gear ratio and input speed. The torque rating/capacity of a gearbox is also used in the selection or sizing of the unit. For a given ratio, torque capacity increases as shaft input speed decreases.

Ratio/Reduction The gearbox ratio is the input speed divided by the output speed. The ratio is determined by the number of teeth on each meshing pinion and gear set, as well as the number of sets in each unit. Standard ratios have been established by AGMA and are listed in manufacturer publications. Most common ratios are nominally 5:1, 10:1, 15:1, 20:1, 30:1, 40:1 and increase by 10 until they reach 100:1.

Gearboxes may have multiple sets of gears within their housing (see Figure 14–44). These sets are also referred to as stages or reductions. If the unit has only one set of mating gears, it is called a single reduction unit. Single reduction units have an input shaft (high-speed shaft) with a high-speed pinion mounted on it that meshes with a low-speed gear mounted on the output shaft (low-speed shaft). A gearbox with two sets (stages) is referred to as a double reduction unit. Double reduction (two stage) units have a high-speed pinion mounted on the input shaft that meshes with a high-speed gear mounted on the low-speed pinion shaft. The shaft is usually named for the pinion on it: The low-speed pinion meshes with the low-speed gear, which is mounted on the output shaft (low-speed shaft). A unit with three sets is a triple reduction gearbox. A triple reduction unit has an intermediate set of gears. The high-speed gear is mounted on the intermediate pinion shaft. The intermediate pinion meshes with the intermediate gear, which is mounted on the low-speed pinion shaft. That low-speed pinion meshes with the low-speed gear on the output shaft. Multiple set reductions, an example of compound gearing, give higher ratios of reduction with each additional set. Generally the more reductions and higher the ratio, the lower the efficiency of the drive.

Most gearboxes are used to decrease the speed delivered to the driven equipment, thus the term "speed reducer." Because of the torque/speed relationship, the torque delivered is inversely proportionate to the speed ratio change. For example, an input speed of 1750 rpm is delivered to a gearbox with a 10:1 ratio. The output speed will be 1/10 of 1750 rpm or 175 rpm. If the torque delivered to the input shaft is 360 ft.-lbs., it will be multiplied approximately ten times, delivering 3600 ft.-lbs. of torque. (Of course efficiencies would have to be taken into consideration.) Thus, not only is a gearbox a speed reducer, it is also a torque multiplier.

Overhung Load An overhung load (OHL) is a right angle load applied to the input and output shafts of the gearbox through a connected component (see Power Transmission Fundamentals). These loads or forces are created by tensions in belt

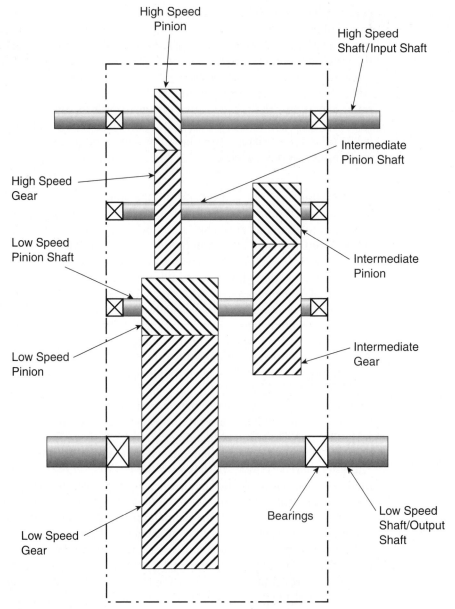

FIGURE 14–44
Triple reduction gearbox configuration listing the names of shafts, pinions, and gears.

and chain drives, or through misaligned couplings or other connecting devices mounted on the shafts of the reducer. The size, mounting location, weights, and tension forces applied by the connecting component also must be considered when calculating OHL. The OHL rating of the gearbox is dictated by the size of the bearings and shafts, and the housing strength. Manufacturers list the overhung load capacities and diameter limits for their units in cataloged tables. Generally, the ratings of a unit represent the maximum force that can be applied to the shaft at a right angle that is one shaft diameter's distance from the outside of the housing. Excessive overhung load forces induced to the bearings and shafts can cause a loss of service life or lead to catastrophic failure.

Thrust Loads Thrust loads are forces imposed on the gearbox shafts parallel to their axes. Internal thrust loads created by the gear types (helical) used in the unit are normally not a concern. The design and selection of supporting bearings used in the reducer can handle these loads. External thrust loads created by driven mechanisms, such as those from fans and mixers or due to thermal expansion, must be considered. The direction and magnitude of the load must be calculated and the results checked against the unit's limits. The manufacturer's published thrust load limits must not be exceeded to prevent bearing failure. When excessive thrust and overhung loads are applied simultaneously, it is recommended that the original drive design engineers be consulted.

Service Factor Part of the selection procedure for gearboxes used in a specific application is to make use of a service factor. The service factor is a number applied to the horsepower or load to modify it to reflect the type of service the unit will see. Essentially the gearbox is oversized from a capacity standpoint to ensure an acceptable service life (see Power Transmission Fundamentals). The service factors established by AGMA vary with the type of service, prime mover, and hours of service. Some manufacturers use a class of service (class 1, 2, or 3) in their selection tables. Class of service is another form of service factor.

Backstops Backstops are mechanisms installed as part of the gearbox to prevent reverse rotation of the shafts. A backstop essentially is an overrunning clutch that allows the transmission of torque and rotation in one direction (see Clutch and Brake chapter). On lower to medium horsepower applications, the backstop will be mounted internally on the end of the high-speed shaft opposite the input connection. External versions are available on applications that have significant amounts of torque being transferred.

Selection of Enclosed Gear Drives

The selection of gearboxes can be done by using manufacturer software or catalogs. All dimensional specifications and ratings are available upon request from a variety of companies that produce enclosed gear drives. The following considerations are general and should be accompanied with product literature.

- Establish all of the application requirements of power, torque, loads, prime mover type, and the driven equipment.
- Calculate the required ratio based on the input and output speeds. Consider if auxiliary drives like belts or chain are needed for subratios.

- Analyze the installation location and space. Note all limitations, weight restrictions, special environmental considerations, and ambient temperatures.
- Determine the needed mounting position and shaft configuration suited to the application. Horizontal or vertical positioning, foot mount, motor mount, and relative location of input and output shafts all need to be considered.
- Choose the type of gearbox and gears that best suit the application and budget constraints.
- Compute all thrust and OHL loads to check against the published limits of the gearbox.
- Analyze the overall system for momentary or peak overloads and potential jam-ups. Additional clutching or soft-start mechanisms might be needed.
- Consider the need to prevent load reversal by use of a backstop. Incline conveyors require rotation in a single direction to prevent the system from unloading backwards when stopped.
- Select the correct service factor from AGMA tables. Consider prime mover, driven equipment, operating time, and type of service.
- Refer to manufacturer torque and horsepower ratings/selection tables to select a gearbox. Make sure the unit selected meets or exceeds the requirements.
- Verify that the selected unit will fit or is a "drop-in" replacement by checking dimensions in the catalog. Important dimensions to cross-check are the overall height, length, shaft sizes, and bolt-down patterns.
- Anticipate maintenance requirements regarding lubrication, replacement parts, and normal upkeep. Special lubricant or sealing mechanisms might be required to operate in a particular environment.

ENCLOSED GEARBOX TYPES

There are many types of enclosed gear drives available in a wide range of mounting methods and shaft configurations. Five of the most common are concentric, worm gear, parallel shaft, shaft mount, and planetary.

Concentric Gearboxes Concentric gearboxes, also known as in-line speed reducers, have the input and output shafts arranged concentrically (see Figure 14–45). Generally the two shafts are in line. Concentric gearboxes are available with a variety of motor-mount and shaft arrangement options. All concentric gearboxes are made with mounting pads (feet) at the base. Standard concentric speed reducers are furnished in single reduction to quadruple reduction. Ratios range from 1.5:1 with single-stage boxes, to greater than 500:1 on multiple-stage units. Helical and planetary gearing are two of the most common types of gears used in concentric reducers.

Worm Gearboxes Worm gearboxes use a worm and worm gear to achieve large reduction of speed in a compact economical package (see Figure 14–46). Most of these drives are put to use in fractional and low horsepower applications such as packaging

conveyors. The great advantages, besides purchase cost, are their compactness and the variety of mounting methods and shaft arrangements available. The box may be foot or shaft mounted, and the shaft configurations (assemblies) are numerous. Standard single-reduction unit ratio ranges are 5:1 to a maximum of 70:1. Multiple-stage units will achieve ratios as high as 3600:1. Worm gear drives can be combined with helical gearing that acts as the primary drive between the motor and the input shaft of the worm gear set. Because of the compactness and sliding contact of the worm gear set, the amount of heat generated is significant. Some means of preventing overheating such as a fan will assist in cooling. Worm gearboxes are not the most efficient means of power transmission, but they are readily available and inexpensive.

Parallel Shaft Gearboxes Parallel Shaft gearboxes are called by that name when the gearing shafts are parallel to each other (see Figure 14–47). These units are frequently used in applications that have high horsepower and torque requirements. Long conveyors handling bulk material and rotating kiln drives are two of the many different high-load applications they are suited for. Multiple reductions and various shaft arrangements make this type of reducer extremely versatile. The ratio range is approximately 2:1 to 300:1. Helical gearing is standard and double-helical gears are used in the larger units. The gearing is typically enclosed in steel, horizontally split housing that simplifies rebuilding of the unit.

FIGURE 14–47
Double reduction parallel shaft
gearbox.

Courtesy of Falk Corporation.

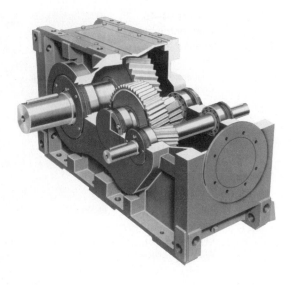

Shaft-Mounted Gearboxes Shaft-mounted gearboxes use a hollow output shaft mounted on the driven mechanism shaft for support and to drive the load. Almost any type of gearbox—whether it is a worm, planetary, or parallel—can be a shaft-mounted drive. The most common is the "shaft-mount" speed reducer with helical gears (see Figure 14–48). Shaft mounts are used frequently on incline conveyors for small-to-medium-sized bulk materials. Because the reducer is mounted and supported on the driven shaft arrangement, no additional base is required. A turnbuckle

FIGURE 14–48
Shaft-mount gearbox with
V-belt drive and motor.

Courtesy of Rockwell Automation.

FIGURE 14–49
Planetary gearbox.
Courtesy of Rexnord.

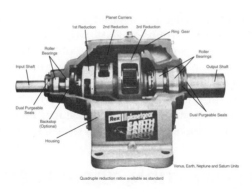

attached between the gearbox and the conveyor framework prevents the unit from spinning on the shaft. A sleeve or tapered bushing of some form fits over the driven shaft and locks the unit in place. There are a variety of single and double bushing systems used on shaft mounts. Ratios range from nominally 5:1 to 25:1. If additional speed reduction is required, a V-belt drive between the motor and reducer is used. The driving electric motor is attached to the unit by a motor-mount plate secured to the top of the gearbox.

Planetary Gearboxes Planetary gearboxes use internal gears to transmit loads in a compact space with shafts in line (see Figure 14–49). Nominal ratios of 3:1 to more than 1000:1 are possible with planetary gearboxes. The two basic types of planetary reducers are the single stage or simple planetary, and the differential planetary.

The simple planetary gear system has a pinion mounted on the input shaft called the sun gear, which drives the three planet gears. These planet gears surround the sun gear and are supported by a carrier that freely rotates about the sun gear (see Figure 14–50). The planet gears are encompassed by a ring gear, which is

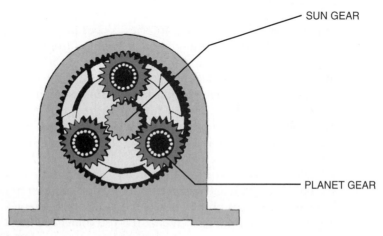

FIGURE 14–50
Planetary gearbox configuration.
Courtesy of Rexnord.

held in the housing to prevent it from turning. The entire planet gear and carrier assembly rotates within the stationary ring gear and is driven by the sun gear. The planet carrier is connected to the output shaft, which makes it turn at the same speed as the planet carrier. The number of teeth on the sun gear and the ring gear determines the ratio in this system.

The differential planetary gear system is a more complex form of the single planetary system. It uses multiple stages and differing numbers of teeth between the primary and additional stages to achieve higher ratios.

GEARBOX INSTALLATION, TROUBLESHOOTING, AND MAINTENANCE

If the gearbox is correctly selected for the application, proper installation and maintenance are the keys to a long service life. Faulty installation or poor maintenance practices will result in the premature failure of the unit and its associated components. Certain procedures are common to all reducers and the basics are covered here. It is recommended that the manufacturer installation and maintenance bulletins relevant to the unit be consulted for specifics.

Nameplate The nameplate, or nametag, is one of the most important parts of the gearbox. It lists the rated horsepower, speed, ratio, and usually a serial number for that unit. The serial number can be used to identify and order spare parts, obtain rebuild information, and determine oil fill amounts. Never paint over or remove the nametag.

Modifications Never weld on or modify the original equipment without prior approval from the manufacturer. Welding on the case may cause distortion of the housing and damage to the internally rotating components. Modifications also void warranties.

Foundation The foundation for the reducer provides support and rigidity for the unit. A rigid and solid foundation is essential to maintain a smooth-running and aligned reducer. The gearbox, location, size and the application dictate the type of foundation. It is preferable to elevate the base of the unit above the floor level. This will facilitate the draining of oil and will help isolate the gearbox from contaminants. Horizontal orientation of the unit is preferred. If it is necessary to tilt the unit, be aware that repositioning vents/plugs and adjusting the oil level might be required. If a concrete foundation is used, make sure that it is set firmly and level before bolting down the unit. Steel mounting pads should be grouted into the concrete base. This will allow the unit to be more accurately aligned and positioned. A steel foundation should be designed and engineered to provide sufficient rigidity to prevent induced loads and the mass of the reducer from distorting the gear case and causing misalignment of the rotating elements. It is recommended that the steel baseplate thickness be equal to or greater than the thickness of the feet. The plate should extend under the entire reducer for complete support. Special fabricated bases that support the motor and gearbox can provide compactness, proper alignment, and ease of installation. A general recommendation is to use SAE grade 5 or ASTM-A449 non-lubricated fasteners to mount units to their bases. Tightening the bolts to the correct torque specifications prevents loosening or failure. If special "torque-locking" fasteners are used or if aluminum housings are being bolted down, the manufacturer must be consulted. This also applies to the fastening of cover plates holding shims/gaskets where specific torque values may differ.

Alignment Alignment of a gearbox with respect to both the prime mover and the driven equipment is required to achieve the maximum life from the gearbox. Alignment tolerances should be as near to zero as practical, but no greater than the shaft-joining components' recommended tolerances. Use a level to assure that the unit is level in all directions with respect to any connected machinery. Because the entire drive may be in a vertical or tilted position, the goal is not necessarily that the unit be perfectly level, but that all subsystems are level within the same plane.

The alignment process generally starts from the driven apparatus and proceeds toward the motor. Place flat stainless steel shims under all mounting pads. Start at the low-speed shaft side and level across the length and the width of the unit. Using a feeler gauge, check to make certain that all feet are supported to prevent warping of the housing when the unit is bolted down. Any "soft foot" should be corrected by placing the appropriate quantity of shims in the air gap (see Flexible Couplings and Alignment). After the unit is aligned with the driven machine and securely fastened down, align the motor to the input shaft of the gearbox. If the gearbox and motor are shipped on a prefabricated base plate, some misalignment of the couplings might have occurred during transit. Verify that all premounted components are within acceptable alignment limits.

Shaft Connections Connected power transmission components must be installed properly to minimize any abnormal loading or stresses. All guards should be OSHA acceptable and be secured in place to prevent damage to components and injury to personnel. Couplings must be heated, fitted, installed, and aligned by following the recommended procedures specific to the device (see Flexible Couplings and Alignment). Never use extreme force or hammer on reducer shafts. This will cause damage to the internal bearings and bend or break the shaft. Any chain or V-belt connections must be aligned and tensioned correctly to prevent excessive overhung loads to the shafts and bearings.

Lubrication Lubrication of all the rotating elements (gears and bearings) is needed to prevent seizing and overheating within the unit. AGMA publishes general lubrication guidelines based on the operating environment of the reducer. Nameplates or manufacturers' bulletins should be consulted for the type and quantity of oil. Some gearboxes are prefilled and sealed for life, but many gearboxes are shipped from the factory without oil and require filling before use. EP (extreme pressure)-type oils can be used if authorized by the producer of the gearbox except in units that incorporate an internal backstop. The backstop may malfunction due to the slipperiness of EP-type lubricants. Selection of the most suitable lubricant for the gearbox depends on the method of application, ambient conditions, and recommendations of the manufacturer.

The temperature around the gearbox and the operating temperature of the lubricating fluid will have a significant effect on the life of the gearbox. The load, lube types and levels, operating environment, and gear conditions will play a role in determining the operating temperature of the unit. Minimum and maximum ambient operating temperatures should be determined for all seasons. If ambient temperatures are exceptionally high or low, a special lubricant formulated to maintain proper viscosity under those conditions should be used. The unit should be sheltered or shielded from extreme heat sources, such as direct sunlight or kilns and ovens. Additional cooling devices like fans or heat exchangers may be required to maintain a safe operating temperature. Oil temperatures should be monitored on a regularly scheduled basis. As a guideline, the oil sump temperatures should not exceed 180°F (82°C) with a temperature of 200°F (93°C) being an indication of a potential problem. Some gearboxes will run hotter than others, based on loads. What should be watched for are changes in temperature over a period of time, which might indicate the beginning stages of failure.

Keeping the housing uncovered and clear of all debris and dirt will assist in maintaining an acceptable running temperature. The oil absorbs the heat that is generated internally by meshing gears and rolling bearings. The heat is transferred from the oil to the reducer housing, where it is dissipated to the surrounding atmosphere. If the unit is covered or enclosed, airflow is restricted and the temperature of the gearbox will rise. The easiest and least expensive step that can be taken to lower gearbox temperatures is to prevent or remove any build-up of foreign materials on or around the unit. Fans and guards should be cleaned to allow adequate airflow. Clean housekeeping not only provides a safe operating environment, but helps the gearbox see its design life.

Lubricant Levels The unit type, mounting position, and manufacturer determine the level of oil. Inadequate lubricant levels, either too much or too little, will cause a rise in temperature. Nameplates will typically provide oil level and quantity information. Make sure all fill/drain plugs are properly positioned for the appropriate mounting style. Sight glasses should be protected, yet visible for viewing. Fill the reducer to the oil level indicated on the dipstick or to the line on the vertical standpipe/sight-glass. Be conscious of the difference between static and operating levels. If the unit is requiring frequent oil fills due to leakage, the oil seals might need replacement. Initially, oil should be drained and replaced after approximately the first 250 hours of operation. Flush the unit thoroughly before refilling it; make sure the sump and all troughs/gutters are clean. Routine gear oil changes under normal operating conditions should take place at least every 6 months or 2500 operating hours, whichever occurs first. Observe the magnetic drain plug for unusual particles like parts of gear teeth. Oil sampling should be done as part of a routine maintenance program. Samples should be taken, analyzed, and recorded to determine if contamination by foreign substances or water has occurred. Oil degradation might require fluid changes at more frequent intervals.

Certain units are furnished with grease-purged seal carrier assemblies. These additional seals provide an extra barrier to abrasive contaminants. The seal housing cavity must be pumped with NLGI #2 grease through the fitting with a low-pressure gun until grease appears on the shaft.

Breathers or vents that are threaded into one of the pipe-plug holes allow the unit to breathe. Pressures from internal heat generation build up and must be allowed to equalize. Certain smaller and newer gearboxes come prelubed and sealed for life, and with a pressure bladder inside that will contract and expand as needed. A breather with a replaceable element can be installed to filter contaminants and keep the unit from ingesting dirt. Some breathers combine a filter element with moisture-absorbing substances into one package.

Storage Most gearboxes are shipped from the factory with rust preventative oil sprayed on the internal parts that will protect the units for a period of 12 months in a dry facility. If a unit is to be stored for an extended period of time—beyond 12 months—one of three procedures may be followed: 1) The unit may be completely filled with oil for the duration of the storage time, 2) a specially formulated fluid that emits rust inhibiting vapors can be added to the oil, and 3) the unit can be completely drained of oil and all internal parts sprayed with a vapor-phase rust inhibitor. Regardless of which method is used, the unit must be sealed from contaminants by removing vents/breathers and replacing them with plugs. The high-speed shaft should be turned periodically to prevent metal-to-metal contact damage in the bearings. Before the unit is put back into service, it must be refilled with the proper gear oil and amount.

Rebuilding Gearboxes

The following steps are general steps applicable to gearbox rebuilding and repair. Specific manuals available from the manufacturers of each package should be consulted and followed.

1. Lockout, tagout, and verify that the unit is electrically shut off. Mechanically neutralize any connected loads to prevent shaft movement. Follow all plant safety practices applicable to the project. Do not attempt to bypass or shortcut safety procedures!

2. Acquire and read all the documents needed to successfully accomplish the job. Service manuals will list replacement parts and procedures that are required. Keep accurate records of what was done to the unit for future reference.

3. Organize and gather the appropriate tools needed. In addition to standard mechanics tools, the following are required: hoists, slings, arbor press, bearing pullers, torque wrenches, feeler gauges, dial indicator, inside and outside micrometers, and a bearing heater.

4. Assemble all of the correct replacement parts that might be needed to rebuild the gearbox. Order or obtain the replacement parts (bearings, shims, etc.) in advance of the scheduled work. Tag and store them in a clean and organized manner until they are needed.

5. Disconnect all attached equipment and guards. Inspect them for damage and replace as needed. Remove any materials, debris, or equipment that might interfere with the process.

6. Drain the oil from the unit. Dispose of the used lubricant in an environmentally safe manner. All plant and government rules and regulations should be adhered to when disposing of used lubricants and their containers.

7. If possible, transport the unit to a clean workshop to better facilitate the work. Properly sling and carefully lift the reducer to avoid damage to the unit and plant personnel. In some instances, because of size constraints, the work will need to be done in place.

8. Clean and inspect the exterior of the gearbox. Visually inspect the case and shaft extensions for damage. Record the nameplate information.

9. Remove any breathers, brackets, auxiliary seal covers, or backstops from the unit. Indicate the correct direction of rotation on the backstop cover plate.

10. Loosen and remove all housing fasteners and tapered dowels. Some types of gearboxes (parallel shaft) require sequential removal of fasteners to prevent a component from falling out of the unit during disassembly. Fasteners should be inspected and replaced if required.

11. Remove all seal cover and bearing plates. As cover plates and housing parts are removed, measure the removed shim pack thicknesses and record and label them to assist in reassembly. They may be tied or wired to the appropriate cover plate. Remove and clean all old gasket and shim materials from metal surfaces. Remove seals and record the part numbers.

12. Separate the case parts, cleaning and inspecting as you proceed. Remove all old shim and gasket materials from metal surfaces.

13. Make sure all machined mating surfaces are free of nicks and burrs that could cut flesh or prevent a proper fit. A fine-tooth file can be used to dress off any high points. Thoroughly clean the housing and inspect it for internal damage. Measure housing bearing bores to make sure they are round and within tolerance. Usually housing bores must allow the outer ring of the bearing to "float" within the case, so they must be accurate and undamaged.

14. A drawing or sketch of the inner workings and parts should be made, labeling all bearings, shims, shaft arrangements, etc. This sketch will act as a reference for present and future repairs.

15. Disassemble the unit per the manufacturer's instructions. Remove the gear/shaft assemblies carefully and place them on a soft surface to avoid damaging the teeth. Inspect the gear teeth for damage and excessive wear and replace them as necessary. Small rises or burrs should be dressed off with a fine-tooth file. Use a press to remove worn gears from shafts, taking care not to gall or damage the shaft during the process.

16. Inspect all shafts for grooving caused by seal wear. "Speedi-sleeves" may be installed to give the seals a new surface to run on. Measure all shafts and keyways to determine bore and key sizes if replacement of connecting components is necessary.

17. Identify and inspect all bearings for obvious signs of damage. Old bearings should be replaced, unless new bearings are unavailable. If the bearings are to be reused, they must be cleaned and free of any damage to the contact surfaces.

18. Follow reassembly instructions provided by the manufacturer. Although many units have common assembly procedures, each gear reducer may require specific steps and settings that are absolutely required to ensure correct assembly.

19. Mount any new gears and bearings prior to assembly. Install new pinions and gears on shafts with a press. Make sure that the gears and shafts are properly fitted and seated. Take measurements of all mounting surfaces and the bore of the component to be mounted. Proper amounts of interference fit are needed to prevent the component from coming loose during operation. Some pinions come as "solid on shaft" and must be replaced as a pinion/shaft assembly. General good practice is to replace both the driving pinion and the driven gear as a set. In some cases the entire gear train may be flipped to drive on the unused face of the teeth.

20. Mount new bearings on the shaft using acceptable mounting practices (see Bearings chapter). Make sure that the bearings have a tight fit and are seated against the shaft and housing shoulders.

21. Replace gear/shaft assemblies in the housing in the sequence recommended by the manual. Adjustment of the running clearances or preloads of the bearings is needed to ensure proper gear mesh and bearing life. Adjustment steps are specific to each type of gearbox. Typically, a cover plate with new shims/gaskets equal to those removed is installed on one side and the fasteners are cross-tightened to the proper torque specifications. The shaft is lightly drifted over with a rubber mallet, then rotated to seat the bearings. The opposite-side cover plate is installed with the prescribed amount of new shims/gaskets and then tightened. Most new shims will compress approximately 10% and this must be taken into account. Follow the recommended tightening and torque specifications to avoid crushing the shims or drawing up the plates unevenly. A dial indicator is set up with the magnetic base mounted on the gear case. The plunger of the indicator is placed on the end of the shaft to measure the total lateral movement (end float) when the shaft assembly is barred back and forth. Adjustment is then made to a shim pack to yield the prescribed amount of end-play or preload. This process is followed sequentially until the unit is assembled. Take care not to pinch or fold gaskets when assembling.

22. Install tapered dowels to align the housing halves. Make sure all dowels—solid and roll pin types—are true and undamaged. Install and correctly tighten all housing and cover bolts as required. Tightening sequence and proper torque specifications for the bolt's specific size must be followed.

23. Install new oil seals properly so as not to damage the case or lips (see chapter on Seals and Sealing Devices). Reinstall all breathers, brackets, and plugs. The threads may be coated with a small amount of Permatex #3. If applicable, return the backstop to the housing along with the key that holds it in place. Check that it has been replaced to ensure the correct direction of rotation.

24. Coat all gears with light oil and prime any lubricant feed gutters. If the unit is to be put back into service, add the correct oil to the proper fill level as indicated on the dipstick or sight glass. If it is to be stored, the unit should be tagged with the information pertinent to the rebuild, such as date, oil level, etc.

25. Rotate the input shaft by hand to turn the gear train, making sure there is no binding or resistance.

26. Reinstall the unit in its operating place and connect all components and guards.

27. Run the unit for a short time with a reduced load. Monitor the temperature and oil levels for any indication of problems. Take vibration and amp draw readings for future reference.

Troubleshooting Quick Guide

Problem	Probable Cause	Corrective Action
Excessive Tooth Wear	Overloaded drive	Check gear ratings; redesign.
	Inadequate lubrication	Analyze lubricant for quantity & quality.
	Contamination from foreign material	Inspect, clean, & seal the drive and lubricant.
	Improper backlash and/or root clearance settings	With open gears, set proper clearances and/or backlash.
	Worn mating pinion/gear	Inspect mating pinion/gear & replace as a set.
Gear Breakage	Shock or overloads	Eliminate overloads & jam-ups.
	Foreign object entering mesh	Inspect, clean debris from mesh; guard from contamination.
Excessive Noise and/or Excessive Vibration	Improper backlash and/or root clearance	With open gears, set proper clearances and/or backlash.
	Misalignment	With open gears, align centers.
		With enclosed drives, check shafts & bearings.
	Worn gears	Replace worn gears.
	Excessive speeds	Monitor speed; reduce if needed.
	Problems with connected components	Align couplings; check motor & driven load.
	Loose or improper base	Check base and foundation for looseness; correct and/or tighten, eliminate soft foot.
	Failed bearings	Inspect & replace failed bearings.
	Incorrect bearing settings	Check and set all bearing clearances and/or preloads to proper settings.
	Inadequate lubrication	Analyze lubricant for quantity & quality.
Excessive Gearbox Temperature	Oil level is too low or too high	Check & correct oil level.
	Incorrect grade & viscosity	Check & replace with correct type.
	Incorrect bearing settings	Check and set all bearing clearances and/or preloads to proper settings.
	Inadequate air flow around unit	Check cooling fans; clean the unit of foreign material.
	Overload	Check ratings of unit, possible causes of temporary overloads.
Gearbox Leaking	Leaking seals	Replace seals & inspect shaft. Speedi-sleeve shaft if required.
	Lubricant level too high	Check level. Adjust to proper level.
	Gasket damaged/leaking	Replace old gaskets.

Open Gear Drive Student Exercise

Calculate and determine gear ratios, rotation direction, and the speed of each gear in the change gear drive (Figure 14–51).

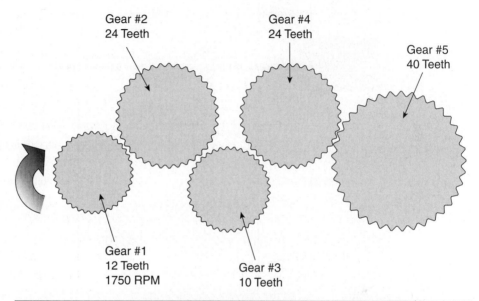

Gear #	Direction	Ratio	Speed (RPM)
1	____		____
2	____	____	____
3	____	____	____
4	____	____	____
5	____	____	____

FIGURE 14–51

Enclosed Gear Drive Student Exercise

Identify the gear and shaft components within the enclosed gear drive and calculate the output speed based on the ratio of the gearbox. Fill out the worksheet in Figure 14–52.

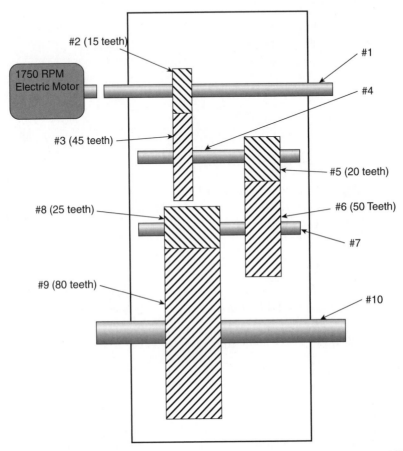

A) Calculate the ratios for each set of gear reductions and the overall ratio.

1st = _____   2nd = _____   3rd = _____

Overall Ratio = _____   Final Output Speed = _____

B) Identify each numbered component.

#1 _____   #2 _____   #3 _____

#4 _____   #5 _____   #6 _____

#7 _____   #8 _____   #9 _____

#10 _____

FIGURE 14–52

Questions

1. What is a gear?
2. Name four different types of gears.
3. What does AGMA stand for?
4. What materials are gears made from?
5. Define backlash.
6. What is the diametral pitch system?
7. What is modular gearing?
8. Name and briefly describe two general methods for aligning open gears.
9. Describe three failure modes for gear teeth.
10. What are the primary components of an enclosed gear drive?
11. How is the ratio of a gearbox determined?
12. What is planetary gearing and how does it work?
13. How does the ratio of a gearbox affect its efficiency?
14. Name three types of gearboxes.
15. What is a C-face motor mount?
16. Describe the advantages and disadvantages of a worm gear drive.
17. Describe the various mounting methods used with hollow shaft gear reducers.
18. Name six installation and maintenance concerns for gearboxes.
19. Why are backstops used on gearboxes?
20. Give three reasons why gearboxes require lubrication.
21. List seven important steps when rebuilding a gearbox.

CHAPTER 15

Linear Motion Technology

This chapter covers the principles, components, and terminology of linear motion technology. Linear motion technology entails those systems and components that move, support, and guide loads that move in linear directions. This chapter will examine factors concerning the selection, application, and operation of linear systems. The various components and their terminology are discussed along with where they are used. Installation and maintenance are also covered.

Objectives

Upon completion of this chapter the student will be able to:
✔ Describe the various components used in linear motion.
✔ Explain operating concerns involved in linear motion.
✔ Identify a variety of linear motion components.
✔ Describe the installation of basic linear motion components.

INTRODUCTION

Linear motion technology is a term used to describe automated and semiautomated mechanical systems that create Cartesian x, y, and z axes of motion. Modern manufacturing processes incorporate the use of linear components to allow for rapid, low-friction precision movement. Machine tools, robots, material handling, packaging, and numerous other types of equipment use linear components such as rails, bearings, and actuators to position and lift loads.

Modern manufacturing processes require precise movements, multidimensional machining, high throughput, flexibility, and integrated technologies. A linear system combined with servomotors, drives, and sensors makes filling these requirements possible. The precise positioning of machined parts is accomplished through the incorporation of guidance and thrust mechanisms in conjunction with various electrical and electronic devices. PLCs, gear motors, and servomotors used as the driving and controlling mechanisms provide feedback, control, and power to the system. Pneumatic or hydraulic cylinders often function as gripping or pushing apparatuses in conjunction with other linear components. Working together, all of these components serve to position and move containers, parts, and product efficiently into an exact location.

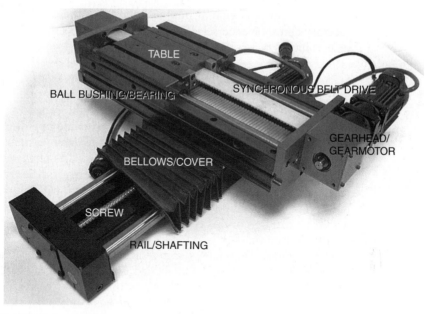

TABLE

SYNCHRONOUS BELT DRIVE

BALL BUSHING/BEARING

GEARHEAD/
GEARMOTOR

BELLOWS/COVER

SCREW

RAIL/SHAFTING

FIGURE 15–1
A compact complete linear motion system.

Linear motion systems can be divided into three basic subsystems: the drive/control, the thrust mechanism/actuator, and the guidance/support components (see Figure 15–1). The drive/control devices include a variety of electric motors such as linear, steppers, and servodrives. Two of the basic mechanical component subsystems of automated and semiautomated linear systems are the guidance and the thrust mechanisms. The guidance mechanisms of the system control the travel direction and linear accuracy, as well as support the load. These components literally guide the load. The thrust mechanism, in conjunction with the drive, provides the thrust and axial positioning accuracy of the load. Together these linear components make up a linear motion positioning system.

LINEAR MOTION OPERATING CONCERNS

When sizing, selecting, and operating a linear system, the loads, velocity, acceleration/deceleration, and duty cycle all must be accommodated. The loads and moments that are imposed on a linear system can be complex and numerous. The positioning of the supporting rails and blocks, along with the number and size of blocks used, is important in carrying these loads. Most linear systems incorporate the use of two rails/shafts and four bearings. The linear bearings are mounted onto a metal saddle.

Improper sizing, positioning, and installation of these components will result in inaccuracy and shortened service life. Design life is expressed in units of linear travel, such as inches or kilometers. Most design engineers select the component by service factors to achieve an extended service life. It is advisable to contact a reputable manufacturer when selecting linear systems.

A basic understanding of the mechanics of a linear motion system is important. The imposed loads are determined by the mass and center of gravity location of the load, as well as any induced moments from machine operations. The life of the linear components is tied to loads imposed on the linear system. Life decreases exponentially with load. The moments can be classified as pitch, yaw, and roll. The loads are classified as radial, reverse-radial, lateral, reverse-lateral, axial, and reverse-axial. Figure 15–2 illustrates a simple linear system with the various loads and moments. The guidance portion of the system handles all of the loads and moments with the exception of the axial and reverse-axial load, which is taken care of by the thrust mechanism.

Many linear components, such as actuators, take into account the "duty cycle" of the item when it is selected and applied. The duty cycle can be defined as the amount of "on time" versus "total time" or as the number of reciprocating motions per minute. Duty cycle for linear components has a relationship with the life of the device. Many actuators are capable of operating at rated load with a duty cycle of 25% "on time." Duty cycle is also defined as "on time" versus "cooling time." Cooling time

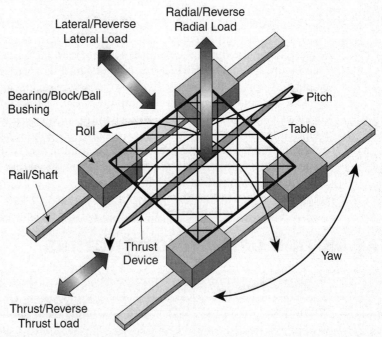

FIGURE 15–2
Linear motion loads and moments.

is when the actuator is not cycling and the component temperature is decreasing. This is generally a consideration of how much heat the device and motor can tolerate during starting, stopping, and reversing. More and faster strokes generally mean less life. The time required to accelerate the mass of the object being moved will affect the power requirements and must be considered. Moving an object at a predetermined speed to an exact location involves many variables. Therefore design and selection of the system is best left to qualified linear engineers.

Accuracy is another of the primary operating concerns with linear systems. Accuracy in a linear system is defined as how close the system is actually capable of coming to a predetermined and commanded position. The overall accuracy in the system is determined by many variables. The accuracy of the assembled system is dependent on the accuracy grade of the components as well as the mounting surfaces. Chief among these variables is the accuracy of the machined base that the assembly is mounted on. The mounting accuracy of the linear guide usually copies the accuracy of the machine base. Mounting errors have an effect on three factors: life, friction, and accuracy.

Holding acceptable parallelism of the linear rails/shafts is imperative to maintain consistent performance and achieve design life. Running parallelism is defined as the error in parallelism between the datum planes of the rail and the block as it traverses over the length of travel. As the table moves, rail parallelism will prevent binding in the bearings and overall inaccuracies in the system. Most manufacturers of linear components publish permissible values of mounting error, mounting surface tolerances, and parallelism standards. These values should be adhered to when installing a system to produce precise motion.

Rigidity is another important issue pertaining to linear systems. One of the goals is to make the system stiff or rigid enough to prevent deformation or unintentional movement in order to maximize static and dynamic rigidity performance. Unwanted deflections of the system during operation can result in manufacturing errors. Preloading of ball screws, properly fastened components, correctly torqued fasteners, and having good end supports on the thrust mechanisms will all enhance the rigidity of the linear system.

Repeatability of a linear system relates to the system's ability to have the components go from one point to another and back again consistently with little error. How much error is acceptable is the question. The answer, in part, lies in being able to control the error and position the table to the precise location electronically. Sensors, limit switches, and encoders of some form can be incorporated into the system to assist positioning. These devices can provide feedback to the controller and drive mechanism—such as a servodrive—to compensate for errors and position the table accurately.

Linear Motion Motors, Drives, and Controls

Linear motors, drives, and controls are the power portion of the linear system. Working together they convert electrical energy into mechanical energy and assist in positioning the mechanism. The mechanical energy can take the form of rotary-to-linear or direct-to-linear motion. Rotary-to-linear drives will convert the rotary

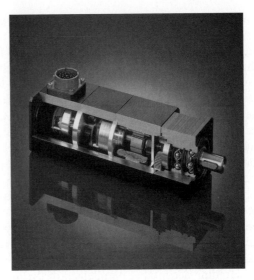

FIGURE 15–3
Typical gear-head/gear-motor used on a linear system.
Courtesy of Danaher Motion.

torque of an electrical motor to axial travel through a gear-head/gear-motor (see Figure 15–3) and/or actuator connection. Electrical linear motors produce direct linear motion without rotary components. Both types of electrical motors are somewhat special in their construction but operate on the basic principle of electromagnetic forces. The electric motor driving the system can be controlled to provide the desired velocity, acceleration, torque, and position to yield the optimum performance with the least amount of detrimental forces on the components.

Two common rotary motors used in linear motion are the "stepper" and "servo" types. Both motors serve the same function but have certain attributes that make them more favorable to use in specific applications, depending on the power, torque, accuracy, and speed requirements.

The stepper motor is a device that provides rotary motion to an actuator, jack, or ball screw to position a load by operating in discrete increments, or steps. The stepping action is accomplished by electronically switching the power to the motor windings so that the motor phases are energized in a specific sequence. It is usually a small, compact brushless permanent magnet motor running in the 0 to 3000 RPM range. It is capable of full or incremental steps at rates of 20,000 steps per second. The motor can be stalled for short periods without damage from overheating. It can be incorporated into open loop (without position feedback) or closed loop systems.

The servomotor is a device that positions a load by operating at a constant speed or torque with a feedback device on the motor. A servomotor is matched to an amplifier that produces power or amps to the motor. The amplifier incorporates a microprocessor that implements torque angle control for maximum torque at high speeds. Servomotors generally have greater torque capabilities than stepper motors and run up to 7500 RPM. Servomotors can help produce very accurate linear systems.

Electric linear motors produce the linear movement directly. There are no rotating components, which eliminates the need for numerous connecting devices and force transformers. There are several different types of motors, including voice coil, linear induction, linear force, and linear step motors. Most of these motors can be thought of—in a rough sense—as rotary motors that have been split and rolled out flat. True linear motors operate on the principle of interacting magnetic fields to produce direct axial force. The big advantage to these motors is the ability to operate at high speeds and at high acceleration rates. They are reliable in that they have few moving parts and are mechanically simple. Most true linear motors are compact but can be used in extended-length applications. The downside to these motors is cost and availability.

Controls, switches, sensors, and optical encoders of various types are incorporated into most linear systems to ensure positioning accuracy. An optical encoder acting as a feedback mechanism can be used to detect stall and/or position. With closed-loop systems, encoders will generate pulses as the motor rotates to indicate the exact position. Controllers and sensors will determine any error between the actual system position and the desired location. Additional sensors can be used to monitor the table position and the shaft rotary position. Control devices are integrated with programmable controls and computers to permit multi-axis movement of the linear system. Control and positioning can be accomplished in a variety of ways and most are discussed in the electrical chapters of this book. The important issue is that control of the exact position of the linear system is predetermined and must be maintained.

LINEAR GUIDANCE COMPONENTS

The guidance portion of a linear system is made of round rails (shafting) or profile rails and bearings mounted in some type of block. The blocks are fastened to a saddle or carriage. The blocks are often referred to as bearings, housings, bushings, or trucks. The saddle is also known as the carriage, platen, or table. The saddle will have the part that is being machined or moved clamped firmly to it. The guidance portion of the system controls the direction of movement by the way it is laid out on a bed. The bed is the overall mounting surface for the system and needs to be flat and true. The accuracy of the system will be determined in part by the condition of the mounting bed. If the rails were fastened to the floor, the movement would be in the horizontal plane. The rails also can be mounted to a wall, which would make the movement vertical or horizontal, depending on the rail orientation. In some systems multiple rail set table assemblies allow for multiple axis movement. The rails also provide support for the table that is mounted to the blocks. Two rails of some type are required for proper support, and they are mounted to a machined support surface (bed). Both round and profile rails, along with the bearing blocks that ride on them, serve the same function of support and directional control.

Round Rail/Shafting

Round rail is essentially steel shafting made to more exacting standards and tolerances than power transmission shafting. Round rail is available in different diameters measured in inches or millimeters. Standard diameters are from 1/8″ to 4″ and 5 mm to 80 mm. Linear shafting can be machined with holes and special ends for mounting purposes (see Figure 15–4). The shaft ends can be machined square to allow for the assembly of multiple lengths. These are known as "butted joints." Shaft rail parts are also available as preassembled units to allow for support and easier installation.

The quality of round rail (shafting) is an important issue to the linear motion world. It is imperative to have a rail that is straight, round, smooth, and relatively hard to be able to deliver the required accuracy of linear systems used in a manufacturing process. Typically the shafting will be made from a high carbon, bearing grade steel. Stainless steel (440C) is also available for applications that are exposed to corrosive elements.

Quality linear shafts should be case hardened to a minimum specification of HRC 60. This is because the bearings that ride on the shaft will be moving back and forth repeatedly across the surface of the rail. The rail acts as a load-bearing surface subjected to repeated stresses. The depth of the shaft hardness will be approximately .080″. The finish of the shaft surface should be in the "Ra" range of 2 micro inch (0.05 μm) maximum. "Ra" is a traditional measurement of peak to valley height. A smooth surface shaft will reduce friction. Straightness is recommended to be .001″ per foot (25 μm/300 mm) cumulative .002″ (50 μm) TIR (total indicator run-out). Shaft roundness should be within .000050″ to .000080″. It should have a maximum taper of .0001″. These exacting specifications are necessary because ultimately the linear system will be only as accurate and true as the rail it uses.

Round rail is available in different tolerance classes. The classes established by certain manufacturers designate the tolerance of the diameter. Four of the class designations are "L," "S," "D," and "N." Following is an example of the specifications for a 1″ nominal diameter shaft for the four classes.

Class L	Class S	Class D	Class N
.9995″/.9990″	.9990″/.9985″	1.0003″/1.0000″	1.0000″/.9998″

FIGURE 15–4
Round rail or linear shafting of assorted sizes and shapes.
Courtesy of Danaher Motion.

FIGURE 15–5
Continuous linear shaft supports with and without shafts.
Courtesy of Danaher Motion.

FIGURE 15–6
End linear shaft supports
used on terminal ends of
round rail.
Courtesy of Danaher Motion.

Round Rail Supports and Ends

Shaft supports hold up the rail/shaft and are fixed by fasteners to the bed. They simplify the mounting of the round rail within the system. Supports are available in standard types: continuous, intermittent, and end. Continuous supports run the entire length of the rail (see Figure 15–5) and offer the most rigidity and load-carrying capacity, intermittent supports are spaced along the rail, and end supports are at each end of the shaft (see Figure 15–6). The supports are predrilled for mounting with holddown bolts to the bed, as well as predrilled for the shaft to be secured to the top of the rail. End and intermittent support blocks are used for light loads and allow clamping of the shaft ends. Shaft deflection can occur when end supports are used with high-load, long-span applications that can result in inaccuracies in the system. Regardless of which type of end support is used, it is recommended that they be shimmed to level the rails and minimize bed inaccuracies. Flanged support blocks are also available for applications that require perpendicular mounting.

Linear Ball Bushings/Bearings

The linear bearing is designed for movement in the lateral direction and not for rotary movement. The linear bearing runs back and forth on the round rail at a predetermined cycle rate. Linear bearings for round rail are available as ball bushings or

FIGURE 15–7
Standard and open linear
ball bushing/bearing.
Courtesy of Danaher Motion.

plane bearings. The bearing will be enclosed and supported in some type of block/housing that is drilled for mounting to a bed.

The ball bushing consists of a shell made from steel, plastic, or a combination of both (see Figure 15–7). The shell will typically have a hardened steel ring around its circumference that is precisely machined to allow for accurate fit within the block. A self-aligning bearing plate will also be part of the shell and is designed to absorb minor amounts of misalignment caused by inaccuracies in the block/housing bore and shaft deflection. Within the shell there is a retainer with multiple grooves in the shape of a loop for captured, recirculating steel balls. The steel balls ride on the round rail/shaft and transfer the loads. These balls are continuously circulating within the retainer. Ball bushings are available in an open configuration that allows the rail to be intermittently or continuously supported. The rail size determines the dimension of the bushing bore. Both inch and metric versions are used in industry. Ball bushings can be sealed to prevent the intrusion of contaminants that will damage the rolling elements and rail surface. The seal is attached to the leading edge of the bushing and acts as a scraper to wipe away any foreign material.

Die-set linear bearing bushings are designed to fit in the mounting holes of the punch holder in standard dies (see Figure 15–8). They are used on end-supported applications.

Plane Linear Bearings

Plane linear bearings are linear bushings without rolling elements (see Figure 15–9). The plane bearing uses a synthetic material that is combined with substances such as Teflon or graphite, which has a low coefficient of friction. Plane linear bushings tend to be self-lubricating and will accept both linear and rotary movement.

FIGURE 15–8
Die-set linear ball
bushing/bearing.
Courtesy of Danaher Motion.

FIGURE 15–9
Plane linear
bushing/bearing.
Courtesy of Danaher Motion.

They are capable of operating in contaminated environments and will accommodate the intrusion of soft contaminants that imbed in the bearing material. However, one of the disadvantages of plane bearings is that they do not accept misalignment. The shell must be crowned if it is used in an application where misalignment is inevitable. Plane linear bushings are available in an oversized bore version that will not bind when subjected to slight amounts of misalignment. These oversized bushings are referred to as "overcompensated," and their degree of accuracy is less.

Round Rail Linear Bearing Blocks/Housings

Linear bearing housings are used to hold the bearings and are mounted to the saddle (see Figure 15–10). The blocks or housings used for containing the bearings are usually made of aluminum. They are precisely machined to accept a slight push fit of the bearing. The bearing is held in place with snap rings that fit into internal grooves in the block bore. Quality housings will have a machined reference edge to assist in the assembly alignment. Housings are available in inch or metric dimensions, and are made in a closed or open style. The closed style is used on end-supported applications. The open style is used in continuously supported applications where rigidity is required. Most linear bearing housings come with a wiper seal that keeps lubrication within the bearing and helps keep contaminants out. An adjustable housing is made that allows for a reduction of the clearance between the shaft and the bearing for a stiffer, more rigid assembly with less play. Twin housings are an extended-length version of a standard block. Twin housings accommodate two bearing bushings for increased load capacity.

FIGURE 15–10
Linear bearing housing/
block used as supports
for rail.
Courtesy of Danaher Motion.

FIGURE 15–11
Profile linear rail
and blocks.
Courtesy of Danaher
Motion.

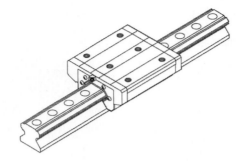

Profile Rail

Profile rail is a shaped version of the round rail (see Figure 15–11). It serves the same purpose as the round rail shafting in providing guidance and support for the system. The design of a profile rail is intended to reduce the naturally occurring sag that is prevalent in round rail. Profile rail is designed for stiff, quiet, high-load applications. Inherent in the shape of the rail is a capacity to handle all of the moments and loads that a linear system is subjected to. Machined into the rail are lateral grooves that act as pathways for the rolling balls. Each manufacturer will have its own uniquely shaped rail based on design criteria. Rail sizes are metric in width, height, and length. All manufacturers of profile linear rail have a unique numbering system. A typical part number might read "LH-25-1200," which is a rail with a 25 mm bottom width that is 1200 mm long. On profile rails a groove will be cut on one side of the rail. This groove is known as the "datum" and is used as a reference for mounting.

Profile Linear Guides/Blocks

Profile linear guides are referred to by numerous names: linear profile blocks, ball slides, cross roller assemblies, trucks, trolleys, and profile bearings describe the combined bearing and block assembly used with profile rail. The guide or ball slide must match the shape and size of the rail that it is riding on. As with rail, the design and shape of the guide is unique to each manufacturer. Typical guides will have caged recirculating ball construction designed for high-speed cycles. Multiple rows of recirculating balls within the guide act as part of the load-bearing component. A flat mounting surface on top of the guide with through or threaded holes allows mounting to the table. The block's flat mounting surface can extend beyond the side of the guide and is referred to as a "flanged" version. Wipers can be attached to the leading edge of the guide to act as an end seal against contaminants and to help retain the lubricant in the bearing. Most systems use a two-rail, four-block assembly.

LINEAR THRUST MECHANISMS

Linear thrust mechanisms are part of the driving portion of a linear system. Linear thrust devices are also known as actuators. They are usually connected to a prime mover in conjunction with other power transmission equipment. Typically a gear motor will provide the power source and be connected to a device by a flexible coupling. The flexible shaft coupling will usually be the type to limit backlash in the system. Many thrust devices are used to convert rotary motion (torque) into linear thrust or single-axis motion. The thrust device can be a ball screw, actuator, lifting jack, cylinder actuator, solenoid, or any other device that provides a single-axis force. These components handle the axial loads imposed on the system. A linear system might have one or multiple thrust mechanisms as part of the system. Several actuators can be incorporated into the system to give multiple-axis movement, or in the case of lifting heavy loads it may use multiple units in parallel. The purpose of the thrust devices is to assist in providing a driving force to move the load to a particular location accurately.

Acme Screws

Acme screws are linear thrust mechanisms that have a thread on a shaft to convert rotary motion to linear motion (see Figure 15–12). The force is transmitted through a matching nut usually made of bronze or plastic. Acme screws are usually found in low cycle, slow speed, or hand-operated applications. The efficiency of acme screws is in the 60% range. The modified square thread rod is available in right-hand or left-hand threads with single or multiple starts. A start is the number of helical thread elements on the screw shaft. Each screw has various diameters, pitch sizes, lengths, and leads, but only a certain number of starts. The land diameter is the outside diameter of the screw. The pitch is defined as the distance along the screw axis from a point on one thread to a corresponding point on the adjacent thread. The lead is the distance the nut advances in one revolution. The threads can be manufactured by being rolled, milled, or ground. Ground threads are the most precise. The nut is a simple device that is threaded to match the rod and flanged for mounting purposes. An antifriction radial bearing must support the ends of the threaded rod. The nut will be mounted to the table that carries the load.

Ball Screws

Ball screws are thrust mechanisms that use screws with flanged nuts that have recirculating steel balls (see Figure 15–13). The use of steel balls in the nuts lowers the friction. These devices are intended for faster speeds and higher-duty-cycle applications than the acme screw device. The operating speeds are limited by the tendency of the screw to vibrate when it reaches a critical speed. Ball screws function by means of steel balls enclosed within a nut that is traveling on a screw. The low coefficient of friction of the

FIGURE 15–12
Acme screw and nut thrust mechanisms.
Courtesy of Nook Industries Inc.

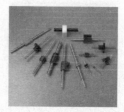

FIGURE 15–13
Linear ball screws. Note return
tubes for balls on the side
of the nut.

Courtesy of Nook Industries Inc.

rolling elements results in an operating efficiency of 90% or better. Because of their
high efficiency, a brake is required to prevent the load from back-driving.

The screws used with these mechanisms will be machined and hardened to Rc
58-62. Both ground and rolled screws are produced for use with ball screw devices.
Rolled screws are sometimes protected with a special black oxide finish to prevent
corrosion. As in acme screws, ball screws will be defined by the various diameters,
number of starts, lead, pitch size, and length (see Figure 15–14). The ball circle diameter is the circle generated by the center of the bearing balls when in contact
with the screw. The root diameter is the diameter measured at the bottom of the
ball groove. The lead is the distance the nut advances in one revolution. The lead is
equal to the number of starts multiplied by the pitch.

Ball screw nuts will have a closed circuit of ball paths within the nut assembly.
The balls will be circulated through the threads and returned by internal or external diagonal transfer tubes. Many ball nuts will be delivered with a transfer tube to
hold the balls in place. Transferring the ball nut from the arbor to the screw is
achieved by locating the arbor against the end of the screw thread and carefully rotating the ball nut onto the screw. Do not remove the arbor from the ball nut prior
to installation or the bearing balls will be lost. When ball screws are operated in a
nonvertical orientation, the transfer tubes should be positioned vertically.

The screw, nut, and flange are all part of the thrust mechanism. Axial and
reverse axial loads generated during operation are transferred back and forth from
the flanged nut to the balls to the screw. The internal threads on the nut will be

FIGURE 15–14
Linear ball screw
terminology.

Courtesy of Nook
Industries Inc.

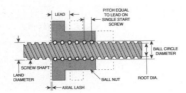

precision-ground to prevent premature failure. The nut will be pinned or setscrewed to a mounting flange. The flange attached to the nut can be drilled with mounting holes in a variety of patterns to allow ease of mounting to a table. The screw can be oriented in a horizontal or vertical position depending on the application.

Because ball screws are often used in precise positioning applications, accuracy and close tolerances are a concern. The lead accuracy of the threaded rod will be one of the determining factors in the overall accuracy of the system. Lead accuracy is defined as the correlation between turns of the screw and the theoretical versus the actual travel of the nut. The lead error is the maximum lead variation in thousandths of inches/foot. For example, the screw will make the correct amount of turns to move one foot. The nut has actually traveled one foot plus the error in the screw. An acceptable lead error tolerance should be .004″/foot or less. Higher accuracy screws are available that will not deviate from nominal lead by more than .0005″/foot. Backlash or axial movement between the nut and the screw can also cause a problem with positioning. In some cases the nut will be preloaded in a variety of ways to prevent backlash. Preloaded ball nuts assist in promoting accurate repeatability and increased stiffness in the system.

Ball screws have low radial rigidity because the supported span is significantly longer compared to its shaft diameter. One of the primary installation and operating concerns of ball screws is rigid mounting. It is important to have a rigid screw mount to prevent unintentional movement and to inhibit vibration. Damaging vibration can occur when the screw is operated at its critical speed or natural vibration frequency. The critical speed of a ball screw is the maximum speed at which a ball screw can rotate without setting up harmonic vibration in the screw. The vibration is partially created from imbalance in the rotating system. The screw will function better when it is mounted with slight tension rather than compression. The screw should be supported on at least one end—but preferably both ends—with radial ball bearings. Oftentimes a pair of angular contact ball bearings is used on one end to compensate for the thrust loads created during operation. Bidirectional thrust bearings at both ends of the screw will also yield excellent support and stiffness.

Roller Screws

A "roller screw" is a thrust device used for converting rotary motion into linear motion in a manner that is similar to acme and ball screws. The basic design of the roller screw uses threaded helical rollers assembled in a planetary arrangement around a threaded shaft (see Figure 15–15). The rollers are held in place within a

FIGURE 15–15
Roller screw.
Courtesy of SKF Industries.

sealed housing. Roller screws are generally capable of higher speeds and loads, size for size, than ball screws.

Lifting Jacks

Lifting jacks, also known as screw jacks, are thrust mechanisms that incorporate the use of ball or acme (machined) screws and a worm gear drive within an enclosed unit (see Figure 15–16). The purpose of a lifting jack is to lift or lower a load. The device will have an input shaft for drive connection and a screw shaft for attachment to the load or a fixture. The assembly is enclosed within a housing that has a mounting flange with through holes. A variety of configurations are available with the screw upright or inverted. The ends of the screw can be provided with clevis brackets or plates for attachment. Mounting conditions are varied (see Figure 15–17). Ends may be fixed, free, or attached to a guided structure by plates or clevis brackets. The jack can be mounted as a stationary unit to allow the shaft to move linearly. The unit may also have an external antirotating device to prevent rotation of the lifting screw. Ball screws are another type of jack and are more efficient than machine-screw jacks. Ball screw jacks have recirculating bearing balls separating the screw and nut. The low coefficient of friction of rolling balls creates less heat and allows higher speeds and cycle rates. Power requirements to drive ball screw lifting jacks are significantly lower than machined/acme screw jacks.

Linear Actuators

Linear actuators are sealed, enclosed thrust devices that are either a rod type or rodless type. The brake motor, gearbox, screw, acme or ball nuts, and housing can all be supplied as a single packaged unit. Their primary purpose is to push/pull or lift/lower a load. Multiple actuators may be installed in a system for load sharing or various movements. Care must be taken to electrically and mechanically synchronize the actuators to prevent binding and uneven movements when traveling in the same axial plane.

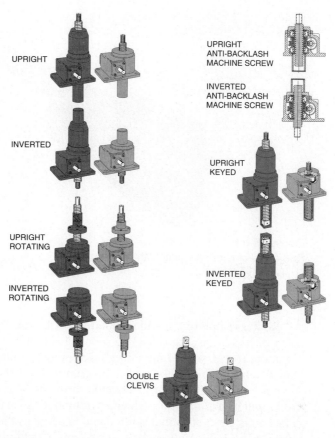

UPRIGHT

INVERTED

UPRIGHT
ROTATING

INVERTED
ROTATING

UPRIGHT
ANTI-BACKLASH
MACHINE SCREW

INVERTED
ANTI-BACKLASH
MACHINE SCREW

UPRIGHT
KEYED

INVERTED
KEYED

DOUBLE
CLEVIS

FIGURE 15–17
Lifting jacks are available in a variety of mounting configurations.
Courtesy of Nook Industries Inc.

The rod-type actuator consists of an enclosed rod connected to a screw driven by a sealed gear motor (see Figure 15–18). The screw can be an acme or ball screw version. A sealed AC or DC motor is connected by direct coupling or other means to a series of gears based on their ratio and configuration. The two basic configurations are in-line and parallel. The entire assembly can be foot mounted or have tapped hole flanges on the ends for mounting. Most quality rod actuators are sealed for wash-down applications. A clevis on the end of the device along the same axis of the screw allows attachment. Tapped, threaded, or universal rod ends are available for connection to the load. Limit switches and brakes can be added to control stroke length and provide back-driving protection. The mechanisms are found everywhere in industry and commerce. Large HVAC systems use actuators to open and close dampers and assist in controlling airflow. Gates, booms, doors, and chutes can all be opened and closed with actuators. They are an inexpensive, compact, and relatively trouble-free linear device.

FIGURE 15–18
Rod-type actuator complete with driving motor.

Courtesy of Danaher Motion.

Rodless actuators provide guided linear travel. They typically use either a ball screw or belt drive connected to a saddle as the thrust mechanism. The saddle rides along the top or side of the extruded aluminum housing and carries the attached load or table. There will be no rod protruding from the end of the housing as in the rod type. The saddle will be connected through a slot to the drive components. The saddle will have either through holes or tapped holes for easy mounting of the load. A magnetic cover strip that is lifted and lowered as the saddle traverses the housing protects the ball screw or belt drive from external contaminants. Figure 15–19 shows a rodless ball screw actuator.

Belt driven actuators use a steel-cable reinforced, low-stretch synchronous belt for accurate positioning (see Figure 15–20). The overall length of the actuator

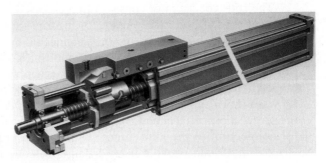

FIGURE 15–19
Rodless ball screw actuator.

Courtesy of Danaher Motion.

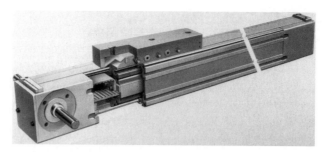

FIGURE 15–20
Belt driven actuator.
Courtesy of Danaher Motion.

remains constant regardless of the position of the saddle and can be anywhere from 2 to 20 feet. The constant length permits a smaller envelope for the actuator. The saddle can be mounted stationary and the housing moves axially, or vice versa. The saddle, which is connected to a ball screw or belt drive, is powered by an electric motor through a gearbox. Rodless actuators are used increasingly in modern manufacturing plants for positioning and processing a variety of materials and parts. The device will be controlled by an electrical variable-speed device and be supervised by a programmable controller.

Short-stroke actuators are rod-type, small, common linear devices used in low power applications that require extremely numerous movements (see Figure 15–21). They are often referred to as linear solenoids. The construction is simple, with an outer ferrous shell enclosing a coil around a ferrous plunger or rod. The energized coil generates a magnetic field that causes the plunger to react in response to the field. Typically the stroke lengths are short, ranging from a fraction of an inch to several inches. They are found in light-duty applications and clean environments such as vending machines and office equipment.

FIGURE 15–21
Short-stroke linear
actuator.
Courtesy of Saia-Burgess.

INSTALLATION, OPERATING, AND MAINTENANCE CONCERNS

Proper installation, operation, and maintenance of linear components are essential to maximize the service life of the system. The following are general guidelines for linear thrust and guidance mechanisms.

- Lockout, tagout, and verify that the system is neutralized prior to any service work.
- Check and do not exceed the limiting speeds of all linear components.
- Mount the components on a clean, level, and solid foundation. Mounting accuracy of linear systems usually will copy the accuracy of the base or bed.
- Position the actuator or screw axis centered and parallel to the load as much as possible to prevent detrimental moments. Side loading and cantilever mounting should always be avoided.
- Make sure the entire linear assembly is as rigidly mounted as practical.
- All rails should be level and parallel with each other. Use dial indicators to determine parallelism.
- Avoid excessive compression loads (column loading) on thrust devices, which may cause the screw to buckle. Heavily loaded vertical applications and high-speed horizontal applications are prone to column loading. Fixed end supports and increased diameter-to-length ratio will help prevent buckling.
- Ensure that the thrust mechanism screw-nut will freely run back and forth during installation to ensure running alignment.
- Minimize spans between drive components. Short, straight connecting shafts are better for the system.
- Use adequate and additional bearing supports for drive shafts whenever possible.
- Use zero-backlash flexible couplings with a high strength-to-bore ratio.
- Incorporate limit switches and sensors into the system to control travel distances and positioning.
- Make sure brakes and antirotation devices are used to prevent unintentional back-driving and effective stops. Only machine screw jacks with a ratio greater than 20:1 are considered self-locking.
- Use a torque-limiting device to prevent overload on the system and to protect the components.
- Use protective bellows, boots, seals, and covers whenever possible to prevent foreign materials from interfering with movement and causing wear of moving parts.
- When mounting profile rails, observe the datum line on the reference rail and place it against the datum plane.
- Tighten all fasteners to the specified amount of torque.

- Many linear components are shipped with an anticorrosion oil coating that must be removed with light oil prior to use.
- Lubricate all steel surfaces with light machine oil.
- Lubrication of linear ball bearings with the proper manufacturer's recommended grease or oil is required to achieve satisfactory service life.

Questions

1. Briefly define the uses of linear motion technology.
2. What are the three main subsystems in a linear system?
3. Name and describe the types of loads and moments a linear system may be subjected to.
4. Name six operating concerns with linear systems.
5. What are three different motors used to power linear systems?
6. What is the difference between round rail and profile rail?
7. How is round rail supported?
8. What are some additional names for profile linear guides?
9. Define and explain linear thrust mechanisms.
10. What is a lifting jack and what is its purpose?
11. What is a linear actuator?
12. Name three forms of linear actuators.

CHAPTER 16

Material Conveying Systems

This chapter covers the systems and components that are used to move materials. Conveying systems are the delivery mechanisms for raw and finished goods. Types of conveyors and how they function will be addressed. Proper maintenance and the importance of basic safety issues concerning all conveying systems will also be examined. The primary purpose of this chapter is to familiarize the student with the basic operating principles, types, terminology, and maintenance of conveyor systems.

Objectives

Upon completion of this chapter, the student will be able to:
✔ Identify the various types of conveying systems.
✔ Identify the various types of conveyor components.
✔ Understand how conveying systems work and how to troubleshoot them if they do not.
✔ Have a basic understanding of how conveyor components are selected.
✔ Describe the various belted conveyer parts and their functions.
✔ Detail the procedures involved in the installation of belted conveyor components.
✔ Know basic safety procedures pertaining to conveyors.

INTRODUCTION

Conveyors move raw material in bulk or finished product from one location to another. Conveyors can move material in horizontal, vertical, incline, or a combination of directions. Certain types of conveyors, such as a roller bed, rely on gravity to move the product. Most conveyors use some form of power transmission device to provide the power. Typically the conveyor system is made up of a drive, belting, framework, and assorted power transmission and conveying components. The conveyor drive is usually an electric motor connected to a gear reducer to provide power and torque to the system. The power transmission devices included in a conveying system are bearings, gear reducers, adjustable speed drives, clutch/brakes, V-belt/sheaves, chain/sprockets, sensors, and assorted pneumatic and hydraulic components. The conveyor components consist of pulleys, idlers, buckets, and screws—or roller beds, depending on the system design.

Conveyor Systems

Conveyor systems are used in a variety of industries; most manufacturing plants will have some type of operating conveyor system. Conveyors reduce the repeated handling of a material. Conveyors save time, money, and lessen the chance of damage to the product or injury to the operator.

There is a wide variety of different conveying systems. Conveyor Equipment Manufacturers Association (CEMA) defines over 100 different forms of conveyors, feeders, and elevators used for moving product. Belt conveyors, bucket elevators, roller conveyors, and screw conveyors are some of the most common types of systems. CEMA publishes numerous bulletins and catalogs detailing different conveying components. CEMA has established standards and specifications for the items that make up a conveyor.

There are two general groups of conveying systems: bulk material and unit conveyors. Bulk material systems are designed to move a continuous flow of material from one processing point to another. Bulk material conveyors will be found in mines, quarries, gravel pits, and agriculture facilities handling large quantities of unfinished product. Unit conveyors, sometimes referred to as discrete item conveyors, are used to move, shift, sort, and handle items of various sizes and shapes, and to handle packages, pallets, and finished or semifinished parts. Unit-handling systems typically are smaller and require less power and space than bulk systems.

Conveyor Types

Belt Conveyors The belt conveyor is one of the dominant forms of conveying systems utilized for the movement of finished goods or transporting bulk material. There are numerous advantages to using belted conveyors as compared to other forms of transportation. Large quantities of product can be moved rapidly, reliably, and economically over varied distances. The cost of energy and labor is low relative to the quantity of material moved.

Belted conveyors can be grouped into two basic types, flat belt and troughed belt. Flat belt conveyors have a flat belt running across rollers or a slider bed (see Figure 16–1). These roller or slider bed conveyors usually are used to move a semifinished or packaged product. The flat belt system is typically used for sorting, feeding, or distributing packages to bins. Distribution warehouses that repeatedly handle numerous items in a short time period benefit from the advantages of a flat belt conveyor. Troughed belt systems use troughing idlers and conveyor pulleys to transport raw materials for processing (see Figure 16–2). Troughed belt conveyors are used extensively in the mining and aggregate industry to move raw material from the pit to the plant, and within the plant for processing. Regardless of which type of conveying system is used, the goal is to accomplish the safe, efficient, and economical movement of materials.

FIGURE 16–1
Flat belt roller
conveyor used for
moving packages
and parts.

Courtesy of Hytrol
Conveyor Company Inc.

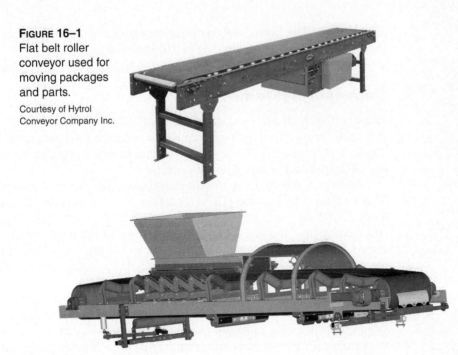

FIGURE 16–2
A short length, troughed belt conveyor system showing various components.

Courtesy of Stephens-Adamson Conveyor Components.

Troughed Belt Conveyors A typical troughed belt conveying system consists of
a drive, belting, pulleys, support structure, idlers, power transmission components,
take-ups, and loading/unloading devices (see Figure 16–3). They are high-capacity
systems designed for transporting material over long distances. Belt conveyors can
be arranged to follow a variety of profiles or belt paths. Many belt conveyors follow
a straight horizontal path. If necessary, an incline conveyor configuration can allow

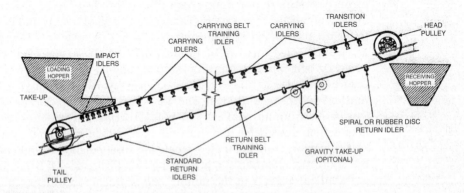

FIGURE 16–3
Typical bulk material conveyor system and components.

Courtesy of Rexnord.

vertical lifting as it is moved horizontally. A belt conveyor can follow a horizontal path for a distance and gradually ascend. Belt conveyors can handle large volumes of material. These systems are capable of being operated 24 hours a day, 7 days a week at capacities measured in tons. The capacity is determined not only by the speed of the belt but the width of the belt. Troughed belt conveyors are available in a variety of widths from eighteen inches to more than six feet. High or low horse-power drive systems can be matched to the requirements of the application. The conveyor components of a troughed system function together to provide an efficient and economical means of material handling.

Roller Conveyors Roller conveyors are made up of many closely spaced rollers that are freewheeling (unpowered) or driven by a belt. Package handling is a typical roller conveyor application.

Bucket Conveyors Bucket elevators use buckets connected to a chain or a belt to lift a product vertically for material, such as grain, that needs to be raised and dumped into a silo for storage.

Screw Conveyors Screw conveyors incorporate the use of a turning screw to move the product in a trough. The volume and rate of feed can be controlled with a screw conveyor.

Pneumatic Conveyors Pneumatic systems rely on air pressure to move the material through an enclosed tube. Pneumatic tube conveyors are often used to transfer lightweight or powdery materials.

CONVEYOR DESIGN AND SELECTION

The selection of the type of conveyor to be used will be based on variables such as the type of product, transport distance, operating environment, and economics. For instance, if the application would require a vertical lift in a short horizontal distance, a bucket elevator might be the chosen method. If the transfer of a coarse rock material over hundreds of feet is required, a troughed belt conveyor would be the best choice. The selected conveyor system must meet or exceed the required expectations within budgetary constraints.

The selection of the right components for a conveyor is an involved process. Manufacturers of conveying systems can be consulted to insure maximum component service life. Most companies that make conveyor components can provide the selection manuals or software that details the process. The following general information is prerequisite to proper selection of conveying components for new installations or replacement on existing systems.

- Material that is being transported. The form it is in: raw, processed, or packaged. The material classification and characteristics: lumpy, wet, coarse, or fine, etc.
- The distance it is being transported. The length and vertical lift distances.

- Conveyor belt width and length. The bucket or screw size.
- Required capacity in cubic feet or tons per hour of the system. The desired volume of material to be moved.
- Density, size, and weight of the transported material.
- Velocity of the system. (feet/minute)
- Required horsepower to move the material. Calculate torque requirements for conveyor drives.
- On belt conveyors, the effective (Te), tight-side (T1), and slack-side (T2) tensions.
- Belt information regarding tension/PIW ratings.
- Loading and unloading conditions: chutes, force-fed, uniformly loaded, etc.
- Any part numbers from previously installed conveyor parts for reference.
- The layouts, profiles, and space limitations of the system.

Conveyor Pulley Design

A conveyor pulley is a cyclically loaded, drum-shaped, fabricated welded device used to transfer power and maintain tension on a belt. The pulley is mounted on a shaft, usually using compression hubs or tapered bushings. The shaft and pulley should act as a unit in supporting the load and transmitting torque if it is connected to the drive. The diameter of the shaft is the basic factor in determining the load capacity of the assembly. Bearing blocks are used to support the shaft and pulley. The three primary purposes of a conveyor pulley are support for directional changes, power transmission, and assistance in belt guidance and tension. Pulleys will be found at the terminal ends of a conveyor system where the belt makes a directional change. Pulleys are used as the driving mechanism of the system, transferring power from the shaft through the pulley rim to the belt. Friction between the surface of the rim and the belt causes the belt to move as the pulley turns. The pulley serving this function is referred to as the drive pulley. Some highly loaded systems have multiple drives. The pulley that is located at the end of the conveyor system to provide a return point for the endless belt is called the tail pulley. Pulleys are also used at various locations in the system to provide tension and snubbing of the belt.

Pulleys are manufactured in a wide range of sizes and shapes. Steel is the most common material used in their construction. The pulley industry, along with CEMA and ANSI, have established dimensional and tolerance standards for pulleys. Manufacturers usually adhere to these recommendations and have additional standards for many of the common pulley sizes and their construction methods.

The common "drum pulley" consists of a rim, end discs, and hubs, and is fitted with bushings for mounting to the shaft (see Figure 16–4). The end discs are welded to the rim. The thickness of the rim and end discs is usually determined by the tension and loading on the pulley. Heavier loads require thicker rims. Certain heavy duty pulleys have additional stiffening discs welded inside the rim.

Pulley design and construction is based on loading considerations and the purpose the pulley will serve. The cyclic and peak stresses that a pulley is routinely subjected to

FIGURE 16–4
Basic drum
pulley assembly.

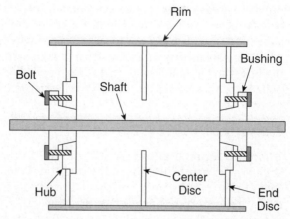

can cause it to fail if not properly sized and constructed. The thickness of the rims and end discs, as well as how they are welded together, will determine its strength. The type of service a pulley will see, as well as its design and construction, will determine the service life of the unit. Pulley end disc to rim thickness ratio can differ between manufacturers. Some manufacturers build a pulley to accommodate the end disc stresses and shaft deflection through a rigid design. Others use a flexible design. Each design has advantages and drawbacks. It is recommended that detailed loading and tension information be provided to the manufacturer to determine the optimum design for the application.

Pulleys are assigned names that define their location within the conveyor system (see Figure 16–5). A head pulley is commonly the unit at the terminal or discharge end of the conveyor. The drive pulley is the powered pulley that is driving the belt. Oftentimes the drive pulley is the head pulley. Conveyors may have a drive pulley that is not in the head pulley position. A snub pulley is fitted adjacent to the drive pulley to

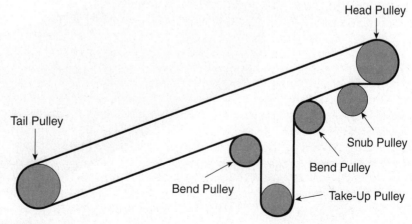

FIGURE 16–5
Conveyor pulley locations on a belted incline conveyor system.

provide a greater degree of wrap around the drive pulley. It also assists in tracking the belt. Bend pulleys are located on the return side of the conveyor where the belt makes a dramatic turn, such as prior to and after a gravity take-up assembly. A take-up pulley is used in conjunction with some form of take-up mechanism to adjust the tension on the belt. The tail pulley is located at the opposite end of the conveyor from the head pulley. It is usually near the feed end or loading point of the belt.

Conveyor Pulley Types

Standard Drum Pulleys A straight-faced drum pulley is a steel cylinder with straight sides (see Figure 16–6). The steel rim should have a minimum rim thickness of 1/2″. Quality pulleys have a machine-turned rim to hold concentricity to certain standards. Industry recommends that the concentricity about the shaft centerline, after machining, shall be that the TIR (total indicator run-out) shall not exceed .060″ for the rim. Standard diameters range between 6 to 60″. Common pulley widths are 20 to 102″. Generally the width of a pulley is 2 to 6″ wider than the flat belt riding on it, but this varies as the belt width increases. The industry typically describes a conveyor pulley by its diameter and width; for example, 20″ diameter × 38″ straight-face steel pulley.

Crowned Drum Pulleys Crowned pulleys have a "crowned" rim (face). On taper-crowned pulleys, the face forms a "V" with the peak in the center of the pulley. Normal crowns of this type vary from 1/16″ to 1/8″ per foot of total face width. Curved crowned pulleys have a long flat surface in the center of the pulley, with the ends curved to a smaller diameter. A crowned face is effective in centering a belt on the pulley if the approach to the pulley is a long, unsupported span and unaffected by the steering action of idlers. On most troughed-belt conveyor systems, the crown is

FIGURE 16–6
Drum pulley with tapered type bushings for mounting to a shaft.
Courtesy of Precision Pulley & Idler.

of little benefit in keeping the belt centered. Crown-faced pulleys should never be used on systems with steel cable belts or multi-ply high tension belt systems. Crown-face pulleys are best suited for use on low tension, return run bend and tail positions where there is a long, unsupported span between the return idler and pulley.

Engineered Pulleys Most manufacturers make an "engineered class" or "mine duty" pulley to meet load conditions that exceed those specified in ANSI standards for drum pulleys. They will be made to more exact tolerances. Generally they are of a heavier duty construction to withstand the extremes of mining applications. The pulley construction and specifications will be determined based on the parameters of the specific application. Pulley design engineers will require detailed data on all aspects of the application to produce an engineered pulley.

Wing Pulleys Another type of pulley is the wing pulley. Wing pulleys have a number of evenly spaced steel wing plates that extend radially about the pulley (see Figure 16–7). They are available in straight or crown-face construction. The primary purpose of a wing pulley is belt cleaning. The vibrating action of the wings hitting the belt knocks off any material that sticks to the belt. A version of the wing pulley is the spiral wing pulley. It offers the advantages of a standard wing pulley while minimizing the pounding action on the belt. With spiral designs, there is more rolling action and less beating of the belt, resulting in less wear.

Split Pulleys Split pulleys are usually die formed and riveted steel construction (see Figure 16–8). They are designed for drive applications using narrow flat belts.

FIGURE 16–7
Wing type pulley used to assist in cleaning of the belt.
Courtesy of Precision Pulley & Idler.

FIGURE 16–8
Split pulleys are used for ease of installation.
Courtesy of Rockwell Automation.

FIGURE 16–9
Elevator pulley.
Courtesy of Rockwell Automation.

FIGURE 16–10
Motorized pulley.
Courtesy of Emerson Power
Transmission.

Elevator Pulleys Elevator pulleys have relatively narrow face widths compared to their diameters. They are use on bucket elevator applications (see Figure 16–9).

Motorized Pulleys Motorized pulleys have a geared motor drive and bearings enclosed within the drum pulley (see Figure 16–10). The bearings are housed within a sealed enclosure that is part of the pulley. The bearings and pulley rotate on stationary shafts that are mounted to brackets on the conveyor framework. The gear motor transmits torque directly to the shell of the pulley through an internal flange connection. They are a compact design that eliminates the need for external driving power transmission components.

Dead Shaft Pulleys Dead shaft pulleys have internally mounted bearings. The pulley rotates on the shaft, which is locked in position. They are often used where space limitations do not allow for externally mounted bearings.

Pulley Lagging

Conveyor pulleys can be covered with some form of material to increase the coefficient of friction between the belt and pulley. The increase in the coefficient of friction will prevent the belt from slipping. Lagging also assists in cleaning the surface of the belt and pulley. Another purpose of the lagging is to reduce the wear on the face of the unit and to protect the steel rim. The thickness of the lagging can vary from a few thousandths of an inch, as with sprayed-on coatings, to inches as with solid rubber lagging. Methods used for attaching the lagging are bolting, cementing, painting, vulcanizing, and mechanical fasteners.

Lagging compounds are varied and selected based on their cost and the parameters of the application. One of the most common types of lagging is a synthetic rubber known as SBR (styrene-butadiene rubber). It has good abrasion resistance qualities and is relatively low cost. The pulley steel surface is prepared and then painted with an adhesive paint. Strips of raw SBR are wrapped around the pulley and pressure wrapped with a special tape. The pulley is placed in an autoclave where it is subjected to a controlled amount of heat and pressure for a calculated time period. This process of vulcanization changes the characteristics of the SBR as well as bonding it firmly to the steel pulley. The hardness of the lagging after vulcanization is approximately 55-65 as measured by a Shore A durometer.

Lagging typically has grooves cut in it (see Figure 16–11). Evenly spaced grooves can be cut into the lagging in a variety of patterns. Generally the grooves are 1/4″ wide by 1/4″ deep, spaced 1 to 2″ apart. The grooves improve the traction between the belt and the pulley. On pulleys that run in a wet and dirty environment, the grooves act as a channel for the removal of the moisture. Some of the classic shapes or patterns cut in the lagging are herringbone, chevron, and diamond (see Figure 16–12). Groove shapes can be cut to facilitate the removal of material from the pulley and reduce the accumulation of foreign material. In both the herringbone and chevron patterns, the apex should point in the direction of belt travel.

FIGURE 16–11
Lagged drum pulley with grooved pattern.
Courtesy of Rockwell Automation.

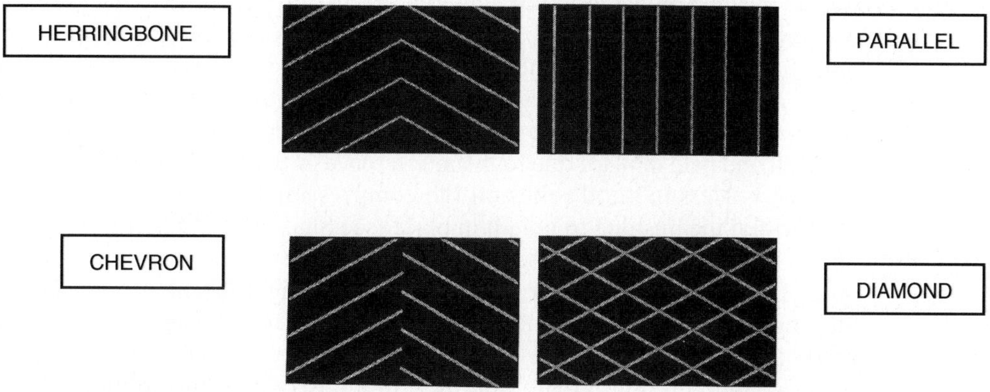

HERRINGBONE

PARALLEL

CHEVRON

DIAMOND

FIGURE 16–12
Various pulley lagging groove patterns.
Courtesy of Precision Pulley & Idler.

FIGURE 16–13
Lagged wing pulley
with shaft.
Courtesy of Rockwell
Automation.

Neoprene lagging is oil and flame resistant. It is also resistant to deterioration caused by ozone and exposure to the sun. Some neoprene lagging can be formulated with additives so that it dissipates static charges.

Ceramic lagging is extremely resistant to abrasion. It is very expensive and delivery times for a ceramic-lagged pulley are extended.

Bolt-on or mechanically fastened lagging is attached to the pulley by fasteners or clamps. Lagged pads formed to match the pulley diameter are bonded to a steel plate. Holding strips or clamps that are welded to the pulley hold the lagged plates in place. This feature allows the lagging plates to be changed out as they wear through. The plates can be replaced without removing the pulley from the conveyor, thus saving time. Lagging can also be bolted directly to the rim with self-tapping screws.

Wing pulleys can be lagged with standard SBR that is bonded and vulcanized. Another method involves installing special replaceable strips that can be slid onto the wing tips of the pulley to allow easy replacement (see Figure 16–13).

Pulley Mounting Systems

Conveyor pulleys must be attached to the shaft. Under operating loads, the shaft and pulley should work together as a common unit. It is important that the pulley be securely fastened to avoid the possibility of coming loose and causing a catastrophic failure in the system. The pulley end disc will be welded to the rim on its outside and welded to a hub on the inside circumference. This hub will be fitted with some form of bushing to clamp onto the shaft. There are essentially three different methods of mounting pulleys to a shaft. The most common method is to use some form of tapered bushing (see Shafting and Shaft Bushing Components chapter). Keyless locking assemblies are another method. Occasionally the pulley hub is mounted directly onto the shaft and held in place due to the interference fit between the shaft and hub.

The keyless method relies on the compression and expansion of concentric rings within the bushing to hold it in place (see Shafting and Shaft Bushing Components chapter). As the cap screws are tightened, the locking assembly clamps down on the shaft and expands into the hub bore for a tight mechanical fit. This type of bushing is the preferred mounting arrangement on high torque and heavily loaded pulley applications.

A common mounting method used in industry is to fit the pulley hub to the shaft with some form of tapered bushing. The tapered bushing is drawn tight with fasteners and clamps on the shaft, wedging it into the hub. The same bushing styles that are used on V-belt sheaves or sprockets can be used on pulleys. The "QD" or

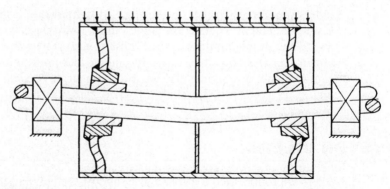

FIGURE 16–14
Stressed end discs of a pulley from mounting tapered bushings, and belt tensions showing the "bellows effect" that can lead to failure.
Courtesy of Rockwell Automation.

"taper-lock" bushings are two of the most prevalent types. They are a reasonably reliable method to lock the pulley to the shaft. Although they are often used for the sake of convenience, they are not the best choice because the taper angle is relatively shallow and the hub of the pulley must travel laterally a considerable distance to lock the assembly in place. During installation, one side of the pulley bushing and hub assembly is locked down onto the shaft initially. When installing the other side, the bushing grips the shaft before it is completely drawn into the hub. This action pulls the hub and end disc out to the bushing. This is sometimes referred to as the "bellows effect" (see Figure 16–14). This movement of the end disc causes bending stress and loading on the welds. The load-carrying capability of a pulley will be reduced and failure of the welds can occur.

To minimize the "bellows effect," manufacturers of conveyor pulleys have specifically designed a group of tapered bushings for use on pulleys. These flanged bushings are similar to standard flanged bushings—but with a few significant differences. A variety of bushings is shown in Figure 16–15. Conveyor pulley bushings have an increased taper angle over standard tapered bushings. Greater bushing

FIGURE 16–15
A variety of conveyor pulley bushings used to mount the pulley to the supporting shaft.

Courtesy of Precision Pulley & Idler.

tapers require less travel on the shaft to tighten the bushing. This decreases the amount of end disc deflection in the pulley and reduces pre-stressing of the pulley due to the installation procedure. These special conveyor bushings have more and larger diameter bolts than standard tapered bushings. The flange is thicker to prevent cracking during installation and removal. It is recommended that conveyor pulleys be mounted to the shaft with these pulley bushings.

CONVEYOR IDLERS

Idlers are rollers used to protect and support the belt and the load being conveyed. T. Edison and T. Hewitt are credited with inventing the conveyor belt idler. Belt conveyor idlers for bulk material handling are available in a wide range of sizes and types. CEMA designates certain specifications and dimensions for the most common forms of idlers. CEMA specifies dimensional standards and load ratings for B, C, D, and E series of idlers. This information can be checked by obtaining a copy of their publication #502-1996.

Most idlers are basically a hollow metal can (roll) that has an internal shaft and bearing assembly. Both roller and ball bearings are used to allow the roll to turn and support the load. The bearing size and type will determine the capacity of the idler. The bearing must be sealed from the surrounding environment to prevent contamination. A seal is usually used to exclude contamination. Idlers come in both sealed bearing or regreasable versions. The regreasable idlers will have a grease line and fitting to allow connection with a grease gun. Certain types of idlers use a solid roller made from urethane or rubber. The idler assembly will typically be composed of the roll with bearings, and shaft mounted into a frame. The roll shaft ends are inserted into a bracket that is attached to a metal frame. The frame is then mounted to the conveyor support structure.

Idler Types

There are two basic types of idlers used on belt conveying systems: the carrying idler and the return idler. Carrying idlers support the belt and load in the section of the conveyor that transports the material. They come in two configurations: troughing idlers and flat belt idlers. Return idlers are used on the return side of the conveyor offering belt support and assisting in training the belt.

Troughing Idlers Troughing idlers are carrying idlers that use multiple rolls, usually three, to form a trough for the belt to ride in (see Figure 16–16). The overall width of the idler assembly will be determined by the width of the belt it is supporting. The rolls will be uniform in diameter, with 4″, 5″, and 6″ being common sizes. The roll length can be equal or the center roll can be longer. Idlers with a longer center roll are referred to as "picking idlers" or "unequal length troughing" and are used in a location on the conveyor where sorting, weighing, or collecting takes place (see Figure 16–17). The center roll carries the majority of the load. The trough

FIGURE 16–16
Three-roll inline
troughing idler with
equal length rolls.
Courtesy of Precision
Pulley & Idler.

FIGURE 16–17
Picking idler with longer
center roller.
Courtesy of Precision
Pulley & Idler.

FIGURE 16–18
Offset roll troughing idler.
Note the position of the
center roll relative to the
outer rolls.
Courtesy of Stephens-Adamson
Conveyor Components.

shape allows more tonnage to be carried than flat rolls. The trough shape also reduces losses of product due to spillage. They are available in 20°, 35°, and 45° angles. The steeper angle troughing idlers will have a greater carrying capacity but
require greater transverse flexibility in the belt. Another version of the troughing
idler has an offset center roll (see Figure 16–18). This version is often used where
space limitations require a narrower frame. Adjustable angle troughing idlers are
used to make a transition to the terminal pulley.

Impact Troughing Idlers Impact troughing idlers have molded urethane or rubber rolls mounted on the shaft rather than hollow cans (see Figure 16–19). They are
designed to prevent damage to the belt at the loading point. Large lumps and heavy
material falling on the belt can seriously damage it unless impact rolls absorb the
energy. It is standard practice to use impact idlers at all loading and transfer points
and to space them close together.

FIGURE 16–19
Impact troughing idler
with rolls made of rubber
or other impact-resistant
material.
Courtesy of Precision
Pulley & Idler.

FIGURE 16–20
Troughing trainer
idler with side guide
rollers used to assist
in training the belt.
Courtesy of Precision
Pulley & Idler.

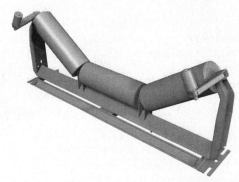

Troughing Training Idlers Troughing training idlers are carrying idlers that assist in the tracking of the belt (see Figure 16–20). A properly aligned and centrally loaded belt will track in the center of the idlers. Occasionally transient and varying conditions will cause the belt to run misaligned. A training idler will help maintain belt alignment in some conveyor applications. A training idler will have the carrying roll frame mounted on a central pivot point. Guide rolls, mounted on the idler frame perpendicular to the belt edge, urge a displaced belt to return to the center. The idler base must be kept free of dirt so it can pivot and does not lock up. The side guide rolls must turn freely or excessive belt edge wear will occur. There are two types of troughing training idlers: the positive and the actuating shoe. The positive type has side mounted guide rolls perpendicular and offset to the roll shaft axis. They should be used for one-direction belt travel. Actuating shoe types have nonrotating guides on the sides of the rolls and can cause wear on the belt edges. The advantage of this type is they can be used for reversing belts. Troughing training idlers should be spaced 100′ apart and not within 50′ of the tail or head pulley.

Suspended Idler Standard troughing idlers with steel frames are available with special brackets to be attached and suspended from wire ropes. Suspended carrying idlers known as "garland" idlers do not have a steel frame. They are directly attached to the wire rope and can have three or five steel rolls linked together or a single solid rubber roll suspended from the conveyor framework (see Figure 16–21). One advantage of this type of idler is its ability to handle off center and uneven loads. It is also tolerant of poor alignment conditions. The supporting structure size and weight can be minimized and the idlers can be suspended from wire ropes.

FIGURE 16–21
Suspended or garland idlers hanging from a frame.
Courtesy of Stephens-Adamson Conveyor Components.

FIGURE 16–22
Flat carrying roll idler used to support the belt on top of the framework.
Courtesy of Precision Pulley & Idler.

FIGURE 16–23
Flat return idler used under the conveyor framework to support the belt on its return run.
Courtesy of Precision Pulley & Idler.

Flat Carrying Roll Idlers Flat carrying or carrier rolls are single horizontal rolls positioned between brackets that are mounted to the conveyor framework (see Figure 16–22). They are available in varied widths depending on the belt width. Impact carrying rolls will have molded rubber or urethane rolls to prevent damage to the belt. They are used at loading points on the belt to absorb shock.

Flat Return Idlers Flat return idlers are used to support the belt on its return run (see Figure 16–23). They are suspended below the flanges of the frame stringers. Standard return rolls are made of steel in different widths and diameters. They are also available with rubber discs.

Self-Cleaning Return Idlers Self-cleaning return idlers are made in two versions: spiral and disc (see Figure 16–24). The spiral version has rubber or neoprene material wound in a spiral in the shape and diameter of standard rolls. Some material will

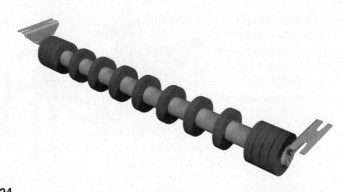

FIGURE 16–24
Disc/impact return idler.
Courtesy of Precision Pulley & Idler.

be removed from the belt when using these idlers, but they are not belt cleaners. The disc return idler has spaced discs in the center and massed discs on the end that prevent the build-up of material on the rolls, whereas the spiral return idler has a spiral of rubber material wound around the idler that assists in removing wet and sticky materials.

CONVEYOR BELTING

The conveyor belt is the part in the system that carries and moves the load; therefore it must be selected, installed, and maintained properly. Various synthetics, rubbers, and elastomeric compounds are used in the construction of conveyor belting. They are bonded and woven together in layers to form the belt. The size and construction of the belt will be determined by the criteria of the application. A conveyor belt is rated in part by its ability to handle tension. Tension ratings can be established on the basis of pounds per inch per ply, or pounds per inch of belt width (PIW). PIW ratings are the industry-accepted system. Ratings, construction, and materials can vary widely among manufacturers.

Conveyor Belt Construction

Conveyor belt construction consists of three elements: top cover, carcass, and bottom cover (see Figure 16–25). The purpose of the covers is to protect the belt carcass against damage from the operating environment or material being transported. The material used for covers is usually a rubber compound. Various elastomers, rubbers, and compounds are mixed to obtain the properties needed for the application. Some important considerations of the belt cover are its ability to withstand abrasion, its tensile strength, and the amount of elongation that can occur. The thickness of the cover can vary from 1/32″ to more than 1″ on heavy-duty applications.

FIGURE 16–25
Basic conveyor belt construction showing its layers.

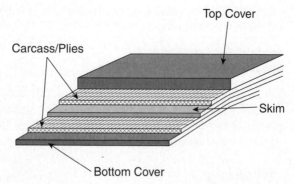

The belt carcass is the tension-carrying portion of the belt. It absorbs the impact energy that occurs during loading. It also provides stability for alignment and load support over the idler rolls. Most belt carcasses have one or more plies of woven fabric. Multiple ply belts have layers of a rubber compound (skims) separating the fabric layers. The fabric yarns are made from materials like rayon, polyester, nylon, and other synthetics. The belt fabric is made of warp yarns that run lengthwise, and weft yarns that run crosswise (transversely). The four basic weave patterns are plain weave, straight-warp weave, solid-woven weave, and woven-cord weave.

There are at least four types of conveyor belt carcass. Multiple-ply belt carcasses are made up of three or more plies (layers) of woven belt fabric that are bonded together. Multiple-ply belts are an older design. Reduced-ply belts consist of a carcass with fewer plies than a comparable multiple-ply belt. They often incorporate special weaves of a combination of high strength synthetic fibers. Steel-cable belts are made with parallel steel cables imbedded into the rubber. They are often used in applications with high operating tensions. Solid-woven belts consist of a single ply of a solid, woven fabric that is usually impregnated and covered with PVC or urethane compounds.

For incline or horizontal service, a conveyor belt can have special "cleats" or ridges molded into the top cover. These cleats prevent material and product from slipping back on the belt (see Figure 16–26). A variety of cleat types and sizes can be bonded to the belt. Belts can also be treated or impregnated with compounds to make them more fire resistant. Special fire-resistant belting is used in potentially explosive environments, such as in a coal mine. The food and beverage industry will

FIGURE 16–26
Conveyor belt with cleats. Cleats prevent bulk material and product from slipping back on the belt.

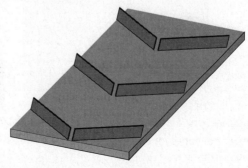

often use a white belt made of nitrile rubber or PVC compounds. It is designed for frequent wash-down duty.

Conveyor Belt Splices

There are basically two methods, mechanical splice or vulcanized splice, for joining the conveyor belt. A mechanical splice will use some type of rivets, fasteners, bolted plates, or lacing to join the two squared ends of the belt (see Figure 16–27). A vulcanized splice involves the careful preparation and joining of overlapped, cemented belt sections in a cold or hot process. Hot vulcanized splicing requires the belt ends to be subjected to heat and pressure from a "Vulcanizer."

The constraints of time, equipment, economics, and type of service the belt will be subjected to determine the type of splice used. The advantages of a mechanical splice are that they are quickly and easily installed. They have a low initial cost and can be done by most maintenance technicians. The advantages of a vulcanized splice are strength and long service life. It is also a cleaner, smoother union of belt ends.

The most common problem with belts used on conveyors is keeping the belt tracking in the center of the frame and idler rolls. If the belt consistently fails to track correctly—even after idler adjustment has been tried—it is probably due to an improperly spliced belt. Belt ends must be "squared" off or the belt will track improperly and fail prematurely. The most precise method is to use a center point as reference.

Following are general belt end squaring guidelines:

- Beginning 15′ to 20′ back from both ends of the belt measure and mark the center points every 1 to 2′.
- Do not use the belt edge as a squaring guide.
- Use a chalk line to draw or mark an average centerline with the points as a guide.
- Use a steel square to mark the transverse line at a point at which the belt is to be cut (see Figure 16–28).

FIGURE 16–27
Mechanical conveyor belt splices.
Courtesy of Flexible Steel Lacing Co.

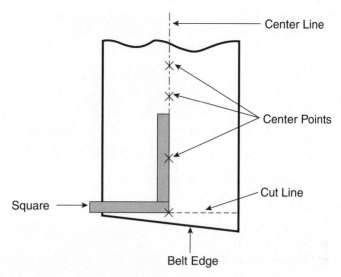

FIGURE 16–28
Correctly squared conveyor belt edge using the belt center line as a reference.

Conveyor Belt Cleaners and Scrapers

Many different types of apparatus can be employed to keep material from building up on the conveyor belt. Debris build-up is detrimental to the life of the belt and the other conveyor components. A precleaner is a scraping device that uses polyurethane blades to scalp the dirt off of the face of the pulley just below the centerline of the pulley (see Figure 16–29). The blades are intended to be in light contact with the belt. The blades are sectional and are clamped or slid into a mounting bracket. A self-adjusting tension spring provides constant pressure on the belt. A "T" cleaner also uses replaceable blades of metal or carbide bonded into a

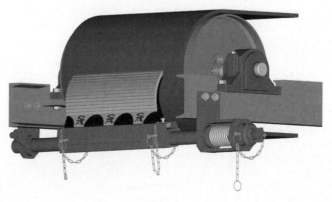

FIGURE 16–29
Belt precleaner.

Courtesy of Stephens-Adamson Conveyor Components.

FIGURE 16–30
Belt T-cleaner.

Courtesy of Stephens-
Adamson Conveyor
Components.

FIGURE 16–31
Arm Cleaner.

Courtesy of Stephens-
Adamson Conveyor
Components.

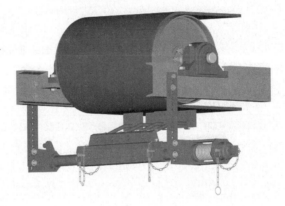

rubber mounting block (see Figure 16–30). It is mounted beneath the pulley and functions regardless of the belt direction. Arm or spring-arm-type cleaners have staggered arms with cleaning blades (see Figure 16–31). An air knife cleaner uses air pressure to remove the debris from the belt surface (see Figure 16–32). An electric motor drives a blower that forces air through a narrow orifice box that spans the width of the underside of the belt. It is often used to remove wet and sticky materials. Brush cleaners have rotating brushes in contact with the underside of the

FIGURE 16–32
Air knife belt cleaner.

Courtesy of Stephens-
Adamson Conveyor
Components.

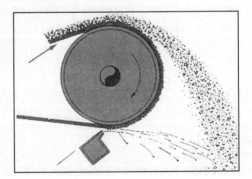

belt at the pulley location. They will work on any type of belt but are especially suited for patterned belts. Regardless of which form of belt cleaner is used, the belt must remain clean to prevent premature wear of all parts of the system.

Belted Conveyor System Installation and Maintenance

The proper installation and maintenance of a conveyor system will help ensure its safe and efficient operation. Assuming that the design and manufacturing steps were properly done, the assembly and upkeep of the system is crucial to trouble-free operation. The conveyor must be thought of as an interconnected system. If one component is performing less than satisfactorily, it will adversely affect numerous other parts of the system. Prior to any assembly or maintenance procedures all local, state, federal, and plant safety rules must be complied with. The following guidelines are general in nature and are not meant to replace shop procedures or documents on the subject.

- Lockout, tagout, and verify that any driving mechanisms are neutralized. Wear proper safety equipment when working around conveyors.

- Prior to installation check all pulleys, idlers, and power transmission components for damage from mishandling. Make sure that the rolls in the idlers turn freely. Replace all worn or damaged components.

- Clean the surrounding work area.

- All conveyor trusses, channels, stringers, and frames must be installed parallel, straight, square, and level to insure proper belt tracking. Frames and stringers should be parallel within a 1/8″ tolerance. The maximum allowable lateral offset in conveyor stringers shall be 1/8″ in 40′. The idler supports should be level within 1/8″ regardless of the belt width. The elevation of the stringer above the supporting structure should be held within plus or minus 1/4″.

- After the supporting structure has been installed and prior to the installation of the belt, install the pulleys. Conveyor pulleys should be set level and the shaft centerline perpendicular to the centerline of the belt. The exception to this would be on special turnover conveyor applications. Alignment of the pulleys is critical to prevent belt wear and training problems. Shim and shift the bearing blocks on the pulley to align it. Using a level and checking both sides of the pulley, set the shaft elevation to within 1/32″. Measure from a line that is perpendicular to the conveyor centerline and adjust the pulley so that it does not deviate greater than plus or minus 1/32″.

- Install all conveyor pulley bushings and bearings by following proper manufacturer-recommended procedures. The use of torque wrenches is recommended when tightening bushing bolts to control the installation and contact pressures.

- Connected drives must be supported by rigid bases and aligned. Power transmission components (couplings, V-belts, etc.) need to be correctly installed and aligned.

Recommended Average Spacing of Idlers (in feet)

| Belt Width Inches | Carrying Idlers | | | | | | Return Idlers |
| | Weight of Material Conveyed Pounds Per Cubic Foot | | | | | | |
	30	50	75	100	150	200	
18	5½	5	5	5	4½	4½	10
20	5½	5	4½	4½	4	4	10
24	5	4½	4½	4	4	4	10
30	5	4½	4½	4	4	4	10
36	5	4½	4	4	3½	3½	10
42	4½	4½	4	3½	3	3	10
48	4½	4	4	3½	3	3	10
54	4½	4	3½	3½	3	3	10
60	4	4	3½	3	3	3	10
72	4	3½	3½	3	2½	2½	8
84	3½	3½	3	2½	2½	2	8
96	3½	3½	3	2½	2	2	8

Note: 1. **Loading Point** (Impact Idlers)—Normally, one half the carrying idler spacing is recommended. For fine materials, place the idlers as close together as possible.

2. **Convex Curves**—One-half of normal spacing unless radius is small, then space closer.

FIGURE 16–33
Idler spacing table. Spacing is determined by the belt width and the material weight.
Courtesy of Rexnord.

■ The idlers should be set from a previously squared and leveled terminal pulley. Spacing of the idlers is determined by the weight of the material, belt weight, idler rating, sag, belt tension, and width of the belt (see Figure 16–33). Loading points require closer spacing than at other locations along the belt. The amount of belt sag between idlers should be limited to prevent spillage and damage to the belt. Place idlers in position by sliding them firmly downstream against the mounting bolts and lightly tightening them (see Figure 16–34). The idlers must be square and in line to the conveyor centerline, and parallel to one another. They should be centered and level across the width of the conveyor (see Figure 16–35). Run a 100″ tight wire on the conveyor centerline to form a reference. The line should be referenced to the starting squared pulley. Place the idlers at their design spacing and squared to the line. After 50′ of idlers have been installed, the 100′ line should be relocated so that there is 50′ of overlap. Reposition the wire

FIGURE 16–34
Positioning idlers against downstream hold-down bolts.
Courtesy of Rexnord.

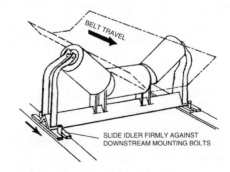

BELT TRAVEL

SLIDE IDLER FIRMLY AGAINST
DOWNSTREAM MOUNTING BOLTS

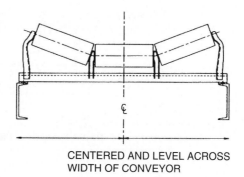

Figure 16–35
Centered and leveled idlers are important for proper belt tracking and even load distribution.

Courtesy of Rexnord.

CENTERED AND LEVEL ACROSS
WIDTH OF CONVEYOR

as needed until the entire conveyor is filled with idlers. Tighten the mounting bolts before start-up.

■ The distance between the last troughing idler to the terminal pulley is critical. The outer edges of the belt are stretched in the transition zone between the idler and pulley, increasing tension and potentially permanently stretching the edges. This will result in belt training problems. If too great a distance is allowed, spillage of the material will result. Idlers of varying or adjustable angles can be used in the transition area. Figure 16–36 shows transition distance recommendations.

■ Comparing diagonal measurements between idlers will check on the squareness of the frame and idler spacing.

■ Install the belt and make sure that there are no defects. Belt splices must be square and secure.

■ Tension the belt so it does not slip on the drive pulley. It should be no tighter than necessary to prevent slipping under the most extreme loading conditions. Take-up counter weights can be predetermined by qualified conveyor engineers to ensure proper tensioning.

■ Start the system empty to determine the belt alignment. The belt should not track over to one side. If one or more segments of the belt run off at all points along the conveyor, the cause is probably due to belt bow, poor belt splice, or uneven material loading. If the belt runs out of line consistently at one location on the conveyor, the cause is probably misaligned idlers. Usually the idlers that require adjustment are located upstream of the point where the belt runs out of line.

■ Shut the system down for alignment. The belt should be trained by adjusting the idlers—not the position of the pulleys. Although it is a common practice on conveyors with screw-type take-up blocks to train the belt by tail pulley adjustment, this is improper. Twisting the tail pulley will result in a temporary fix because the tension on the system and components will be increased. Ideally, a level and square frame along with a straight belt that has been correctly spliced will track the belt. Changing the frame or belt is not always an option; in those instances when the belt and frame cannot be adjusted, shifting or "knocking" idlers might be required. Train the belt by loosening the bolts of several idlers on the upstream side and shifting the idler axis (see Figure 16–37). Never shift

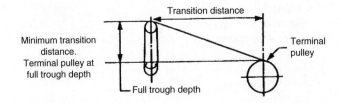

Idler Angle	% Rated Tension	Fabric Belts	Steel Cable Belts
20°	Over 90 60 to 90 Less than 60	1.8b 1.6b 1.2b	4.0b 3.2b 2.8b
35°	Over 90 60 to 90 Less than 60	3.2b 2.4b 1.8b	6.8b 5.2b 3.6b
45°	Over 90 60 to 90 Less than 60	4.0b 3.2b 2.4b	8.0b 6.4b 4.4b

(a)

Idler Angle	% Rated Tension	Fabric Belts	Steel Cable Belts
20°	Over 90 60 to 90 Less than 60	.9b .8b .6b	2.0b 1.6b 1.0b
35°	Over 90 60 to 90 Less than 60	1.6b 1.3b 1.0b	3.4b 2.6b 1.8b
45°	Over 90 60 to 90 Less than 60	2.0b 1.6b 1.3b	4.0b 3.2b 2,3b

b = Belt width

(b)

FIGURE 16–36
Chart showing the proper idler transition distances for terminal pulley at (a) full trough depth and (b) at or near one-half trough depth. Idler angle and tension must be considered.
Courtesy of Rexnord.

an idler more than 1/4″ in any direction from square. Do not shift idlers on reversing conveyors. Care must be taken to not overcompensate by shifting an excessive number of idlers. Minimize the number of idlers that are adjusted and mark them as being adjusted from center. In the case of erratic running belts, the troughing idler can be tilted forward no more than 2° in the direction

FIGURE 16–37
Shifting idlers to assist belt tracking.

Courtesy of Rexnord.

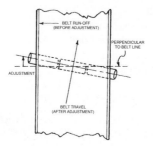

FIGURE 16–38
Tilting idlers to assist belt tracking.

Courtesy of Rexnord.

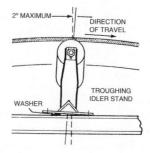

of belt travel (see Figure 16–38). Training the belt onto the tail pulley can also be accomplished by shifting the axis of the return idlers. Some industry experts recommend that the belt be aligned from the return side, due to lower tensions. The best practical alignment should be accomplished with minimal adjustment.

■ Start the system and inch it along, if possible. Check the belt for training and to see if it is running properly. Gradually load the belt until its full operational load is achieved. If the belt runs for 8 continuous hours within the width of the pulley faces, it is considered aligned.

■ Proper loading of the belt is necessary to assure alignment and achieve maximum service life from the conveyor components. The loading area of the belt is one of the most critical points on the conveyor. Check the chutes to make sure that the material is being directed onto the center of the belt (see Figure 16–39). Off-center loading will adversely affect belt alignment and is harmful to the belt, idlers, and frame. Ideally the material should pass from the chute to the belt at the same speed and direction of belt travel. Grizzly or spaced bars that are part of the chute will allow fine materials to drop on the belt ahead of any lumps to cushion impact.

■ The skirting must be adjusted to minimize spillage of material and to assist in centering the load. The recommended maximum distance between skirt boards is 2/3 the width of the troughed belt. The length of the skirting should be at the point where the material is still and not tumbling. Uniform pressure from the skirting, scrapers, and plows is required. If the skirting has too much contact pressure on the belt, both the skirt and belt will wear. Old conveyor belting is not the best skirting material. Special resilient skirt-board compounds with a low coefficient of friction are made and designed for that purpose.

FIGURE 16–39

Proper loading onto a conveyor belt will assist in straight belt tracking and prevent uneven wear of conveyor system components.

Courtesy of Rexnord.

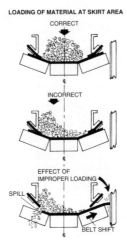

- Many idlers have sealed-for-life bearings and require no relubrication. Idlers that require relubrication should be done at least every 4000 to 6000 hours of operation. The lubricating cycle should be adjusted according to the severity of the application. In a clean environment the time can be lengthened; in a dirty and wet environment it might need to be shortened. An NLGI #2 lithium based mineral oil grease with an operating temperature range of $-10°F$ to $+225°F$ is recommended. Do not use high-pressure lubricating equipment because the seals may become damaged. All fittings should be wiped clean before delivering the grease. Seven to ten strokes, or until a small amount of grease is visible around the seal, is all that is required.

Conveyor Troubleshooting

See Figure 16–40.

Screw Conveyors

Screw conveyors have been in use for thousands of years. The first screw conveyors were developed for the purpose of lifting water. During the early stages of the industrial revolution, an inventor named Oliver Evans constructed a mechanized flour mill that incorporated the use of a bucket elevator, belt conveyors, and a crude screw conveyor. The mill, built in 1785, was powered by a water wheel. The screw conveyor was made from wood. In 1898 Frank Caldwell patented the one-piece continuous helicoid flight (screw) which marked the beginning of the modern screw conveyor system.

The operating principle behind the screw conveyor is based on the mechanical advantage realized by wrapping an inclined plane around a rotating core. As the helix (flighting) is turned, any material that is resting on the carrying side of the helix will move. The advantages of a screw conveyor system include the following: It is completely enclosed, compact, blends the material, lifts in elevation, has

BELT CONVEYOR TROUBLE-SHOOTING

Any belt conveyor installation can be subject to a wide variety of problems which may become costly in terms of replacement and plant downtime unless quickly diagnosed and corrected. This guide is intended to point out the majority of belt conveyor problems and to set forth their probable causes and solutions.

Locate the specific problem in the "problem" column below and note the numbers to the right. They represent the most likely causes for the problem, in order of probable occurrence, and how to correct them. The list at the bottom of the page details the causes and solutions by number.

COMMON BELT CONVEYOR PROBLEMS

PROBLEM	CAUSE AND SOLUTION In Order of Probable Occurrence					
Belt runs off at tail pulley.	7	15	14	17	21	—
Entire belt runs off at all points of the conveyor.	26	17	15	21	4	16
One belt section runs off at all points of the conveyor.	2	11	1	—	—	—
Belt runs off at head pulley.	15	22	21	16	—	—
Belt runs to one side throughout entire length at specified idlers.	15	16	21	—	—	—
Belt slip.	19	7	21	14	22	—
Belt slip on starting.	19	7	22	10	—	—
Excessive belt stretch.	13	10	21	6	9	8
Belt breaks at or behind fasteners; fasteners tear loose.	2	23	13	22	20	10
Vulcanized splice separation.	13	23	10	20	2	9

PROBLEM	CAUSE AND SOLUTION In Order of Probable Occurrence					
Excessive wear, including rips, gouges, ruptures and tears.	12	25	17	21	8	5
Excessive bottom cover wear.	21	14	5	19	20	22
Excessive edge wear, broken edges.	26	4	17	8	1	21
Cover swells in spots or streaks.	8	—	—	—	—	—
Belt hardens or cracks.	8	23	22	18	—	—
Covers become checked or brittle.	8	18	—	—	—	—
Longitudinal grooving or cracking of top cover.	27	14	21	12	—	—
Longitudinal grooving or cracking of bottom cover.	14	21	22	—	—	—
Fabric decay, carcass cracks, ruptures, gouges (soft spots in belt).	12	20	5	10	8	24
Ply separation.	13	23	11	8	3	—

PROBABLE CAUSES AND SOLUTIONS

1. **Belt bowed** — Avoid telescoping belt rolls or storing them in damp locations. A new belt should straighten out when "broken in" or it must be replaced.

2. **Belt improperly spliced or wrong fasteners** — Use correct fasteners. Retighten after running for a short while. If improperly spliced, remove belt splice and make new splice. Set up regular inspection schedule.

3. **Belt speed too fast** — Reduce belt speed.

4. **Belt strained on one side** — Allow time for new belt to "break in." If belt does not break in properly or is not new, remove strained section and splice in a new piece.

5. **Breaker strip missing or inadequate** — When service is lost, install belt with proper breaker strip.

6. **Counterweight too heavy** — Recalculate weight and adjust counterweight accordingly. (If using screw takeups, reduce takeup tension to point of slip, then tighten slightly.)

7. **Counterweight too light** — Recalculate weight required and adjust counterweight accordingly. (If using screw takeups, increase tension.)

8. **Damage by abrasives, acid, chemicals, heat, mildew, oil** — Use belt designed for specific condition. For abrasive materials working into cuts and between plies, make spot repairs with cold patch or with Permanent Repair Patch. Seal metal fasteners or replace with vulcanized step splice. Enclose conveyor for protection against rain, snow and sun. Don't over-lubricate idlers.

9. **Differential speed wrong on dual pulleys** — Make necessary adjustment.

10. **Drive underbelted** — Recalculate maximum belt tensions and select correct belt. If conveyor is over-extended, consider using two-flight system with transfer point. If carcass is not rigid enough for load, install belt with proper flexibility when service is lost.

11. **Edge worn or broken** — Repair belt edge. Remove badly worn or out-of-square section and splice in a new piece.

12. **Excessive impact of material on belt or fasteners** — Use correctly designed chutes and baffles. Make vulcanized splices. Install impact idlers. Where possible, load fines first. Where material is trapped under skirts, adjust skirtboards to minimum clearance.

13. **Excessive tension** — Recalculate and adjust tension. Use vulcanized splice within recommended limits.

14. **Idler rolls not turning** — Correct or replace stalled rolls. Lubricate. Improve maintenance. (Don't over-lubricate.)

15. **Idlers or pulleys out-of-square with center line of conveyor** — Realign. Install limit switches for greater safety.

16. **Idlers improperly placed** — Relocate idlers or insert additional idlers spaced to support belt.

17. **Improper loading, spillage** — Feed should be in direction of belt travel and at belt speed, centered on the belt. Control flow with feeders, chutes and skirtboards.

18. **Improper storage or handling** — Refer to your belt supplier for storage and handling tips.

19. **Insufficient traction between belt and pulley** — Increase wrap with snub pulleys. Lag drive pulley. In wet conditions, use grooved lagging. Install correct cleaning devices on belt. Install centrifugal switch for safety.

20. **Material between belt and pulley** — Use skirtboards properly. Remove accumulation. Improve maintenance.

21. **Material build-up** — Remove accumulation. Install cleaning devices, scrapers, and inverted "V" decking. Improve housekeeping.

22. **Pulley lagging worn** — Replace worn pulley lagging. Use grooved lagging for wet conditions.

23. **Pulleys too small** — Use larger-diameter pulleys.

24. **Radius of convex vertical curve too small** — Increase radius by vertical realignment of idlers to prevent excessive edge tension.

25. **Relative loading velocity too high or too low** — Adjust chutes or correct belt speed. Consider use of impact idlers.

26. **Side loading** — Load in direction of belt travel, in center of conveyor.

27. **Skirts improperly placed** — Install skirtboards so that they do not rub against belt.

FIGURE 16–40

Conveyor troubleshooting guide.

Courtesy of Rexnord.

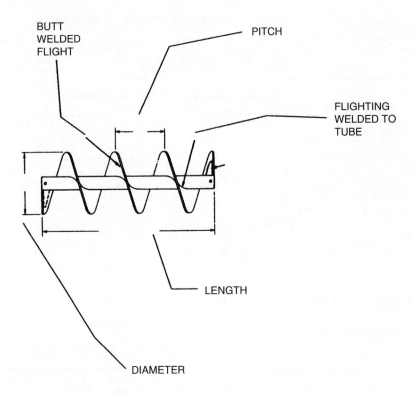

BUTT
WELDED
FLIGHT

PITCH

FLIGHTING
WELDED TO
TUBE

LENGTH

DIAMETER

FIGURE 16–41
Standard screw conveyor flight terminology and construction.
Courtesy of Martin Sprocket & Gear.

metered flow, and heats or cools the product. The chief disadvantages include limited volume and material degradation, and that they are limited by the torsional strength of their components. Screw conveyor systems can be made to move the product horizontally as well as vertically.

The screw flights are available in a variety of sizes, materials, and configurations. The critical dimensions of the screw are the diameter, pitch, and the flight thickness (see Figure 16–41). The pitch of the screw is the measurement between the centerline of two adjacent flights and is usually equal to the diameter of the screw. Screw flights can be made in left-hand or right-hand helix forms (see Figure 16–42). Some screw flights are made in sections (sectional) and joined together with plates and rivets, or they are welded. Screw flights can also be made in one continuous piece and are referred to as "helicoid flighting." The helicoid screw flight cross section can be made to be tapered, with the thickness at the inner edge approximately twice the thickness of the outer edge. The choice of screw shape, sizes, and types is based on the volume and the material being moved.

The screw helix is usually mounted on a central pipe or shaft that is contained within a trough and connected to a driving mechanism. Newer design systems can incorporate the use of a shaftless screw. The screw and shaft will be supported on one

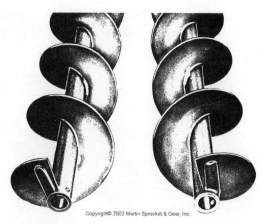

Copyright© 2002 Martin Sprocket & Gear, Inc.

FIGURE 16–42
Left-hand and right-hand screw flights.

Courtesy of Martin Sprocket & Gear.

FIGURE 16–43
Screw conveyor hanger
bearing.

Courtesy of Martin
Sprocket & Gear.

Copyright© 2002 Martin Sprocket & Gear, Inc.

end—along its length on long sections—by a hanger bearing (see Figure 16–43). The hanger bearing is contained in a "T"-shaped bracket that is bolted to the trough top. Various bearing materials are available, such as bronze or plastic. The other end may have a trough-end bearing block. The hollow screw shaft will have a stub shaft inserted and pinned into it. This drive shaft passes through a seal of some type and protrudes from the trough end. A drive is connected to the shaft. The drive apparatus is typically a motor and reducer combination (see Figure 16–44). It is preferred to have the drive at the discharge end of the trough to keep the components in tension, not compression. The trough is a variable length, metal "U", square or round-shaped tube that is slightly wider than the screw. The material being moved along with the flight is contained in the trough or tube.

A feeder is a different type of screw conveyer. It operates on the same principle as a standard screw conveyor but with some significant differences. A screw feeder will usually have the inlet flood loaded. It is designed to regulate the material flow from the hopper to the bin.

Screw conveyors are designed to transport a wide variety of bulk materials. The material can be anything from grains to glass. There is almost no limit to what can be moved with a screw conveyor. Each material has its own characteristics such as density, particle size, flowability, abrasiveness, weight, etc. When designing screw conveyor systems, it is important to know how the material will react when it is being conveyed. CEMA publishes charts and tables that define these important characteristics for different types of materials. The design and selection process for screw conveyors is complex.

Bucket Elevators

Bucket elevators are designed to convey bulk materials in a vertical path. These elevators have buckets attached to chains or an endless belt enclosed in a casing (see Figure 16–45). The buckets are available in steel, cast malleable iron, and plastics of various compositions. Material is fed to the buckets at the bottom portion of the system called the "boot." It is then elevated to the top of the structure and discharged at the "head end." A drive mechanism, which is usually a gear motor, is connected to the driving pulley or drive sprockets.

There are four general classifications of bucket elevators. The centrifugal discharge elevator has buckets mounted on a belt or chains that are spaced at intervals. The material is fed into the buckets at the boot and picked up by more buckets as they pass around the foot wheel. When the buckets pass over the head wheel at the top of the elevator, the material is discharged by centrifugal force. The operating speeds for these types of elevators are in the range of 160 feet per minute (FPM) to 325 FPM. The continuous discharge elevator has nonspaced buckets mounted continuously on chains or a belt. The operating range for these systems is approximately 100 FPM to 150 FPM. Gravity causes the material to be discharged. They are used with coarse, lumpy materials in the 2 to 4″ size range. Super-capacity elevators are a version of the continuous type and are designed for very coarse

FIGURE 16–45
Enclosed bucket elevator system.
Courtesy of Martin Sprocket & Gear.

materials. The buckets are mounted between two chains that can support large buckets. They operate in the 80 to 120 FPM range. Positive discharge elevators have spaced buckets that are turned over by idler wheels. Operating speeds are generally slow: 120 FPM or less.

Elevator chain is designed for the attachment of buckets. Pitches of the chain links range from 2.6″ to 9″. Chain is used on applications of severe duty where the environment or temperature would be destructive to belted elevators. It is important to have matched chain lengths for both sides. Do not mix old chain strands with new strands on bucket elevators.

ROLLER BED, SLIDER BED, AND GRAVITY CONVEYORS

The conveyors that make up this group are generally considered "package," "sorting," or "unit handling" conveyors. They are designed to handle packages and finished product, or are used as light-duty assembly line conveyors. Anywhere packages and parts are sorted, handled, shipped, and distributed you will find these conveyors. They generally are lighter in construction and have a shorter span and smaller horsepower drives than bulk material-handling conveyors. The capacity ratings of these types of systems are usually measured in pounds. Linear speeds or velocities are measured in feet per minute (FPM). In the past the velocity of these systems was usually below 100 FPM. With the advent of high-speed drives and sophisticated sensors, the operating speeds are increasing.

The design and construction of these conveyors allow them to move in a horizontal or incline direction, or to turn corners. Unit handling conveyors are made in a variety of different types such as roller bed, slider bed, gravity roller, and others. Several of these conveyor types can be integrated into a computerized network for central control of gates and switches at intersections. These conveyors are referred to as sorting conveyors. Sorting conveyors can be very complex, involving the use of various electrical variable speed drives, encoders, sensors, PLCs, gates, switches, and other devices in the system. One way that sensors are used in sorting conveying systems is to give each unit or package a label (bar code). This label can contain such information as the composition, location, address, route, etc., of the unit. When the sensor reads the unit label, a signal is sent to the brains of the system. Within a fraction of a second a command is given to automated and mechanized sorting devices to direct the path of the package. Sliding shoes, pushers, grippers, gates, and a variety of diverting mechanisms can all be triggered to divert the package onto a spur or into a bin. This eliminates the repeated manhandling of the unit and increases the speed of the process significantly.

Roller Bed Conveyors

The roller bed conveyor is made up of multiple closely spaced rollers supported by a steel or aluminum frame (see Figure 16–46). The frame sections are typically sectional to allow custom flow arrangements. Some systems might require straight runs or numerous diverging and merging intersections. Roller bed sections are available in curved sections also. The system can be either powered—referred to as "live" conveyors—or rely on gravity to move the product. Often the two types are combined within a plant to facilitate the flow of products or packages as they are sorted or assembled.

The rollers support the product and can be driven or freewheeling. The tubular rollers contain internally mounted ball bearings that allow them to rotate freely. The roller ends are secured into a channel frame that stands on legs. The frame legs should be adjustable to allow the bed to be leveled. Most of the rollers are made so

FIGURE 16–46
Roller bed conveyor.

Courtesy of Hytrol Conveyor
Company Inc.

they can be removed easily for replacement if they should fail or become damaged. The positioning of the rollers, relative to the side height of the channels, can be either above or below. If the rollers are mounted lower than the side channels, the channels assist in containing the package. If the rollers are mounted slightly higher than the frame, oversize packages can be moved. The size and weight of the units loaded onto the conveyor dictate the length, diameter, and spacing of the rollers.

Powered or live-roller conveyors can be driven from the end of the bed or they may have the motor-reducer drive positioned in the center of the conveyor. Center drive systems on roller bed conveyors are recommended for belt travel in two directions and are used on medium to heavy loads. The belt can be positioned in one of two ways. The flat belt can be positioned on top of the rollers to move the units and be driven from a drive roller (see Figure 16–1). With this arrangement it is necessary to have some type of take-up mechanism to provide tension to the belt. The take-up assembly is an adjustable roller arrangement that compensates for changes in belt length due to wear or stretch. The units are placed on top of the belt and transported along the belt as it moves. Snubbing and take-up rollers assist in maintaining the tension on the belt. Return rollers attached to the frame support the belt on its return run. Irregular-shaped units can be safely handled without jamming with roller conveyors that have the belt on top.

Another form of belt-powered roller conveyor has the belt positioned underneath the rollers and driving each roller. They are called belt-driven live-roller conveyors. The load-carrying rollers are called "tread rolls." Tension on the belt is provided by snubbing rollers and the take-up arrangement. Pressure rollers maintain friction between the rollers and belt. The pressure rollers should be adjustable in height in relation to the bottom of the carrying rollers. Normally pressure rollers are spaced every second or third carrying roller.

Chain-driven live-roller conveyors use a roller bed for the carrying surface. The conveyed units ride on the surface of the rollers, which are driven by standard roller chain and sprockets. The sprockets are mounted on the roller shaft ends. There are two types of chain-driven roller conveyors: roll-to-roll and continuous chain. With the roll-to-roll type, power is transmitted from roll to roll via short chain loops encircling the sprocket, which is attached to each adjacent roller (see Figure 16–47). Nonpowered idler rolls can be spaced every other roller position. The driven rollers

FIGURE 16–47
Drive-chain-driven roller conveyor.

Courtesy of
Hytrol Conveyor
Company Inc.

FIGURE 16–48
Skate wheels gravity conveyor.
Courtesy of Hytrol Conveyor Company Inc.

are required to have two sprockets mounted on them. Sprocket and chain-driven rollers may also use one continuous strand or loop of chain to engage a single sprocket mounted on the side of the roller.

Live-roll light-duty roller conveyors for small packages and cartons can use a round urethane belt to power the rollers individually from a line shaft running beneath the bed, with some of the rollers connected to each. Some of the rollers, referred to as slaves, are driven off of the master rollers, which are connected to the line shaft.

Gravity or nonpowered roller conveyors rely on the force of gravity to move the packages. The same types of roller and frame configurations that are used on a driven system are used on gravity types, but without any driving mechanisms. They are simple, economical, and require little maintenance. Transition points between powered conveyors and on curves are places where the nonpowered conveyor is often used. Another version of the gravity conveyor uses a series of skate-type wheel bearings mounted on rods and fixed to the frame (see Figure 16–48).

Slider Bed Conveyors

Slider bed conveyors use a belt riding on a flat bed to move the units or packages (see Figure 16–49). Rollers are used only for driving, snubbing, tensioning, and supporting the belt on its return run. A flat bed of some material with low coefficient of friction supports the belt and load. Slider bed systems must consider the weight of the unit to ensure adequate drive power to move the belt and product. The belt sliding on the bed will create friction. Slow moving systems of this type may use a metal or wood bed. Newer and better designed systems will use a special plastic compound with a low coefficient of friction to reduce the power requirements and wear.

Optimum performance of slider bed conveyors is obtained when speeds are in the 10 to 90 FPM range. The unit loads should be kept to less than 100 pounds. At higher speeds and loads, power consumption and wear will become excessive.

FIGURE 16–49
Slider bed conveyor system complete with supporting steel frame and drive package mounted underneath.

Courtesy of Hytrol Conveyor Company Inc.

Apron/Pan Conveyors

There are many different apron conveyor arrangements, but three of the most common are the chain roller supported, frame roller supported, and the outboard roller supported. Chain roller supported apron conveyors consist of two or more strands of heavy-duty chain connected to the sides or bottom of formed steel "aprons" or "pans." Frame roller supported conveyors have carrying rollers mounted on the frame to support the chain on the sidebars, not on the chain. Outboard roller supported types have rollers mounted on chain strands outside of the overlapping pans and run on a set of rails.

Apron conveyors are sometimes referred to as pan conveyors (see Figure 16–50). They are common in the mining and aggregate industries. Apron conveyors are designed to move large heavy materials for short distances. Materials such as coal, rock, clinkers, and ores of various types are transferred on apron conveyors. Low-speed

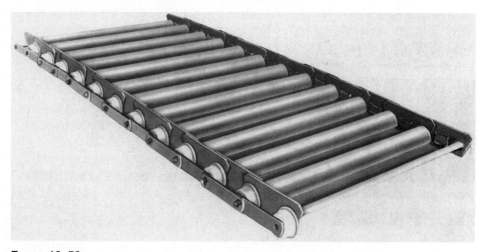

FIGURE 16–50
Frame roller supported apron conveyor often used to move raw, heavy materials for short distances.

Courtesy of Webster Industries.

operation and high horsepower drives make them a natural choice for use as a feeder to crushers and grinders. The driving mechanism is usually a gear motor combination connected to a roller chain drive to yield high ratios and generate significant amounts of torque. Because of their all-steel and iron composition, they are capable of operating in hostile environments where abrasives and high temperatures are present.

Overhead/Trolley Conveyors

Overhead or trolley conveyors incorporate rails supported from overhead structural iron to move material on a trolley mechanism. The rails connected to overhead supports are in the shape of I-beams or rolled tracks. Rails can be singular or two parallel tracks when supporting heavy loads. Trolleys run along the length of the track on multiple bearing wheels. The trolley will have some form of hook or attachment to lift and hold the load. The overhead rails are joined together in a loop so that the conveyor can run in a straight or curved path. The system will be powered by a motor-reducer drive connected to a caterpillar drive mechanism. A caterpillar drive consists of a loop of drive chain and sprockets operating on idlers and wheels. Conveyor and drive chains of numerous configurations are connected to the trolleys. Take-ups and tensioning devices assist in removing slack within the system and help move the components smoothly. Certain overhead systems incorporate both free and powered trolleys. Powered trolleys are connected to the drive or power chain. Free trolleys are used for load support. Overhead systems are found in the mobile equipment and automobile manufacturing plants where large parts are handled and processed.

Conveyor Safety

Conveyor safety begins with a sound design and proper installation. Most often, faulty design or component failure does not cause accidents that occur around conveying systems. Personal injury accidents are usually the result of negligence or ignorance. Operators and maintenance personnel must be instructed in safe operating procedures, recognizing hazards, and proper maintenance, and continuously encouraged to be alert. The safe operation of conveying systems can be summed up in three words: awareness, training, and inspection.

Organizations such as ANSI, OSHA, MSHA, and CEMA publish numerous bulletins on the safe operation of different types of conveying systems. It is suggested that ANSI B20.1-1996 and ANSI B15.1-1972 be obtained and consulted for detailed information on conveyor safety.

The following general safety guidelines are not intended to replace ANSI, MSHA, nor OSHA publications and regulations. Specific plant safety regulations pertinent to conveyors should be followed. The list is to illustrate the various general safety considerations that can be applied to the installation and maintenance of most conveyor systems.

- All personnel must be trained on the safe installation, operation, and maintenance of the conveyor. A formal, documented safety program will help reduce the potential for accidents.

- Lockout, tagout, and verification procedures must be followed at all times.

- Only trained personnel should operate or work on maintaining the conveyor. Never field-modify the conveyor unless the appropriate authorities have given prior approval.

- Have a detailed checklist available for the start-up, shutdown, and maintenance process. Provide it to those individuals who need it to safely and satisfactorily work on and operate the conveyor. Always operate the system in accordance with the plant guidelines.

- Qualified individuals should do routine inspections of the system on a regularly scheduled basis. A thorough inspection of the system must be performed prior to start-up.

- Good housekeeping is a prerequisite to safe conditions around conveyors. Always clean up before, during, and after any conveyor work. Remove all obstacles and debris from around the system.

- Conveyors should handle only the material that they were designed to transport. Regulate the feeding of material at a uniform rate.

- Never touch a moving conveyor with a tool, hand, or feet. Shut down the conveyor before working on it. Never poke or prod the material with anything while it is operating.

- Emergency shutoff and safety device locations must be known to all staff and the devices must be in good working order.

- All moving parts and connected power transmission components should be properly guarded. The guards must be in place before operating the system. Never place any part of your body in a potential pinch-point. Adequate fencing and railing should be installed around the conveyor system.

- Cross over or under conveyors only at safe crossover points that were intended for that purpose. Avoid areas with falling debris. Under no circumstances should an individual ride on a conveyor.

- Good lighting is imperative to prevent accidents.

- Any pertinent caution/warning signs must be legible and posted in any potentially dangerous area.

Questions

1. Name four different types of conveyor systems.

2. How are conveyor systems powered?

3. What is a roller conveyor?

4. What function does a conveyor idler serve?

5. Name four functions and positions of a conveyor pulley.

6. What are some of the important issues regarding conveyor system design and component selection?

7. What is a wing pulley and where should it be used?

8. What are the differences between a motorized pulley and a drum pulley?

9. Name two different methods used to mount conveyor pulleys to the shaft.

10. Lagging a pulley serves what three purposes?

11. Describe the differences between a troughing idler and a flat carrying idler.

12. How are conveyor belts constructed?

13. Name two types of joining conveyor belts.

14. List six important steps for installing and maintaining belted conveyors.

15. What is a bucket elevator conveyor?

16. What is a pan conveyor?

17. What is a screw conveyor?

18. What is a slider bed conveyor?

19. How is a live roller bed conveyor driven?

20. What is a gravity conveyor?

21. List ten safety procedures that must be followed when working with conveyors.

Introduction to Fluid Power Actuation

Pneumatics and hydraulics are very prevalent in industry. The basic understanding of such systems is crucial in design, troubleshooting, and maintenance of automated systems.

Objectives

Upon completion of this chapter you will be able to:

✔ Understand the basic principles of fluid power.
✔ Define the terms solenoid, spring-return, double-acting, and flow valves.
✔ Design simple fluid power applications, including calculating sizes, and choose appropriate equipment.
✔ Explain the types and operation of simple valves.

FLUID POWER ACTUATION

Fluids such as air and oil have long been used to do work. Pascal's law, which is the basis of fluid power, states that pressure applied on a confined fluid is transmitted in all directions, and acts with equal force on equal areas and at right angles to them (see Figure 17–1). Consider a container similar to the one in Figure 17–1. If the cork is pushed far enough into the bottle, the bottle will break. This happens because the liquid is incompressible and transmits the force from the stopper through the whole container. This causes a much higher force on a much larger area than the cork. Thus, the glass container can break with relatively small pressure.

A simple example of this law applied to a practical application involves a simple hydraulic press (see Figure 17–2). The left part of the press has a small input area and force is applied to this side. This causes the same pressure to be applied throughout the press. The forces are proportional to the area of the pistons. Thus, a 5-pound force can lift a 50-pound load if the output piston has 10 times the area of the input piston. Figure 17–3 shows the comparison between volume and distance.

FIGURE 17–1
Illustration of Pascal's law. The pressure is exerted in all directions and with equal force at 90 degrees to the surfaces.

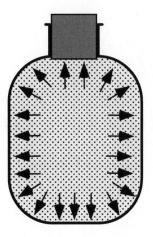

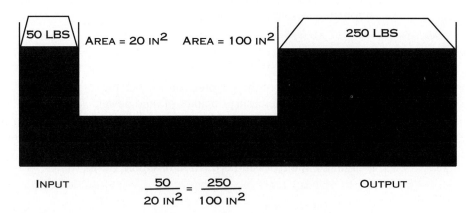

INPUT $$\frac{50}{20 \text{ IN}^2} = \frac{250}{100 \text{ IN}^2}$$ OUTPUT

FIGURE 17–2
Some basics of hydraulic leverage. The area of the input side of the press is 20 square inches, and the area of the output side is 5 times larger (100 square inches). This means that we can lift 5 times as much weight as the amount of pressure we apply at the input.

Pressure

Pressure is expressed as atmospheric, gauge, and absolute. *Atmospheric pressure* is the force exerted by the weight of the atmosphere on the earth's surface. The weight of the atmosphere at sea level is 14.7 pounds per square inch. Atmospheric pressure can also be expressed in inches of mercury. A mercury barometer calibrated in inches of mercury can be used to measure atmospheric pressure. The 14.7 pounds at sea level represent one atmosphere; it is equal to 29.92 inches of mercury. *Gauge pressure* is the pressure above atmospheric pressure. Most pressures are gauge pressure. *Absolute pressure* is the pressure above a perfect vacuum. It is the sum of gauge

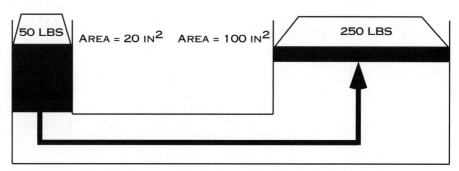

FIGURE 17–3
If we move a given volume of fluid on the input side, the same volume is moved on the output side, resulting in a smaller move. In this case while we could lift 5 times our input pressure, we lose distance. The output in this case only moves one-fifth of the distance.

pressure and atmospheric pressure. Absolute pressure is expressed in pounds per square inch absolute (psia).

Work and Power

Work = Force · Distance

Work is expressed in foot pounds. If 5 pounds are lifted 10 feet, the work actually done is 50 foot pounds (5 pounds · 10 feet). Imagine walking up a flight of stairs. The work done equals your weight multiplied by the height of the stairs. If you run up the stairs, you are doing the same amount of work but at a faster rate, called *power*. It takes much more power to run up a flight of steps than to walk up them even though the same amount of work is done.

$$\text{Power} = \frac{\text{Force} \cdot \text{Distance}}{\text{Time}}$$

Power is usually measured in horsepower or watts. One horsepower equals 33,000 pounds lifted in one minute. One watt equals 1 newton lifted one meter in one second. One horsepower is equal to 746 watts. Pressure equals the force of the load divided by the piston area (see Figure 17–4). The flow in fluid power systems is measured by the flow rate of gallons per minute (GPM). Speed and distance are indicated by the flow. Force is indicated by pressure.

FIGURE 17–4
Formulas for pressure.

$$P = \frac{F}{A} \qquad 1W = \frac{1N\,m}{\text{sec.}}$$

$$A = \frac{F}{P}$$

$$F = P * A$$

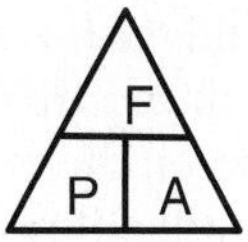

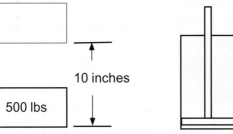

FIGURE 17–5
Simple application
of lifting the load
10 inches.

10 inches

500 lbs

A Simple Application Let's try a simple application. We need to lift a 500-pound load 10 inches. First we must select an actuator. In this case, a simple linear actuator would do the job. The load must be moved 10 inches, so we must choose a cylinder that has at least a 10-inch stroke. The area of the cylinder depends on the weight of the load and the pressure that will be used in the system. For this example, assume that our maximum operating pressure is 100 pounds. A 5-square-inch cylinder would lift 500 pounds. Our pressure of 100 multiplied by the 5-square-inch cylinder would be 500 pounds. This would not be a good idea, however. We should provide a margin of error. An 8-square-inch cylinder would lift the load at 62.5 pounds of pressure and provide a margin of error. See Figure 17–5.

TYPES OF ACTUATORS

Linear Actuators

Linear actuators (cylinders) are classified according to their construction and method of operation. The single-acting and double-acting cylinders are the basis for all other cylinders. A *single-acting cylinder* has a spring to return the piston rod (see Figure 17–6a). Air or oil is used to extend the cylinder and when the pressure is released, the spring returns the piston. A *double-acting cylinder* uses air to extend and retract the piston (see Figure 17–6b).

Cushioning can be added to decelerate the piston at one or both ends of travel. The amount of cushioning is adjustable with a needle valve that controls the flow rate of air or oil as it escapes the cylinder. The cushioning is designed to work at the end of the travel. The cylinder can move full speed between the endpoints of travel. This is especially useful for moving heavy loads. The cushioning allows controlled deceleration before a stop. See Figure 17–6c for a double-acting cylinder with adjustable cushioning at both ends.

Both single-ended and double-ended cylinders are available. A *single-ended cylinder* produces unequal forces on the extend and retract stroke. On the extend stroke, the fluid pushes against the whole surface of the piston. On the retract stroke, the area is reduced by the diameter of the piston rod. A double-ended cylinder has equal pressure on the extend and retract stroke. Cylinders are also available with piston rods that will not rotate.

FIGURE 17–6
Various cylinder symbols.

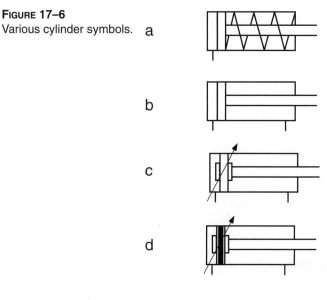

FIGURE 17–7
Rodless cylinder.

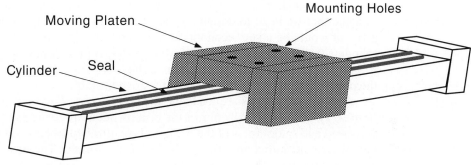

Rodless cylinders are also available (see Figure 17–7). These cylinders have a housing on the outside attached to the piston. The outside of the cylinder has a seal so that air fluid does not leak. Rodless cylinders are particularly useful where space is tight. Instead of having a rod attached to the piston that extends out of the cylinder, they have a platen that is attached through one side of the cylinder (see Figure 17–8).

FIGURE 17–8
Principle of a
rodless cylinder.

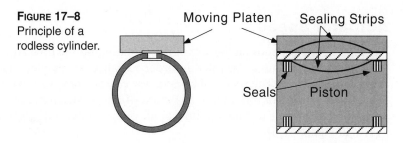

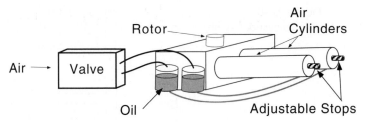

FIGURE 17–9
Air-over-oil system with a rotary actuator. Rotary actuators are available for pneumatic
or oil systems.

The platen moves along the outside of the cylinder. The mounting holes on the platen
attach to the object that needs to move.

Rotary Actuators

A rotary actuator is a motor with a rotor that can be pushed by air to a stop in either
direction. A single-rotation type can only make a revolution of 360 degrees or less.
Imagine that we need to design a rotary table that stops in two positions. A rotary
actuator would be a very economical way to do this. A rotary actuator uses two air
cylinders (see Figure 17–9), each of which moves a rack. A pinion is attached to the
rotor. The rotor is moved by the action of the cylinders pushing the rack and moving
the pinion. The stops on a rotary actuator are adjustable.

Air-over-oil systems are available if smooth, controlled movement is required.
In these systems oil actually moves the pistons, but air pressurizes the oil. This allows
the speed and smoothness of the move to be controlled because oil is noncompress-
ible. Rotary actuators are also available for pneumatics. The two types are: motor
and single rotation. The motor type is an air-operated motor.

Feedback on Cylinder Position

It is sometimes necessary to have feedback as to the actual position of the piston.
Magnetic sensing can be added to give feedback on whether the cylinder is in the
extended or retracted position. See Figure 17–6d for the symbol for a double-acting
cylinder with adjustable cushioning and magnetic sensing.

Actuator Speed

Actuator speed depends on the size of the actuator and the flow into it. To calculate
the speed, the volume to be filled in order to cause the needed amount of travel
must be considered. The speed of an actuator can be controlled with an adjustable
flow valve, which can restrict the flow rate of air. Restrictors are installed at the
outlet of the cylinder to slow the flow of air out of the cylinder. The relationship is
shown in Figure 17–10.

FIGURE 17–10
Formulas used to calculate speed.

$$\text{volume / time} = \text{speed} * \text{area}$$

$$\text{speed} = \frac{\text{volume / time}}{\text{area}}$$

$$\text{area} = \frac{\text{volume / time}}{\text{speed}}$$

$$\text{v/t} = \text{in}^3/\text{minute}$$
$$\text{a} = \text{in}^2$$
$$\text{s} = \text{in/minute}$$

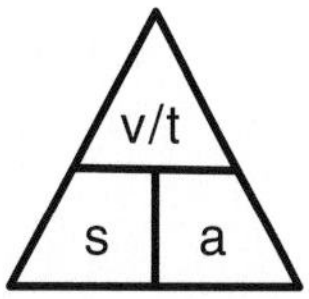

Air Pressure

Figure 17–11 shows a variety of symbols used for air supply and control devices. Pressure is usually shown with the first symbol. A variety of filter/regulator units are available to regulate the air pressure as well as clean and lubricate the air. Some units remove oil from the air for certain applications.

Directional Valves

Directional valves are used to stop, start, and control the direction of fluid flow. They can be classified according to the following characteristics:

The method of actuation: manually, mechanically, or by pneumatic, hydraulic, electrical power, and a combination of actuation such as pneumatic and manual.

The number of flow paths they provide: two-way, three-way, and four-way valves.

The type of connection: pipe thread, straight thread, flanged and subplate, or manifold-type mounting. Valves can also be classified by the internal type of valve that is used: poppet (piston or ball), sliding spool, and rotary spool.

Valve Symbols Valve positions are represented by squares (see Figure 17–12). The number of squares represents the number of switching positions. The one in Figure 17–12b has two switching positions. Lines in the boxes indicate flow paths, and arrows represent the direction of the flow (see Figure 17–12c). Valve shut-off positions are represented by lines at right angles (see Figure 17–12d). Inlet and outlet ports are shown by exterior lines (see Figure 17–12e).

Directional control valves are represented by the number of ports and the number of control positions that they have (see Figure 17–13). Normally open and normally closed are the opposite in fluid power. In electrical terms, a normally open valve does not pass current. In fluid power, a normally open valve does pass fluid.

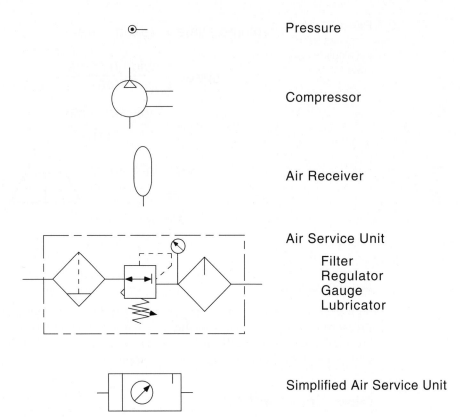

Pressure

Compressor

Air Receiver

Air Service Unit
 Filter
 Regulator
 Gauge
 Lubricator

Simplified Air Service Unit

FIGURE 17–11
Air supply and control device symbols.

FIGURE 17–12
Valve symbols. (a) Each rectangle represents a valve position. (b) This drawing represents a two-position valve. (c) A line in a box with an arrow represents flow and its direction. (d) A line at a right angle represents a shut-off position. (e) Inlet and outlet ports are represented by lines on the outside of the box. These lines are drawn in the initial valve position.

a

b

c

d

e

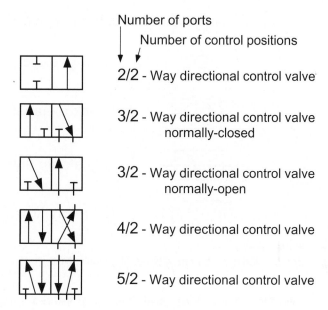

Number of ports

Number of control positions

2/2 - Way directional control valve

3/2 - Way directional control valve
normally-closed

3/2 - Way directional control valve
normally-open

4/2 - Way directional control valve

5/2 - Way directional control valve

FIGURE 17–13
The most common directional valve configurations.

FIGURE 17–14
Standard numbering and
lettering system for ports.

Pressure Port - 1 or P
Exhaust Port - 3 or R
Exhaust Ports - 5, 3 or R, S
Signal Outputs - 2, 4 or B, A

Port Numbering A numbering system identifies the ports on directional control
valves. In the past, a lettering system was used. Both are shown in Figure 17–14. The
pressure port is 1 or P. Exhaust ports are 3 and 5 (R and S). Signal outputs are 2 and
4 (B and A). Directional control valves are represented by the number of ports and
the number of control positions that they have.

The simplest type of directional valve is a check valve, which allows flow in one
direction but blocks flow in the other direction. It is made of a ball, a seat, and a
spring (see Figure 17–15). A light spring pushes the ball against the seat to stop
flow in one direction. If the pressure on the other side is higher than the light spring
pressure, the ball moves away from the seat and allows flow in that direction.

Pilot-Operated Check Valve A pilot-operated check valve is controlled by a pilot
signal that is electrical, mechanical, or fluid operated. The pilot-operated check valve
operates like a simple check valve until pressure is applied through the pilot port.
When pilot pressure is applied, reverse flow is permitted through the valve.

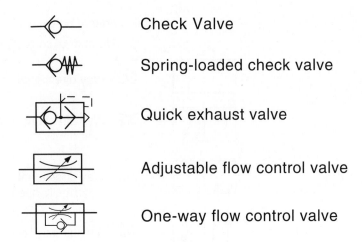

FIGURE 17–15
Some of the more common types of check valves.

Two-Way, Three-Way, and Four-Way Valves Imagine that we need to make a cylinder extend and retract. We need a device to control and change the direction of fluid flow. These devices are called *valves*. They can be controlled manually or by pneumatics or electricity. A valve consists of a body that has internal flow passages and ports to connect to. Valves are all designed to direct flow from the inlet port (pressure port) to either of two output ports. The number of ports to and from which fluid flows determines whether the valve is a two-way, three-way, or four-way valve.

Two-Way Valves A two-way directional control valve has two ports connected to each other with passages. These passages can be opened or blocked by a spool (see Figure 17–16). The figure shows the body of the valve, the spool, and the ports. Figure 17–17 shows the two possible valve positions. In the first position, the flow path is open between the pressure port and the actuator output port. The graphic

FIGURE 17–16
Two-way valve
showing spool.

FIGURE 17–17
Two possible valve
positions.

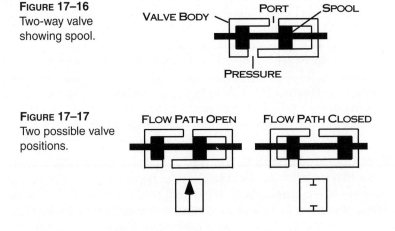

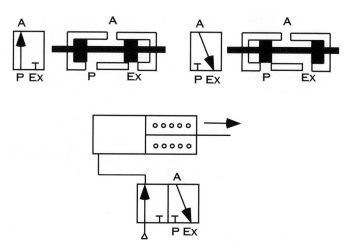

FIGURE 17–18
Two possible spool positions.

below this valve is the representation of this valve position. The arrow shows that the ports are connected, and the direction of the flow. Note that this would be a 2/2 valve because there are two ports and the valve has two possible positions.

The second position shows the valve closed. No flow can move from the pressure port to the actuator output port. The graphic below the valve is the representation of this valve position with the ports blocked by the spool. A two-way valve provides an on/off function only.

Three-Way Valves A three-way valve has three ports that can be connected by passages in the body of the valve. Figure 17–18 shows the two possible valve spool positions. Note that port P is the supply pressure, port A is connected to the cylinder, and port EX is the exhaust port. The graphical representation on the left shows that the supply port and the actuator port are connected and the exhaust port is blocked. The graphic on the right shows that the supply port is now blocked and the cylinder port is connected to the exhaust. This valve can be used to apply or exhaust pressure to one port. This valve is a 3/2 valve because it has three ports and two positions.

Four-Way Directional Valves The four-way valve is probably the most commonly used valve in pneumatic systems. It has a pressure port (supply pressure), two actuator ports, and one or more exhaust ports. Figure 17–19 shows a four-way

FIGURE 17–19
Four-way valve.

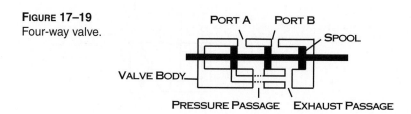

FIGURE 17–20
Two possible spool positions and a graphical representation of each valve.

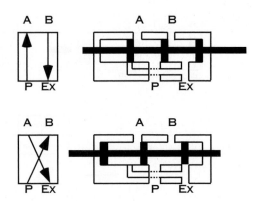

valve. In this figure the spool has connected the pressure port to the A port, and the B port is connected to the exhaust port.

Figure 17–20 shows the two possible positions of the spool and the graphical representation of each valve position. The left view shows the spool in the left-most position. In this position port A is connected to the exhaust port. The pressure port is connected to Port B. The symbolic representation shows the same thing. The arrows represent the connections and the direction. This is a 4/2 valve because it has four ports and two positions.

The top graphic in Figure 17–20 shows the spool in the right-most position in which the pressure port is connected to the A port, and the B port is connected to the exhaust. The symbolic representation is clearer. The arrows show that the A port is connected to the exhaust and the pressure port is connected to the B port.

We could construct a simple but very useful circuit with this valve (see Figure 17–21). This figure shows a symbolic representation of a four-way valve connected to a double-acting air cylinder. The spool is in the left position. The pressure port is connected to the A port, and the B port is connected to the exhaust. As pressure is applied to the rear of the cylinder, the pressure in front of the position can go through the B port to the exhaust. If we move the spool to the right position, the pressure is applied to the B port (front of the cylinder), and the back of the cylinder could exhaust the pressure through the connection of the A port to the exhaust.

FIGURE 17–21
Simple circuit that utilizes a four-way valve.

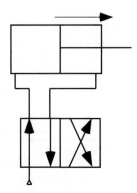

BLOCKED CENTER   **EXHAUST CENTER**   **PRESSURE CENTER**

FIGURE 17–22
Three examples of three-position valves.

Three-Position Valves The valves discussed to this point were all two-position valves. Three-position valves are also available in four- and five-port models. Figure 17–22 shows three examples of three-position five-port valves. On the left is a blocked center valve. The center position in this valve blocks all of the working ports. The left position extends the piston cylinder, and the right position retracts the piston cylinder. The center graphic in Figure 17–22 is an exhaust center valve. When the valve is in the center position, both ports are opened to the exhaust. In this case, the piston in a cylinder could float in either direction. In the left position in this type of valve, the piston in the cylinder extends. In the right position, the piston cylinder retracts. The right graphic in Figure 17–22 is a pressure center valve. When the valve is in the center position, both ports are opened to pressure. In the left position, the piston in the cylinder extends. In the right position, the piston cylinder retracts.

Spool Position Valves *Spool position* classification refers to the normal or deenergized valve condition. There are several types. We consider two of the types available.

FIGURE 17–23
Symbolic representation of four spring-return valves.

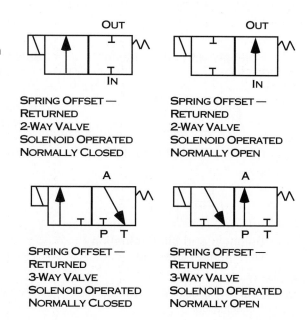

SPRING OFFSET —
RETURNED
2-WAY VALVE
SOLENOID OPERATED
NORMALLY CLOSED

SPRING OFFSET —
RETURNED
2-WAY VALVE
SOLENOID OPERATED
NORMALLY OPEN

SPRING OFFSET —
RETURNED
3-WAY VALVE
SOLENOID OPERATED
NORMALLY CLOSED

SPRING OFFSET —
RETURNED
3-WAY VALVE
SOLENOID OPERATED
NORMALLY OPEN

One type is spring centered. In a spring-centered valve, the spool is returned to the center position by spring force. When the actuation pressure is gone, the spring returns the spool to the centered position. The spring-return type is valuable in some applications if power is lost, because the valve can return the cylinder to a safe position.

Spring-return valves can be purchased as either normally open or normally closed, which is determined by the condition when the valve is not energized. Figure 17–23 shows a symbolic representation of four spring-return valves. The valves in Figure 17–23 are solenoid operated. A solenoid is a coil used to generate a magnetic field to move the spool in one direction. When the coil is deenergized, the spring returns the spool. The symbolic representation of spring-return valves always shows the valve in the deenergized condition.

A second type is a two-position spring offset valve. Its spool is normally offset to one end by a spring. It has one actuator that shifts the spool to the other location. When the actuation force is released, the spool returns to the end position.

VALVE ACTUATION

There are several ways in which valves are actuated. One way is to manually activate the valve. It is activated by the mechanical movement of a lever or a plunger. A mechanical example is a handle (lever) that an operator must move to make something happen. The plunger style is used for automatic operation. Some part of the machine

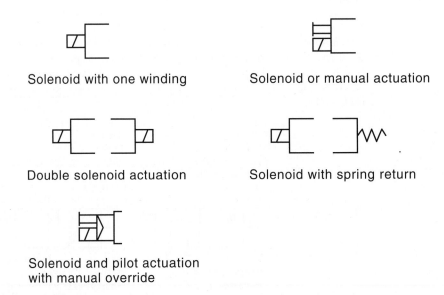

Solenoid with one winding

Solenoid or manual actuation

Double solenoid actuation

Solenoid with spring return

Solenoid and pilot actuation
with manual override

FIGURE 17–24
Diagram of several common valve actuation types. A valve can have more than one method of actuation.

pushes the plunger, which changes the valve spool position. Figure 17–24 shows examples of several types of activation.

Mechanical Actuation

Though very similar to manual actuation, mechanical actuation does not involve a person. This type of valve has a plunger that is activated when it is moved by a cylinder or machine element; this movement moves the spool.

Fluid Power Actuation

Fluid shifts the valve spool in these valves. A pilot is another port on the valve. If pressure is present on the pilot valve, the piston shifts and allows flow. When the piston shifts, it moves the spool, which changes the flow direction.

Electrical Actuation

Valves that provide electrical actuation are usually called *solenoid valves*. A solenoid consists of a coil and an armature. When electrical current is applied to the coil, it generates an electrical field. The field causes the armature to move, and as it is pulled into the magnetic field, it moves the spool.

AC/DC

Solenoids can use either AC or DC. An AC solenoid has a large current draw when it is first turned on (in-rush current) but has a low draw after that (holding current). If an AC sensor cannot complete the shift, it continues to draw high current and burns out. This can also happen if a solenoid is turned on and off at a high frequency.

A DC solenoid draws constant current when it is energized. It is designed to handle continuous and high current. This prevents burnout if the shifting is incomplete or high cycle rates are required. DC solenoids generally operate at lower voltages and so are usually safer.

DESIGN PRACTICE

Let's complete the simple application we started earlier. We chose a 10-inch minimum stroke and an 8-inch squared area. Next we need to design the rest of the application.

We need to control the movement of the cylinder in both directions, so we use a double-acting cylinder. We need to control the speed of the cylinder on the retract stroke, so we will add a flow valve. To control speed, we need to control the fluid leaving the cylinder, so the flow valve must always be placed where the air will leave the cylinder on the controlled movement. We chose a 4/2 control valve to switch the direction of the cylinder. The valve we chose has one solenoid and a spring return. We also added an air service unit to clean the air and control the pressure (see Figure 17–25).

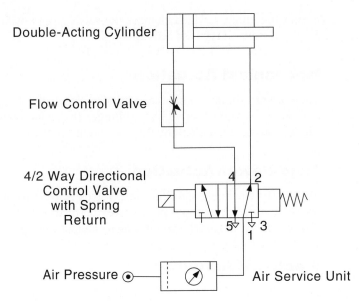

FIGURE 17–25
A simple double-acting cylinder application.

0	Air Supply Unit
1.0, 2.0, 3.0, etc.	Actuator
.1	Control Element
.01, .02, .03, etc.	Elements Between the Control Element and the Working Element

FIGURE 17–26
Numbering system for pneumatic components.

Actuators are labeled with integers 1, 2, 3, and so on. For example, the two cylinders in an application are numbered 1 and 2 (see Figure 17–26). Control elements are numbered .1, .2, .3, and so on. The regulator after the air supply is numbered 0.1. A directional control valve controlling cylinder 1 is numbered 1.1. Elements between a control element and an actuator are labeled .01, .02, .03, and so on. A flow valve between directional control valve 1.1 and actuator 1 is labeled 1.01 (see Figure 17–27).

VACUUM

A vacuum is often used in industrial applications. When a small vacuum is needed, for example, to pick up small parts with a suction cup, a venturi vacuum generator system is often used. A venturi system uses positive air pressure to create a flow through the venturi, which then draws air in through the port attached to the suction cup(s). When more vacuum is required, a vacuum pump is used.

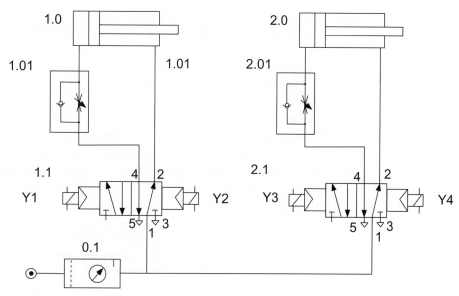

FIGURE 17–27
Numbering of components in a pneumatic schematic.

Questions

1. Label each component in the following diagram and describe its function. Then number each component according to the standard numbering scheme.

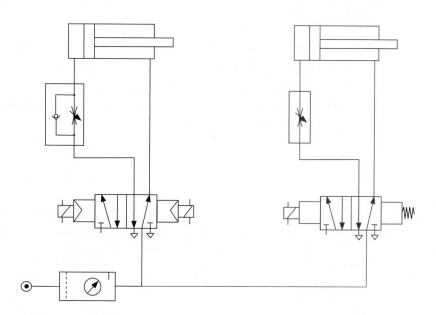

2. You are asked to design a pneumatic clamp for a system that must clamp with 750 pounds of pressure. You have 80 PSI available. What size cylinder should be used?

3. Draw the following valves and explain their operation:
   a.   3/2 solenoid valve with spring return
   b.   4/2 solenoid valve with spring return
   c.   4/2 solenoid valve with dual solenoid

4. Explain how to control the speed of a cylinder.

5. Design the following. We have a 300-pound load to move 5 inches. The speed must be controlled on the extend stroke. We have 100 pounds of air pressure available. Calculate the cylinder size, choose the appropriate components, and draw the circuit.

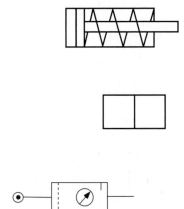

6. Draw and describe a double-acting cylinder.

7. Draw and describe the operation of a single-acting, spring-return cylinder.

8. Describe the operation of a 4/2 valve.

9. Describe the operation of a 5/2 valve.

10. Draw a circuit with an appropriate valve to control a double-acting cylinder.

Abrasive wear Wear that occurs when a hard surface rubs against a softer surface or by small, hard particles rubbing against a softer surface.

Acceleration The change in velocity indicated by the motion within the cycle; also an increase in speed.

Action Refers to the action of a controller. Defines what is done to regulate the final control element to effect control. Types include on-off, proportional, integral and derivative.

Actual mechanical advantage (AMA) Measurement that takes into account efficiency and friction losses of a machine.

Actuator 1. A device that transforms fluid or electrical energy into linear or rotary mechanical force; 2. Output device normally connected to an output module. An example would be an air valve and cylinder.

Addendum The portion of the tooth above the pitch circle.

Additive A substance added to a lubricating fluid to enhance a desired characteristic.

Address Number used to specify a storage location in memory.

Alignment The location (within tolerance) of one axis of a coupled rotating machine shaft relative to that of another rotating machine.

Angle The intersection of two lines or sides.

Angular-contact bearing A rolling element bearing designed to carry both axial (thrust) loads and radial loads.

Angular misalignment A condition where two shafts are at different angles within the horizontal or vertical plane.

Angular soft foot A condition that exists when one or more of a machine's feet are bent up or down, and/or not on the same plane as the other feet.

ANSI American National Standards Institute.

Antifriction bearing A bearing that has rolling elements, such as balls or rollers, which provide a low-friction support surface for rotating surfaces.

Arc A portion of the circumference.

Area The number of unit squares equal to the surface of an object.

Asperities Microscopic peaks and valleys on any material surface as a result of the machining process.

Asymmetrical load A load in which one-half of the load is not equal to that of the other half.

Atmospheric pressure The pressure exerted on a body by the air. It is equal to 14.7 pounds per square inch at sea level.

Axial float Axial movement along the shaft axis of a bearing and shaft arrangement due to thrust loads, thermal growth, and bearing and bearing housing clearances.

Axial load A load applied parallel to the rotating shaft axis.

Axial thrust Force, or push, along the axis of the shaft. Also called thrust load.

Axial-type friction clutches Clutches or brakes that have contact pressure applied perpendicular to the rotating shaft.

Babbitt metal Lead- or tin-based alloys with copper and antimony used for sleeve type bearings.

Backlash The amount of movement or play between meshing gear teeth. The measurable space between meshing teeth measured at the pitch circle.

Ball bearing An antifriction bearing that allows for reduced friction motion between a moving part and a fixed part by means of balls separated by a retainer, confined between inner and outer machined rings.

Base plate A rigid steel support for mounting, coupling, and aligning two or more rotating machines, such as a pump and motor.

Bearing A machine component used to reduce friction and maintain clearance between stationary and moving parts.

Belt A device used to transfer power between two pulleys that are mounted on shafts.

Belt and sheave groove gauge A gauge that has a male form to determine the size and angle of a sheave and a female form to determine the size and type of a V-belt.

Belt creep The movement of the belt on the face of the pulley or within the groove of the sheave when changes in tension occur.

Belt deflection tension method A method used to tension a belt in which the tension is adjusted by measuring the deflection of the belt by applying a specific amount of force.

Belt drive A mechanical drive system that uses a belt and pulleys/sheaves to transfer power and torque between two rotating shafts.

Belt pitch length Circumference of the tensile cord of the V-belt. Also called the "datum."

Belt pitch line A line located on the same plane as the belt tension member.

Belt tension tester A device used to properly tension a belt by measuring the amout of force required to deflect a belt a specified amount.

Bending Stress caused by forces acting perpendicular to the horizontal axis of an object.

Bevel gear A gear with straight tapered teeth cut on a cone used in applications where shaft axes intersect.

Body An object with mass.

Bolt bound A condition that prevents the horizontal movement of a machine due to the contacting of the machine mounting bolts to the sides of the machine anchor holes.

Boundary lubrication Takes place when only an extremely thin film of lubricant is present to separate mating surfaces.

Breakaway torque The initial energy required to get a nonmoving load to turn.

Bulk material conveyor systems Conveyor systems designed to move a continuous flow of material from one process point to another.

Cage (separator) An internal component of an antifriction bearing used to maintain the position and alignment of the balls or rollers.

Calibration Procedure used for determining, correcting, or checking the absolute values corresponding to the graduations on a measuring instrument.

Cascade Programming technique used to extend the range of timers and counters.

Center distance The shortest distance between the axes of two mating or connected rotating components.

Centrifugal force The outward force produced by a rotating object.

Chain 1. A series of metal rings connected to one another and used for support, restraint, or lifting. 2. A series of connected machined articulating components used to transmit power or convey material.

Chain coupling A mechanical-flexing coupling that includes two identical sprocketed hubs connected with a chain.

Chain drive A mechanical drive system that uses a chain connecting two sprockets mounted on separate shafts to transfer power and torque.

Chamfer To bevel the edge of a fastener or shaft.

Chordal action The vibratory motion caused by the rise and fall of the chain as it goes over a sprocket.

Circular pitch 1. In synchronous belts, the distance from the center of one tooth to the center of the next tooth, measured along the pitch line. 2. In gears, the distance from a point on a gear tooth to the corresponding point on the next gear tooth, measured along the pitch circle.

Circumference The boundary of a circle.

Clearance fit Fit in which the mechanical component such as a coupling hub slides onto the shaft without forcing or heating.

Coefficient of friction Resistance to sliding or rolling expressed as a ratio of the force holding the surfaces together divided by the force that resists sliding or rolling.

Combination sheave A V-sheave belt that can be used with different cross-section belts.

Compound gear train Two or more sets of meshing gears where a pinion and gear are mounted and rotate on one common shaft.

Connecting link A chain part used to assemble and connect the two ends of a chain.

Conrad bearing A single-row, deep-groove ball bearing without loading slots.

Conveyor belt An extra wide, flat belt made from a combination of rubber and synthetic material used for the conveying of material on a conveyor system.

Conveyor pulley A long cylinder mounted on a shaft and bearing arrangement used to provide tension, change direction, and support a flat conveyor belt within a conveyor system.

Conveyor system An electromechanical system used to transport material by means of moving components such as belting, rollers, screws, or buckets.

Corrosion The combining of metals with elements in the surrounding environment that leads to the deterioration or wasting away of a material.

Corrosive wear Wear that occurs when acid or moisture deteriorates the surface of metal materials such as gears or bearings.

Coupling A device used to connect a mechanical drive to a prime mover. A device that connects the ends of rotating shafts.

Coupling unbalance An unequal radial weight distribution where the center of mass and the geometric center do not coincide.

CSA (Canadian Standards Organization) Organization that develops standards, tests products, and provides certification for a wide variety of products.

Cylindrical roller bearing A bearing with cylinder-shaped rolling elements.

Dedendum The portion of the gear tooth below the pitch circle.

Deformation A change in the shape of a material.

Dial indicator A device that measures small movements.

Diameter The distance from circumference to circumference through the centerpoint.

Diametral pitch Ratio of the number of teeth for each inch of pitch diameter of the gear.

Die A circular-shaped, hardened steel device with internal threads used to cut or clean threads on a fastener.

Documentation Descriptive paperwork that explains a system or program. It describes the system so that the technician can understand, install, troubleshoot, maintain, or change the system.

Dowel pin A solid or rolled rod-shaped pin used to set and align mating machined parts.

Downtime The time a system is not available for production or operation, which can be caused by breakdowns in systems.

Drip system A lubrication system that uses gravity to provide drop-by-drop lubrication from an orifice, typically controlled by a needle valve, from a manually-filled container.

Drive system A combination of electrical and mechanical components that transfer power and torque from one location to another.

Dropping point of grease The temperature at which the oil in grease separates from the thickening agent.

Dry lubricant A lubricant in nonliquid form, such as graphite.

Dynamic seal A sealing device used between moving parts that helps prevent fluid leakage or contamination entering.

Eccentric Out-of-round or not concentric.

Efficiency A measure of a mechanical component's useful output energy compared to its input energy.

Elastic deformation The ability of a material to return or spring back to its original size and shape after having a load applied then removed.

Elastomeric shaft coupling A flexible shaft coupling that includes two flanged hubs connected and separated by a center member of elastomeric material.

End-float The inclination of shafts to move back and forth across their bearings. Also called *end-play*.

End-play The total amount of axial movement of a shaft and/or bearing.

Energy The capacity to do work.

Face width The length of gear teeth along the axial plane.

False Brinell damage Bearing damage caused by forces passing from one ring to the other through the rolling elements, often associated with vibration.

Fatigue Breakage caused by repeated bending stresses that exceed the material limits.

Fatigue crack A crack that occurs due to bending, mechanical stress, thermal stress, or material flaws.

Fatigue life The theoretical maximum useful life of a bearing prior to the onset of failure.

Fatigue wear Wear created by repeated stresses below the tensile strength of a material.

Fault Failure in a system that prevents normal operation of a system.

Feeler gauge (thickness gauge) A narrow, thin steel leaf machined to a specific thickness and marked as such.

Filter A cartridge-like device containing a semi-porous substance through which air or fluid is allowed to pass but particulate matter cannot.

Fire point The temperature at which oil ignites when touched with a flame.

Firmware A series of instructions contained in read-only memory (ROM) that are used for the operating system functions. Some manufacturers offer upgrades for PLCs. This is often done by replacing a ROM chip. The combination of software and hardware lead to it being called firmware.

First class lever A lever that has the fulcrum located between the resistance and the effort.

Flash point The temperature at which oil gives off enough gas vapor to ignite briefly when touched with a flame.

Flat belt A relatively thin belt that has a rectangular cross-section, used to transmit power or convey material.

Flexible shaft coupling A shaft coupling designed to join two rotating shafts and to allow some misalignment.

Flow chart A diagram using geometric shapes and arrows showing a logical sequence of troubleshooting steps for a given set of conditions.

Fluting The elongated grooves or marks on the rolling elements or the rings of an antifriction bearing due to improper grounding during welding or the passage of current through the bearing.

Force Anything that changes or tends to change the state of rest or motion of a body.

Fractional horsepower belt A V-belt designed for light-duty applications.

Fracture A small crack in metal caused by the stress or fatigue of cyclic pulling or bending forces.

Frequency The number of cycles per minute (cpm), cycles per second (cps).

Fretting corrosion The rusted surface that results when two ill-fitting metal surfaces are in contact and are moved around, breaking loose minute metal particles that become oxidized.

Friction bearing A bearing, without rolling elements, made of a stationary surface, such as babbitt metal or pressed-in bushings, that provides a reduced-friction support surface for rotating or sliding machine components.

Friction disc A device that transmits power or is used as a stopping mechanism, through contact between two discs or plates.

Fulcrum A support on which a lever turns or pivots and which is located somewhere between the effort force and the resistance force.

Galling (adhesive wear) A bonding, shearing, and tearing away of material from the surface of two contacting, sliding or rotating metals.

Garter spring Used to provide force to help maintain contact of the lip to the shaft with a radial lip seal.

Gasket A pliable material placed between mating surfaces, such as a pipe flange, to prevent leakage.

Gear A toothed machine component mounted on a shaft used to transmit power, and torque between rotating shafts.

Gearbox An enclosed fixed center gear drive that has input and output shafts and houses mating gears.

Gear coupling A flexible shaft coupling that includes two hubs with external gear teeth and a sleeve(s) with mating internal gear teeth.

Gear drive A mechanical drive system that uses the meshing of two or more machined gears to transfer power and torque from one shaft to another.

Gear train A combination of two or more gears in mesh used to transmit motion between two rotating shafts.

Grease A semisolid lubricant created by combining oils with thickeners, such as soap or other finely dispersed solids and additives, into one substance.

Grease dropping point The maximum temperature a grease withstands before it softens enough to flow through a testing orifice.

Grease fitting A hollow tubular orifice threaded or pushed into a machine housing, used to direct grease to moving components.

Grease gun A hand-operated lubrication device filled with grease that is attached to a grease fitting to provide grease to a bearing or gear set.

Grid coupling A flexible shaft coupling that includes two hubs with axially-cut slots along the perimeter of a flange portion of the hub and joined by a steel grid.

Ground Direct connection between equipment (chassis) and earth ground.

Helical gear A gear with teeth that are not parallel to the shaft axis that are cut at an angle.

Helix The curve traced on a cylinder or cone by a point rotating at a right angle to the axis.

Herringbone gear A double helical gear that contains a right- and left-hand helix-cut teeth.

Horsepower A unit of power equal to 746 W or 33,000 foot-pounds per minute (550 foot-pounds per second).

Hypoid gear A spiral bevel gear with curved, non-symmetrical teeth that are used to connect shafts at right angles.

Idler gear A gear that transfers motion and direction in a gear train, but does not change speeds.

IEC (International Electrotechnical Commission) Commission that develops and distributes recommended safety and performance standards.

IEEE Institute of Electrical and Electronic Engineers.

Image table Area used to store the status of input and output bits.

Imbalance Lack of balance.

Inclined plane A simple machine, such as a ramp, that allows force to be applied over a long distance to move loads.

Incremental This term typically refers to encoders. Encoders provide logic states of 0 and 1 for each successive cycle of resolution.

Induction The process of causing electrons to align or uniformly join to create a magnetic or electrical force.

Inertia The property of matter by which any physical body persists in its state of rest or uniform motion until acted upon by an external force.

Instruction set Instructions that are available to program the PLC.

Intelligent I/O PLC modules that have a microprocessor built in. An example would be a module that would control closed-loop positioning.

Interference fit Fit in which the mating components, such as a shaft and a bore of a hub, are of different diameters so that there is always an actual interference of metal in which the components must be heated or pressed on so it will slide together.

Involute The curved line produced by a point of a stretched string when it is unwrapped from a given cylinder.

Involute form A tooth form that is curled or curved.

IP rating Rating system established by the IEC that defines the protection offered by electrical enclosures. It is similar to the NEMA rating system.

ISO International Standards Organization.

Isolation Used to segregate real-world inputs and outputs from the central processing unit. Isolation assures that even if there is a major problem with real-world inputs or outputs (such as a short), the CPU will be protected. This isolation is normally provided by optical isolation.

Jack screw A bolt inserted through a block or plate that is attached to a machine base plate, allowing for horizontal machine movement.

Journal The part of a shaft, such as an axle or spindle, that mates with a bearing.

Key A removable part that fits in a slot or seat, providing a positive means of transmitting torque between a shaft and a hub.

Keyseat A groove or slot along the axis of a shaft or hub.

Kinetic energy Energy in motion.

Labyrinth seal Seal that has a series of grooves cut in mating parts with a small clearance between them allowing for rotation.

Lagging Rubber or similar material bonded onto a conveyor pulley to increase the coefficient of friction between the belt and pulley, assist in cleaning the surface, and to reduce the wear on the face of the pulley.

Linear bearing A bearing designed to provide low-friction movement of a mechanical device that moves in a straight line rather than rotates.

Linear motion The movement of an object in a straight line.

Lip seal A grease or oil seal that is made up of a case or shell with a bonded resilient material acting as a sealing edge that is formed into a lip.

Load Any device that current flows through and produces a voltage drop.

Load classifications Loads imposed by machines can be put into three classes. Constant torque loads where the horsepower requirements of the load vary directly with speed. Constant horsepower loads require higher values of torque at lower speeds and lower values of torque at higher speeds. Variable torque loads or those subject to "fan laws" describe the characteristics of fans and centrifugal pumps operating at various speeds. The torque required varies as the square of its speed.

Loading slot A groove or notch on the inside wall of each bearing ring to allow insertion of balls.

Lockout The process of preventing the flow of energy from a power source to a piece of equipment.

Lockout device A device that utilizes a positive means such as a lock, either key or combination type, to hold an energy isolating device in the safe position and prevent the energizing of a machine or equipment.

Lubricant A substance that separates moving (bearing) surfaces to reduce the friction and/or wear between them.

Lubricating grease A semisolid lubricant made of a mixture of oil, additives, and thickening agents.

Lubricating oil A liquid lubricant having a mineral, synthetic, vegetable, or animal origin.

Lubricating oil analysis A predictive maintenance technique used to detect and analyze the presence of contaminants and wear particles in lubricating oil and assist in the prediction of machine failure.

Lubrication The process of delivering and maintaining a fluid film between metal surfaces to prevent the surfaces from contacting each other.

Machine A mechanical device that is used to transfer force, motion, or energy.

Malfunction The failure of a system, equipment, or part to operate as expected.

Master control relay (MCR) Hardwired relay that can be deenergized by any hardwired series-connected switch. Used to deenergize all devices. If one emergency switch is hit, it must cause the master control relay to drop power to all devices. There is also a master control relay available in most PLCs. The master control relay in the PLC is not sufficient to meet safety requirements.

Mechanical advantage The ratio of the output force of a device to the input force.

Mechanical drive A system made up of manufactured components by which power, motion, and/or torque is transmitted.

Memory map Drawing showing the areas, sizes, and uses of memory in a particular PLC.

Metal fatigue The fracture of worked metal due to normal operating conditions or overload situations.

Micron (μ) A unit of length equal to one millionth of a meter (.000039″).

Microsecond A microsecond is one millionth (0.000001) of a second.

Millisecond A millisecond is one thousandth (.001) of a second.

Misalignment The condition where the centers of rotation of two machines are not aligned within specified tolerances.

Miter gear A gear used at right angles to transmit horsepower between two intersecting shafts at a 1:1 ratio.

Mnemonic codes Symbols designated to represent a specific set of instructions for use in a control program. An abbreviation given to an instruction, usually an acronym that is made by combining the initial letters or parts of words.

Needle bearing An antifriction roller-type bearing with long rollers of small diameter.

NEMA (National Electrical Manufacturers Association) Develops standards that define a product, process, or procedure pertaining to electrical apparatus.

Network System that is connected to devices or computers for communication purposes.

Node Point on the network that allows access.

Noise Unwanted electrical interference in a programmable controller or network. It can be caused by motors, coils, high voltages, welders, and so on. It can disrupt communications and control.

Nonvolatile memory Memory in a controller that does not require power to retain its contents.

Occupational Safety and Health Administration A federal government agency established under the Occupational Safety and Health Act of 1970 that requires employers to provide a safe environment for employees.

Offset link A chain link consisting of only one roller/bushing portion and side plates that is used to shorten or lengthen a chain and to connect the chain.

Offset misalignment 1. In flexible belt drives, a condition where two shafts are parallel but the sheaves are not on the same plane. 2. In shaft couplings, a condition where two shaft axes are parallel but are not co-linear.

Oil analysis A predictive maintenance technique used to detect and analyze the presence of contaminants and wear particles in lubricating oil and assist in the prediction of machine failure.

Oil level sight glass A tubular device with a transparent window mounted on a gearbox that indicates the level of oil inside.

Oil seal A device used to retain fluid inside a housing and exclude contaminants.

Oil wick An absorbent material, such as felt, that serves as a conduit for oil from the reservoir to the bearing surface.

Operating system The fundamental software for a system that defines how it will store and transmit information.

O-ring A molded synthetic rubber seal having a round cross-section.

Overhung load Force applied at right angles at the end of a shaft that may cause bending of the shaft or early bearing and belt failure.

Oxidation The combining of oxygen with elements in oil that break down the basic oil composition.

Oxidation inhibitors Additives that slow the rate of a lubricant's natural tendency to oxidize.

Packing A bulk deformable material used as a seal.

Parallel lines Two or more lines that remain the same distance apart.

Parallel soft foot A condition where one or two machine feet are higher than the others but still parallel to the base plate.

Peak The absolute value from a zero point (neutral) to the maximum travel on a waveform.

Perpendicular line A line that makes a 90° angle with another line.

PID (Proportional, integral, derivative) control Control algorithm used to closely control processes such as temperature, mixture, position, and velocity. The proportional portion takes care of the magnitude of the error. The integral takes care of small errors over time. The derivative compensates for the rate of error change.

Pinion The smaller and/or driving gear of a matching set of gears.

Pintle chain An offset-type chain usually made of malleable iron.

Pitch 1. In chain, the distance from the center of one roller to the center of the next roller. 2. In synchronous belts, the distance from the center of one tooth to the center of the next tooth, measured along the pitch line.

Pitch circle The circle that contains the operational pitch point.

Pitch diameter The gear diameter defined by the point on the teeth where force is applied to rotate the gear. Diameter of the pitch circle.

Pitch length The total length of the synchronous belt measured at the belt pitch line.

Pitting Localized corrosion that has the appearance of small pits.

Plain or journal bearing A bearing, without rolling elements, made of a stationary surface, such as babbitt metal or pressed-in bronze bushings, that provides a reduced friction support surface for rotating or sliding machine components.

Plant survey A complete inventory of machines, components, and equipment of a facility.

Plastic deformation The failure of a material to return to its original size and shape after being loaded and unloaded.

Plastic flow The movement of gear tooth material.

Positive belt drive A type of belt drive that transmits power by the positive engagement of belt teeth with pulley teeth, which has no slip.

Pour point The point at which the lubricant becomes so thick that it no longer flows.

Power The rate of doing work.

Power transmission The transmitting of power and torque between rotating machine shafts.

Power transmission components Mechanical devices such as gears, belts, chain, etc., used to transfer power and torque from one rotating shaft to another.

Predictive maintenance The monitoring of the condition of machine characteristics against a predetermined tolerance to predict possible malfunctions or failures.

Preloading A pressure or clamping force placed on a bearing to remove the initial internal clearance.

Pressure angle The angle of contact of a gear tooth. It is usually 14-1/2° or 20°.

Preventive maintenance A combination of unscheduled and scheduled work required to maintain equipment in peak operating condition.

Prime mover A device that changes one form of energy into another, or an electric motor or engine that supplies rotational force at a constant speed.

Puller A device used to remove gears, pulleys, sprockets, bearings, and couplings from a shaft or housing.

QD bushing Type of tapered bushing with a characteristic flange used to mount power transmission components onto a shaft.

Quadrature Two output channels out of phase with each other by 90 degrees.

Race The pathway on which the balls or rollers of a bearing move.

Rack gear A gear with teeth spaced along a straight line.

Radial bearing A rolling element bearing in which the load is transmitted perpendicular to the axis of shaft rotation.

Radial load A load in which the applied force is perpendicular to the axis of rotation. Force is exerted at 90° from the shaft axis.

Radial-type friction clutches Clutches or brakes that have contact pressure applied to the peripheral of a drum or rim.

Radius The distance from the centerpoint to the circumference.

Radius of gyration (WR² or WK²) The radius of gyration is the distance from the axis of rotation to a theoretical point where all of the mass can be considered concentrated.

Ratio The relationship between two quantities of terms. Usually with mechanical components, it is a size comparison.

Repeatability The ability to repeat movements or readings. For a robot, it would be how accurately the robot would return to a position time after time. Repeatability is unrelated to resolution and is usually 3 to 10 times better than accuracy.

Reservoir A container for storing fluid in a hydraulic system.

Resolution A measure of how closely a device can measure or divide a quantity. For example, in an encoder resolution would be defined as counts per turn. For an analog to digital card it would be the number of bits of resolution; for example, for a 12-bit card the resolution would be 4096.

Retaining ring A stamped circular ring used to keep parts from slipping or sliding.

Reverse dial indicator method An alignment method that uses two dial indicators to take readings off of opposing sides of coupling rims, giving angular and parallel alignment readings.

Rigid coupling A mechanical component that couples two shafts that allows no flexibility within the coupling.

Rim-and-face alignment method An alignment method in which the parallel and angular offset of two shafts is determined using two dial indicators that measure the rim and face of a coupling.

Rippling Plastic flow that occurs from heavy loads, vibration, or improper lubrication.

Roller bearing An antifriction bearing that has parallel or tapered steel rollers retained between inner and outer rings.

Roller chain A synchronous drive chain that contains roller, pin, bushing, and side plates connected in a series, used to transmit power.

Roller link A chain link that consists of two bushings placed inside two rollers that are pressed into two side plates.

Rolling-contact (anti-friction) bearing A bearing composed of rolling elements between an outer and inner ring.

Rolling element bearing A machined, cylinder-shaped component containing an inner ring, outer ring, steel balls, or rollers seperated by a retainer/cage.

Rolling friction Friction that occurs when a rolling roller or ball moves on a stationary surface.

Root The bottom section or valley of a thread or gear tooth.

Root clearance Distance between the top of the tooth of one gear and the bottom of the meshing space.

Running torque The energy that a motor develops to keep a load turning. A measure of roundness.

Run-out A radial variation from a true circle, both radial and axial.

Saybolt Seconds Universal (SSU) Unit of measure for industrial lubricant viscosity. May also be referred to as *Saybolt Universal Seconds* (SUS).

Screw An inclined plane placed on a cylinder.

Scuffing The marring of metal surfaces.

Seal A device intended to contain pressure, prevent leakage, and exclude contaminants.

Sealed bearing An antifriction bearing that is completely sealed so the lubricant stays inside the bearing and contaminants are kept out.

Serpentine belt (double-V or hex belt) A belt designed to transmit power from the top and bottom of the belt.

Service factor A number applied to the load, power, or torque to reflect the overall operating parameters and conditions the power transmission component will function in.

Service life The length of service received from a bearing or power transmission component.

Shackle A steel horseshoe-shaped attchment with a removable pin used to make connections between rigging and hoisting devices.

Shear Stress caused by two equal and parallel forces acting upon an object from opposite directions.

Shim A thin precision piece of material, usually metal, used to elevate and align a machine.

Shim stock Steel, brass, or plastic material manufactured in various thicknesses, ranging from .0005″ to .125″.

Silent chain A synchronous chain that consists of a series of links joined together with a bushing and pin.

Sintered bronze bearings Sleeve-type bearings made from a powdered bronze composite and impregnated with oil.

Sleeve bearing A bearing in which the shaft turns and is lubricated by a sleeve.

Sliding friction Friction that occurs when one surface moves across another or both surfaces move in opposite directions.

Sling A wire, chain, or synthetic webbed device designed into a configuration for hoisting or lifting.

Soft foot A condition that occurs when one or more machine feet do not make complete and even contact with the mounting base plate.

Solid lubricant A non-liquid material such as graphite, molybdenum disulfide, or polytetrafluoroethylene (PTFE) that is used to lubricate mating surfaces.

Spalling The flaking off of the surface of a gear tooth or bearing, often due to metal fatigue.

Spectrum A representative combination of the amplitude (total movement) and frequency (time span) of a waveform.

Spiral bevel gear A bevel gear that has curved teeth cut on a cone, which provide smooth operation at high speeds.

Spring coupling A shaft coupling that uses the flexing of a spring to accommodate shaft misalignment.

Sprocket A wheel with evenly spaced, uniformly shaped teeth located around the perimeter of the wheel.

Spur gear A gear with straight teeth cut parallel to the shaft axis.

Starting torque The energy required to start a load turning after it has been broken away from a standstill.

Static friction The resistance to begin movement between two contacting surfaces.

Static load A load that remains constant.

Static seals Designed for service where there is little or no movement between mating surfaces. Gaskets and o-rings are the most common examples of static seals.

Straightedge alignment method A method of flexible shaft coupling alignment in which a straight steel rule is used to align the couplings or shafts.

Synchronous drive system A drive system that provides a positive engagement between the drive and driven sides of the system.

System A combination of components, units, or modules that are intended to move, transform, or alter objects or energy as well as perform work.

Tagout The process of placing a tag on a power source that warns others not to restore energy.

Tagout device A prominent warning device, such as a tag and a means of attachment, which can be securely fastened to an energy isolating device in accordance with an established procedure, to indicate that the energy isolating device and the equipment being controlled may not be operated until the tagout device is removed.

Tap A hardened steel thread-cutting and cleaning tool with external threads in the shape of a tapered rod.

Taper gauge A flat, tapered strip of metal with graduations in thousandths of an inch or millimeters marked along its length.

Taper-lock bushing Type of bushing that is tapered and has no flange. Used to mount power transmission components to a shaft. It is assembled with setscrews that cause the bushing to squeeze the shaft as they are tightened.

Tapered bore bearing A bearing whose bore varies in diameter from the face to the back of the bearing.

Tapered roller bearing A roller bearing having tapered rolling elements.

Temperature A measurement of the intensity of heat.

Tensile member Cording material that is usually braided and runs the entire length of the belt, that increases the tensile strength of the belt.

Tensile strength A measure of the greatest amount of straight-pull stress that metal can bear without tearing apart.

Tension Stress caused by two equal forces acting on the same axial line to pull an object apart.

Third-class lever A lever that has the fulcrum at the end of the rigid bar and the resistance at the opposite end.

Thrust damage Bearing damage due to axial force.

Thrust failure Damage due to excessive axial forces.

Tight-side tension The tension on a belt when it is approaching the drive pulley/sheave.

Timing belt A synchronous flat belt with trapezoidal or parabolic-shaped gear teeth.

Tolerances The amount of size differences allowed from a specific traget measurement. An allowable deviation from a specified number.

Tooth breakage The removal of a gear tooth or part of a tooth from a gear.

Tooth form The shape or geometric form of a tooth in a gear when seen from its profile.

Torque A force acting on a perpendicular radial distance from a point of rotation. A twisting force around an axis producing rotating motion.

Troubleshooting The systematic examination and diagnosis of a system, circuit, or process to locate a malfunctioning part and determine a corrective action.

Troughed belt conveyor Conveyor system using troughing idlers and conveyor pulleys to transport raw materials.

UL (Underwriters Laboratory) Organization that operates laboratories to investigate systems with respect to safety.

Universal joint A form of a flexible coupling that can include two yokes connected by a spider allowing for extensive misalignment.

Variable-speed belt drive A belted mechanism that transmits motion from one shaft to another and allows the speed of the shafts to be varied by changing the diameter of an adjustable pulley.

V-belt An endless drive belt made from rubber or synthetic material that has a cross-section in the shape of a V.

V-belt sheave A grooved pulley with a V-shape.

Vector A force with a magnitude and direction.

Vibration Motion in response to a force represented by upper and lower limits over a given time period. A continuous periodic change in displacement with respect to a fixed reference point.

Vibration amplitude The extent of vibration movement measured from a starting point to an extreme point.

Vibration analysis The monitoring of individual component or the complete machine's vibration characteristics to determine the component condition.

Vibration analyzer A black-box device containing microprocessors, circuitry, and meters, that assists in collecting, pinpointing, and analyzing a machine by identifying its unique vibration characteristics.

Vibration cycle Motion from a neutral position to the upper limit, from the upper limit to the lower limit, and from the lower limit back to the neutral position.

Vibration frequency The number of completed vibration cycles within a specified period of time.

Vibration signature A set of vibration readings inherent and particular in a machine resulting from tolerances and looseness within that machine.

Vibration velocity The speed or time required to travel from the highest peak of the vibration displacement to the neutral position, expressed in inches per second.

Viscosity The measure of the resistance of a fluid's molecules to move. A measure of fluid flow.

Viscosity index A scale used to show the magnitude of viscosity changes in lubrication oils with changes in temperature.

V-ring seal A lip seal shaped like the letter V.

Waveform A graphic presentation of an amplitude as a function of time.

Wedge A simple machine that is a variation of the inclined plane.

Wheel and axle A simple machine consisting of a wheel and an axle held together along the same axis.

Work The energy used when a force is exerted over a distance.

Work order A document that details work required to complete a specific maintenance task.

Worm A shank having at least one complete tooth/start winding around its circumference.

Worm gear A gear that mates with a worm.

English Standard Measures

Long Measure
1 mile = 1760 yards = 5280 feet.
1 yard = 3 feet = 36 inches.
1 foot = 12 inches.

Surveyor's Measure
1 mile = 8 furlongs = 80 chains.
1 furlong = 10 chains = 220 yards.
1 chain = 4 rods = 22 yards = 66 feet = 100 links.
1 link = 7.92 inches.

Square Measure
1 square mile = 640 acres = 6400 square chains.
1 acre = 10 square chains = 4840 square yards = 43,560 square feet.
1 square chain = 16 square rods = 484 square yards = 4356 square feet.
1 square rod = 30.25 square yards = 272.25 square feet = 625 square links.
1 square yard = 9 square feet.
1 square foot = 144 square inches.
An acre is equal to a square, the side of which is 208.7 feet.

Dry Measure
1 bushel (U.S. or Winchester struck bushel) = 1.2445 cubic foot = 2150.42 cubic inches.
1 bushel = 4 pecks = 32 quarts = 64 pints.
1 peck = 8 quarts = 16 pints.
1 quart = 2 pints.
1 heaped bushel = 1 1/4 struck bushel.
1 cubic foot = 0.8036 struck bushel.
1 British Imperial bushel=8 Imperial gallons=1.2837 cubic foot = 2218.19 cubic inches.

Liquid Measure
1 U.S. gallon = 0.1337 cubic foot = 231 cubic inches = 4 quarts = 8 pints.
1 quart = 2 pints = 8 gills.
1 pint = 4 gills.
1 British Imperial gallon=1.2003 U.S. gallon=277.27 cubic inches.
1 cubic foot = 7.48 U.S. gallons.
1 barrel = 31.5 gallons

Circular and Angular Measure
60 seconds (″)	=	1 minute (′).
60 minutes	=	1 degree (°).
360 degrees	=	1 circumference (C).
57.3 degrees	=	1 radian.
2 π radians	=	1 circumference (C).

Specific Gravity
The specific gravity of a substance is its weight as compared with the weight of an equal bulk of pure water.
For making specific gravity determinations the temperature of the water is usually taken at 62° F. when 1 cubic foot of water weighs 62.355 lbs. Water is at its greatest density at 39.20° F. or 4° Centigrade.

Temperature
The following equation will be found convenient for transforming temperature from one system to another:
Let F = degrees Fahrenheit; C = degrees Centigrade; R = degrees Reamur.
$$\frac{F-32}{180} = \frac{C}{100} = \frac{R}{80}$$

Avoirdupois or Commercial Weight
1 gross or long ton = 2240 pounds.
1 net or short ton = 2000 pounds.
1 pound = 16 ounces = 7000 grains.
1 ounce = 16 drams = 437.5 grains.

Measures of Pressure
1 pound per square inch = 144 pounds per square foot = 0.068 atmosphere = 2.042 inches of mercury at 62 degrees F. = 27.7 inches of water at 62 degrees F. = 2.31 feet of water at 62 degrees F.
1 atmosphere = 30 inches of mercury at 62 degrees F. = 14.7 pounds per square inch = 2116.3 pounds per square foot = 33.95 feet of water at 62 degrees F.
1 foot of water at 62 degrees F. = 62.355 pounds per square foot = 0.433 pound per square inch.
1 inch of mercury at 62 degrees F. = 1.132 foot of water = 13.58 inches of water = 0.491 pound per square inch.
Column of water 12 in. high, 1 in. dia. = .341 lbs.

Cubic Measure
1 cubic yard = 27 cubic feet.
1 cubic foot = 1728 cubic inches.
The following measures are also used for wood and masonry:
1 cord of wood = 4 X 4 X 8 feet = 128 cubic feet.
1 perch of masonry = 16-1/2 X 1-1/2 X 1 foot = 24-3/4 cubic feet.

Shipping Measure
For measuring entire internal capacity of a vessel: 1 register ton = 100 cubic feet.
For measurement of cargo:
1 U.S. shipping ton = 40 cubic feet = 32.143 U.S. bushels = 31.16 Imperial bushels.
British shipping ton = 42 cubic feet = 33.75 U.S. bushels = 32.72 Imperial bushels.

Troy Weight, Used for Weighing Gold and Silver
1 pound = 12 ounces = 5760 grains.
1 ounce = 20 pennyweights = 480 grains.
1 pennyweight = 24 grains.
1 carat (used in weighing diamonds) = 3.086 grains.
1 grain Troy = 1 grain avoirdupois = 1 grain apothecaries' weight.

Measure Used for Diameters and Areas of Electric Wires
1 circular inch = area of circle 1 inch in diameter = 0.7854 square inch.
1 circular inch = 1,000,000 circular mils.
1 square inch = 1.2732 circular inch = 1,273,239 circular mils.
A circular mil is the area of a circle 0.001 inch in diameter.

Board Measure
One foot board measure is a piece of wood 12 inches square by 1 inch thick, or 144 cubic inches. 1 cubic foot therefore equals 12 feet board measure.

Metric System of Measurements

Measures of Length

10	millimeters (mm.)	=	1 centimeter (cm.)
10	centimeters	=	1 decimeter (dm.)
10	decimeters	=	1 meter (m.)
1000	meter	=	1 kilometer (km.)

Measures of Weight

10	milligrams (mg.)	=	1 centigram (cg.)
10	centigrams	=	1 decigram (dg.)
10	decigrams	=	1 gram (g.)
10	grams	=	1 decagram (Dg.)
10	decagrams	=	1 hectogram (Hg.)
10	hectograms	=	1 Kilogram (Kg.)
1000	kilograms	=	1 (metric) ton (T.)

Surveyor's Square Measure

100	square meters (m.^{2})	=	1 are (ar.)
100	acres	=	1 hectare (har.)
100	hectares	=	1 sq. kilometer (Km.^{2})

Square Measure

100	sq. millimeters (mm.^{2})	=	1 sq. centimeter (cm.^{2})
100	sq. centimeters	=	1 sq. decimeter (dm.^{2})
100	sq. decimeters	=	1 sq. meter (m.^{2})

Cubic Measure

1000	cu. millimeters (mm.^{3})	=	1 cu. centimeter (cm.^{3})
1000	cu. centimeters	=	1 cu. decimeter (dm.^{3})
1000	cu. decimeters	=	1 cu. meter (m.^{3})

Dry and Liquid Measure

10	milliliters (ml.)	=	1 centiliter (cl.)
10	centiliters	=	1 deciliter (dl.)
10	deciliters	=	1 liter (l.)
100	liters	=	1 hectoliter (Hl.)

1 liter = 1 cubic decimeter = the volume of 1 kilogram of pure water at a temperature of 39.2 degrees F.

Length Conversion Constants for Metric and U.S. Units

Millimeters X .039370 = inches.
Meters x 39.370 = inches.
Meters X 3.2808 = feet.
Meters X 1.09361 = yards.
Kilometers X 3,280.8 = feet.
Kilometers X .62137 = Statute Miles.
Kilometers x .53959 = Nautical Miles.

Inches X 25.4001 = millimeters.
Inches X .0254 = meters.
Feet x .30480 = meters.
Yards X .91440 = meters.
Feet x .0003048 = kilometers.
Statute Miles X 1.60935 = kilometers.
Nautical Miles x 1.85325 = kilometers.

Weight Conversion Constants for Metric and U.S. Units

Grams X 981 = dynes.
Grams X 15.432 = grains.
Grams X .03527 = ounces (Avd.).
Grams x .033818 = fluid ounces (water).
Kilograms X 35.27 = ounces (Avd.).
Kilograms X 2.20462 = pounds (Avd.).
Metric Tons (1000 Kg.) X 1.10231 = Net Ton (2000 lbs.).
Metric Tons (1000 Kg.) X .98421 = Gross Ton (2240 lbs.).

Dynes X .0010193 = grams.
Grains X .0648 = grams.
Ounces (Avd.) X 28.35 = grams.
Fluid Ounces (Water) X 29.57 = grams.
Ounces (Avd.) X .02835 = kilograms.
Pounds (Avd.) X .45359 = kilograms.
Net Ton (2000 lbs.) X .90719 = Metric Tons (1000 Kg.).
Gross Ton (2240 lbs.) X 1.01605 = Metric Tons (1000 Kg.).

Area Conversion Constants for Metric and U.S. Units

Square Millimeters X .00155 = square inches.
Square centimeters X .155 = square inches.
Square Meters X 10.76387 = square feet.
Square Meters X 1.19599 = square yards.
Hectares X 2.47104 = acres.
Square Kilometers X 247.104 = acres.
Square Kilometers X .3861 = square miles.

Square Inches X 645.163 = square millimeters.
Square Inches x 6.45163 = square centimeters.
Square Feet x .0929 = square meters.
Square Yards X .83613 = square meters.
Acres X .40469 = hectares.
Acres X .0040469 = square kilometers.
Square Miles X 2.5899 = square kilometers.

Volume Conversion Constants for Metric and U.S. Units

Cubic centimeters X .033818 = fluid ounces.
Cubic centimeters X .061023 = cubic inches.
Cubic centimeters X .271 = fluid drams.
Liters X 61.023 = cubic inches.
Liters X 1.05668 = quarts.
Liters X .26417 = gallons.
Liters X .035317 = cubic feet.
Hectoliters X 26.417 = gallons.
Hectoliters X 3.5317 = cubic feet.
Hectoliters X 2.83794 = bushel (2150.42 cu. in.).
Hectoliters X .1308 = cubic yards.
Cubic Meters x 264.17 = gallons.
Cubic Meters X 35.317 = cubic feet.
Cubic Meters X 1.308 = cubic yards.

Fluid Ounces X 29.57 = cubic centimeters.
Cubic Inches X 16.387 = cubic centimeters.
Fluid Drams x 3.69 = cubic centimeters.
Cubic Inches X .016387 = liters.
Quarts x .94636 = liters.
Gallons x 3.78543 = liters.
Cubic Feet X 28.316 = liters.
Gallons x .0378543 = hectoliters.
Cubic Feet X .28316 = hectoliters.
Bushels (2150.42 cu. in.) X .352379 = hectoliters.
Cubic Yards x 7.645 = hectoliters.
Gallons x .00378543 = cubic meters.
Cubic Feet x .028316 = cubic meters.
Cubic Yards x .7645 = cubic meters.

Power and Heat Conversion Constants for Metric and U.S. Units

Calorie x 0.003968 = B.T.U.
Joules X .7373 = pound-feet.
Newton-Meters X 8.851 = pound-inches.
Cheval Vapeur X .9863 = Horsepower.
Kilowatts X 1.34 = Horsepower.
Kilowatt Hours X 3415 = B.T.U.
(Degrees Cent. X 1.8) +32 = degrees Fahr.
(Degrees Reamur X 2.25) + 32 = degrees Fahr.

B.T.U. X 252 = calories.
Pound-Feet X 1.3563 = joules.
Pound-inches X .11298 = Newton-meters.
Horsepower X 1.014 = Cheval Vapeur.
Horsepower X .746 = kilowatts.
B.T.U. X .00029282 = kilowatt hours.
(Degrees Fahr. - 32) x .555 = degrees Cent.
(Degrees Fahr. - 32) x .444 = degrees Reamur.

Torque and Horsepower Equivalents

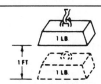

A foot-pound is the amount of energy expended in lifting a one-pound mass a distance of one foot against the pull of gravity.

**FOOT-POUNDS
INDICATE ENERGY**

$$\text{Torque (in Pound-Inches)} = \frac{63{,}025 \times HP}{RPM}$$
$$= \text{Force} \times \text{Lever Arm (In Inches)}$$

$$\text{Torque (in Pound-Feet)} = \frac{5{,}252 \times HP}{RPM}$$
$$= \text{Force} \times \text{Lever Arm (In Feet)}$$

Force = Working Load in Pounds.
FPM = Feet Per Minute.
RPM = Revolutions Per Minute.
Lever Arm = Distance from the Force to the center of rotation in Inches or Feet.

TORQUE

**It is: a turning moment or twisting effort.
Is it expressed in foot-pounds? or pound-feet?**

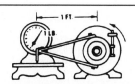

A pound-foot is the moment created by a force of one pound applied to the end of a lever arm one foot long.

**POUND-FEET
INDICATE TORQUE**

Example:—
25 HP at 150 RPM = 10504 Pound-Inches Torque
2.5 HP at 150 RPM = 1050.4 Pound-Inches Torque

For other values of RPM move decimal point in RPM values to the left or right as desired, and in Torque values move to the right or left (opposite way) the same number of places.

Example:—
25 HP at 150 RPM = 10504 Pound-Inches Torque
25 HP at 1.50 RPM = 1050400 Pound-Inches Torque
2.5 HP at 1.50 RPM = 105040 Pound-Inches Torque

HORSEPOWER

**Common Unit of Mechanical power. (HP)
One HP is the rate of work required to raise 33,000 pounds one foot in one minute.**

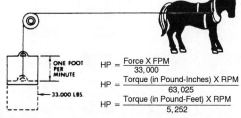

$$HP = \frac{\text{Force} \times FPM}{33{,}000}$$

$$HP = \frac{\text{Torque (in Pound-Inches)} \times RPM}{63{,}025}$$

$$HP = \frac{\text{Torque (in Pound-Feet)} \times RPM}{5{,}252}$$

Overhung Loads

An overhung load is a bending force imposed on a shaft due to the torque transmitted by V-drives, chain drives and other power transmission devices, other than flexible couplings.

Most motor and reducer manufacturers list the maximum values allowable for overhung loads. It is desirable that these figures be compared with the load actually imposed by the connected drive.

Overhung loads may be calculated as follows:

$$\text{O.H.L.} = \frac{63{,}000 \times HP \times F}{N \times R}$$

Where HP = Transmitted hp $\times$ service factor.
N = RPM of shaft.
R = Radius of sprocket, pulley, etc.
F = Factor.

Weights of the drive components are usually negligible. The formula is based on the assumption that the load is applied at a point equal to one shaft diameter from the bearing face. Factor F depends on the type of drive used:

$$F = \begin{cases} 1.00 \text{ for single chain drives.} \\ 1.3 \text{ for TIMING Belt Drives and HTD belt Drives.} \\ 1.25 \text{ for spur or helical gear or double chain drives.} \\ 1.50 \text{ for V-belt drives.} \\ 2.50 \text{ for flat belt drives.} \end{cases}$$

Example: Find the overhung load imposed on a reducer by a double chain drive transmitting 7 hp @ 30 RPM. The pitch diameter of the sprocket is 10"; service factor is 1.3.

Solution:

$$\text{O.H.L.} = \frac{(63{,}000)(7 \times 1.3)(1.25)}{(30)(5)} = 4{,}780 \text{ lbs.}$$

Viscosity Classification Equivalents

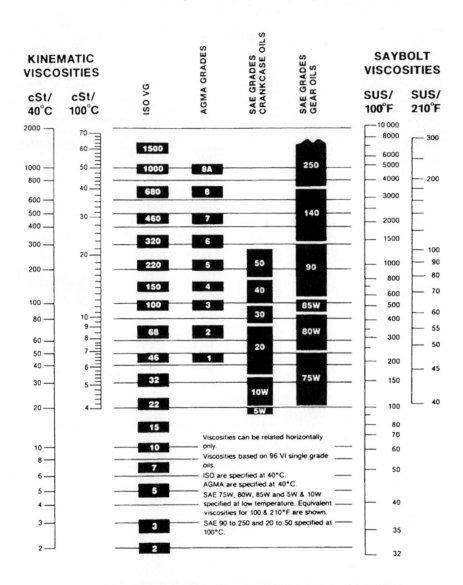

ISO VISCOSITY CLASSIFICATION SYSTEM

All industrial oils are graded according to the ISO Viscosity Classification System, approved by the International Standards Organizations (ISO). Each ISO viscosity grade number corresponds to the mid-point of viscosity range expressed in centistokes (cSt) at 40°C. For example, a lubricant with an ISO grade of 32 has a viscosity within the range of 28.80-35.2, the midpoint of which is 32.

Rule-of-Thumb: The comparable ISO grade of a competitive product whose viscosity in SUS at 100°F is known can be determined by using the following conversion formula:

$$\text{SUS @ 100°F} \div 5 = \text{cSt @ 40°C}$$

Electrical

Table 1—Electrical Formulas

To Find	Alternating Current		To Find	Alternating or Direct Current
	Single-Phase	Three-Phase		
Amperes when horsepower is known	$\dfrac{Hp \times 746}{E \times Eff \times pf}$	$\dfrac{Hp \times 746}{1.73 \times E \times Eff \times pf}$	Amperes when voltage and resistance is known	$\dfrac{E}{R}$
Amperes when kilowatts are known	$\dfrac{Kw \times 1000}{E \times pf}$	$\dfrac{Kw \times 1000}{1.73 \times E \times pf}$	Voltage when resistance and current are known	IR
Amperes when Kva are known	$\dfrac{Kva \times 1000}{E}$	$\dfrac{Kva \times 1000}{1.73 \times E}$	Resistance when voltage and current are known	$\dfrac{E}{I}$
Kilowatts	$\dfrac{I \times E \times pf}{1000}$	$\dfrac{1.73 \times I \times E \times pf}{1000}$	**General Information (Approximation)**	
Kva	$\dfrac{I \times E}{1000}$	$\dfrac{1.73 \times I \times E}{1000}$	At 1800 rpm, a motor develops 36 lb.-in per hp At 1200 rpm, a motor develops 54 lb.-in per hp At 575 volts, a 3-phase motor draws 1 amp per hp At 460 volts, a 3-phase motor draws 1.25 amp per hp At 230 volts, a 3-phase motor draws 2.5 amp per hp	
Horsepower = (Output)	$\dfrac{I \times E \times Eff \times pf}{746}$	$\dfrac{1.73 \times I \times E \times Eff \times pf}{746}$	At 230 volts, a single-phase motor draws 5 amp per hp At 115 volts, a single-phase motor draws 10 amp per hp	

I - Amperes; E = Volts; Eff = Efficiency; pf = power factor; Kva = Kilovolt amperes; Kw = Kilowatts; R = Ohms.

Temperature Conversion:
Deg C = (Deg F - 32) X 5/9
Deg F = (Deg C X 9/5) + 32

(General Information column note: All Values At 100% Load)

Table 2—AC Motor Recommended Wire Size

Volts	Motor Horsepower																					
	1-3	5	7¹/₂	10	15	20	25	30	40	50	60	75	100	125	150	200	250	300	350	400	450	500
230	14	12	10	8	6	4	3	1	0	000	000	300	500	...	...	...	...	...	...	...	...	...
460	14	14	14	12	10	8	6	6	4	3	2	0	000	0000	300	500	700	900	1500	600*	750*	900*
575	14	14	14	14	12	10	8	6	6	4	3	2	0	000	0000	250	500	600	800	1000	1500	600*

Insure that the requirements of the National Electric Code are fully met in all installations. This table is included as a guide only and is based on 3 phase, continuous duty, design B, standard efficiency motors using 600 volt Insulation, Type THW, with individual cooper conductors run in rigid conduit as defined in the 1987 NEC.

Table 3—Motor Amps @ Full Load†

HP	Alternating Current		DC	HP	Alternating Current		DC	HP	Alternating Current		DC	HP	Alternating Current		DC
	Single-phase	3-phase			Single-phase	3-phase			Single-phase	3-phase			Single-phase	3-phase	
1/2	4.9	2.0	2.7	5	28	14.4	20	25	....	60	92	75	....	180	268
1	8.0	3.4	4.8	7-1/2	40	21.0	29	30	....	75	110	100	....	240	355
1-1/2	10.0	4.8	6.6	10	50	26.0	38	40	....	100	146	125	....	300	443
2	12.0	6.2	8.5	15	....	38.0	56	50	....	120	180	150	....	360	534
3	17.0	8.6	12.5	20	....	50.0	74	60	....	150	215	200	....	480	712

† Values are for all speeds and frequencies @ 230 volts. Amperage other than 230 volts can be figured:

$$A = \frac{230 \times Amp\ from\ Table}{New\ Voltage}$$

Example:

For 60 hp, 3 phase @ 550 volts : $\dfrac{(230 \times 150)}{550} = 62$ amps.

Power Factor estimated @ 80% for most motors. Efficiency is usually 80-90%.

Table 4—NEMA Electrical Enclosure Types

Type	Description	Type	Description
NEMA Type 1 (General Purposes)	For indoor use wherever oil, dust or water is not a problem.	NEMA Type 5 Dust Tight (Non-Hazardous)	Used for excluding dust. (All NEMA 12 enclosures are usually suitable for NEMA 5 use.)
NEMA Type 2 (Driptight)	Used indoors to exclude falling moisture and dirt.	NEMA Type 9 Dust Tight (Hazardous) ‡	For locations where combustible dusts are present.
NEMA Type 3 (Weatherproof)	Provides protection against rain, sleet and snow.		
NEMA Type 4 (Watertight) ◆	Needed when subject to great amounts of water from any angle—such as areas which are repeatedly hosed down.	NEMA Type 12 (Industrial Use)	Used for excluding oil, coolant, flying dust, lint, etc.

◆ Not designed to be submerged.
‡ Class II Groups E, F and G.

A-C MOTOR INFORMATION

Table 5—Frame Assignments

HP	Motor Speed, rpm				HP	Motor Speed, rpm			
	3600	1800	1200	900		3600	1800	1200	900
1/8	...	48	...	...	15	254T,256U	254T,284U	284T,284TS,324U	286T,326U
1/6	...	48	...	...	20	256T,286U	256T,286U	286T,286TS,326U	324T,364U
1/4	48	48	48	56	25	284TS,324US	284T,284TS,324U	324T,324TS,364U	326T,365U
1/3	48	48,56	56	56	30	286TS,326US	286T,284TS,326U	326T,326TS,365U	364T,404U
1/2	48,56	48,56	56	56	40	324T,364US	324T,324TS,364U	354T,404U	365T,405U
3/4	56	56	56,143T,182U	56,145T	50	326TS,365US	326T,326TS,365U	365T,405U	404T
1	56,143T,182U	56,143T,182U	56,143T,184U	182T	60	364TS	364T,364TS	404T	405T
1-1/2	56,143T,182U	56,145T,184U	145T,184U	184T	75	365TS	365T,365TS	405T	444T
2	56,145T,184U	56,145T,184U	184T,213U	213T	100	405TS	405T,405TS	444T	445T
3	56,145T,182T,184U	182T,213U	213T,215U	215T,254U	125	444TS	444T,444TS	445T	...
5	184T,213U	184T,215U	215T,254U	254T,256U	150	445TS	445T,445TS	...	...
7-1/2	213T,215U	213T,254U	254T,256U	256T,284U	200	445TS	445T,445TS	...	...
10	215T,254U	215T,256U	256T,284U	284T,286U	250	447TS	447T,447TS	...	...

Table 6—Motor Frame Dimensions

Frame Size	D	E	2F	H Dia. (4) Holes	U Dia.	BA	V Min.	Key
48	3	2-1/8	2-3/4	11/32	1/2	2-1/2	...	3/64 Flat
56	3-1/2	2-7/16	3	11/32	5/8	2-3/4	...	3/16 x 3/16 x 1-3/8
143T	3-1/2	2-3/4	4	11/32	7/8	2-1/4	2	3/16 x 3/16 x 1-3/8
145T	3-1/2	2-3/4	5	11/32	7/8	2-1/4	2	3/16 x 3/16 x 1-3/8
182T	4-1/2	3-3/4	4-1/2	13/32	1-1/8	2-3/4	2-1/2	1/4 x 1/4 x 1-3/4
184T	4-1/2	3-3/4	5-1/2	13/32	1-1/8	2-3/4	2-1/2	1/4 x 1/4 x 1-3/4
213T	5-1/4	4-1/4	5-1/2	13/32	1-3/8	3-1/2	3-1/8	5/16 x 5/16 x 2-3/8
215T	5-1/4	4-1/4	7	13/32	1-3/8	3-1/2	3-1/8	5/16 x 5/16 x 2-3/8
254U	6-1/4	5	8-1/4	17/32	1-3/8	4-1/4	3-1/2	5/16 x 5/16 x 2-3/4
254T	6-1/4	5	8-1/4	17/32	1-5/8	4-1/4	3-3/4	3/8 x 3/8 x 2-7/8
256U	6-1/4	5	10	17/32	1-3/8	4-1/4	3-1/2	5/16 x 5/16 x 2-3/4
256T	6-1/4	5	10	17/32	1-5/8	4-1/4	3-3/4	3/8 x 3/8 x 2-7/8
284U	7	5-1/2	9-1/2	17/32	1-5/8	4-3/4	4-5/8	3/8 x 3/8 x 3-3/4
284T	7	5-1/2	9-1/2	17/32	1-7/8	4-3/4	4-3/8	1/2 x 1/2 x 3-1/4
284TS	7	5-1/2	9-1/2	17/32	1-5/8	4-3/4	3	3/8 x 3/8 x 1-7/8
286U	7	5-1/2	11	17/32	1-5/8	4-3/4	4-5/8	3/8 x 3/8 x 3-3/4
286T	7	5-1/2	11	17/32	1-7/8	4-3/4	4-3/8	1/2 x 1/2 x 3-1/4
286TS	7	5-1/2	11	17/32	1-5/8	4-3/4	3	3/8 x 3/8 x 1-7/8
324U	8	6-1/4	10-1/2	21/32	1-7/8	5-1/4	5-3/8	1/2 x 1/2 x 4-1/4
324T	8	6-1/4	10-1/2	21/32	2-1/8	5-1/4	5	1/2 x 1/2 x 3-7/8
324TS	8	6-1/4	10-1/2	21/32	1-7/8	5-1/4	3-1/2	1/2 x 1/2 x 2
326U	8	6-1/4	12	21/32	1-7/8	5-1/4	5-3/8	1/2 x 1/2 x 4-1/4
326T	8	6-1/4	12	21/32	2-1/8	5-1/4	5	1/2 x 1/2 x 3-7/8
326TS	8	6-1/4	12	21/32	1-7/8	5-1/4	3-1/2	1/2 x 1/2 x 2
364U	9	7	11-1/4	21/32	2-1/8	5-7/8	6-1/8	1/2 x 1/2 x 5
364US	9	7	11-1/4	21/32	1-7/8	5-7/8	3-1/2	1/2 x 1/2 x 2
364T	9	7	11-1/4	21/32	2-3/8	5-7/8	5-5/8	5/8 x 5/8 x 4-1/4
364TS	9	7	11-1/4	21/32	1-7/8	5-7/8	3-1/2	1/2 x 1/2 x 2
365U	9	7	12-1/4	21/32	2-1/8	5-7/8	6-1/8	1/2 x 1/2 x 5
365US	9	7	12-1/4	21/32	1-7/8	5-7/8	3-1/2	1/2 x 1/2 x 2
365T	9	7	12-1/4	21/32	2-3/8	5-7/8	5-5/8	5/8 x 5/8 x 4-1/4
365TS	9	7	12-1/4	21/32	1-7/8	5-7/8	3-1/2	1/2 x 1/2 x 2
404U	10	8	12-1/4	13/16	2-3/8	6-5/8	6-7/8	1/2 x 1/2 x 5-1/2
404US	10	8	12-1/4	13/16	2-1/8	6-5/8	4	1/2 x 1/2 x 2-3/4
404T	10	8	12-1/4	13/16	2-7/8	6-5/8	7	3/4 x 3/4 x 5-5/8
404TS	10	8	12-1/4	13/16	2-1/8	6-5/8	4	1/2 x 1/2 x 2-3/4
405U	10	8	13-3/4	13/16	2-3/8	6-5/8	6-7/8	5/8 x 5/8 x 5-1/2
405US	10	8	13-3/4	13/16	2-1/8	6-5/8	4	1/2 x 1/2 x 2-3/4
405T	10	8	13-3/4	13/16	2-7/8	6-5/8	7	3/4 x 3/4 x 5-5/8
405TS	10	8	13-3/4	13/16	2-1/8	6-5/8	4	1/2 x 1/2 x 2-3/4
444U	11	9	14-1/2	13/16	2-7/8	7-1/2	8-3/8	3/4 x 3/4 x 7
444US	11	9	14-1/2	13/16	2-1/8	7-1/2	4	1/2 x 1/2 x 2-3/4
444T	11	9	14-1/2	13/16	3-3/8	7-1/2	8-1/4	7/8 x 7/8 x 6-7/8
444TS	11	9	14-1/2	13/16	2-3/8	7-1/2	4-1/2	5/8 x 5/8 x 3
445U	11	9	16-1/2	13/16	2-7/8	7-1/2	8-3/8	3/4 x 3/4 x 7
445US	11	9	16-1/2	13/16	2-1/8	7-1/2	4	1/2 x 1/2 x 2-3/4
445T	11	9	16-1/2	13/16	3-3/8	7-1/2	8-1/4	7/8 x 7/8 x 6-7/8
445TS	11	9	16-1/2	13/16	2-3/8	7-1/2	4-1/2	5/8 x 5/8 x 3

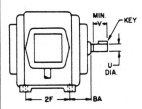

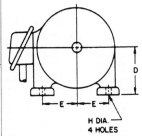

Table 1—Decimal and Millimeter Equivalents of Fractions

Inches (Fractions)	Decimals	Milli-meters	Inches (Fractions)	Decimals	Milli-meters	Inches (Fractions)	Decimals	Milli-meters
1/64	.015625	.397	11/32	.34375	8.731	11/16	.6875	17.463
1/32	.03125	.794	23/64	.359375	9.128	45/64	.703125	17.859
3/64	.046875	1.191	3/8	.375	9.525	23/32	.71875	18.256
1/16	.0625	1.588	25/64	.390625	9.922	47/64	.734375	18.653
5/64	.078125	1.984	13/32	.40625	10.319	3/4	.750	19.050
3/32	.09375	2.381	27/64	.421875	10.716	49/64	.765625	19.447
7/64	.109375	2.778	7/16	.4375	11.113	25/32	.78125	19.844
1/8	.125	3.175	29/64	.453125	11.509	51/64	.796875	20.241
9/64	.140625	3.572	15/32	.46875	11.906	13/16	.8125	20.638
5/32	.15625	3.969	31/64	.484375	12.303	53/64	.828125	21.034
11/64	.171875	4.366	1/2	.500	12.700	27/32	.84375	21.431
3/16	.1875	4.763	33/64	.515625	13.097	55/64	.859375	21.828
13/64	.203125	5.159	17/32	.53125	13.494	7/8	.875	22.225
7/32	.21875	5.556	35/64	.546875	13.891	57/64	.890625	22.622
15/64	.234375	5.953	9/16	.5625	14.288	29/32	.90625	23.019
1/4	.250	6.350	37/64	.578125	14.684	59/64	.921875	23.416
17/64	.265625	6.747	19/32	.59375	15.081	15/16	.9375	23.813
9/32	.28125	7.144	39/64	.609375	15.478	61/64	.953125	24.209
19/64	.296875	7.541	5/8	.625	15.875	31/32	.96875	24.606
5/16	.3125	7.938	41/64	.640625	16.272	63/64	.984375	25.003
21/64	.328125	8.334	21/32	.65625	16.669	1	1.000	25.400
			43/64	.671875	17.066			

Table 2—Millimeter-Inch Equivalents; 1 —254MM (.03937″—10.0″)

Milli-meter	Decimal	Milli-meter	Decimal	Milli-meter	Decimal	Milli-meter	Decimal	Milli-meter	Decimal
1	.03937	52	2.04724	103	4.05511	154	6.06299	205	8.07086
2	.07874	53	2.08661	104	4.09448	155	6.10236	206	8.11023
3	.11811	54	2.12598	105	4.13385	156	6.14173	207	8.14960
4	.15748	55	2.16535	106	4.17322	157	6.18110	208	8.18897
5	.19685	56	2.20472	107	4.21259	158	6.22047	209	8.22834
6	.23622	57	2.24409	108	4.25196	159	6.25984	210	8.26771
7	.27559	58	2.28346	109	4.29133	160	6.29921	211	8.30708
8	.31496	59	2.32283	110	4.33070	161	6.33858	212	8.34645
9	.35433	60	2.36220	111	4.37007	162	6.37795	213	8.38582
10	.39370	61	2.40157	112	4.40944	163	6.41732	214	8.42519
11	.43307	62	2.44094	113	4.44881	164	6.45669	215	8.46456
12	.47244	63	2.48031	114	4.48818	165	6.49606	216	8.50393
13	.51181	64	2.51968	115	4.52755	166	6.53543	217	8.54330
14	.55118	65	2.55905	116	4.56692	167	6.57480	218	8.58267
15	.59055	66	2.59842	117	4.60629	168	6.61417	219	8.62204
16	.62992	67	2.63779	118	4.64566	169	6.65354	220	8.66141
17	.66929	68	2.67716	119	4.68503	170	6.69291	221	8.70078
18	.70866	69	2.71653	120	4.72440	171	6.73228	222	8.74015
19	.74803	70	2.75590	121	4.76378	172	6.77165	223	8.77952
20	.78740	71	2.79527	122	4.80315	173	6.81102	224	8.81889
21	.82677	72	2.83464	123	4.84252	174	6.85039	225	8.85826
22	.86614	73	2.87401	124	4.88189	175	6.88976	226	8.89763
23	.90551	74	2.91338	125	4.92126	176	6.92913	227	8.93700
24	.94488	75	2.95275	126	4.96063	177	6.96850	228	8.97637
25	.98425	76	2.99212	127	5.00000	178	7.00787	229	9.01574
26	1.02362	77	3.03149	128	5.03937	179	7.04724	230	9.05511
27	1.06299	78	3.07086	129	5.07874	180	7.08661	231	9.09448
28	1.10236	79	3.11023	130	5.11811	181	7.12598	232	9.13385
29	1.14173	80	3.14960	131	5.15748	182	7.16535	233	9.17322
30	1.18110	81	3.18897	132	5.19685	183	7.20472	234	9.21259
31	1.22047	82	3.22834	133	5.23622	184	7.24409	235	9.25196
32	1.25984	83	3.26771	134	5.27559	185	7.28346	236	9.29133
33	1.29921	84	3.30708	135	5.31496	186	7.32283	237	9.33070
34	1.33858	85	3.34645	136	5.35433	187	7.36220	238	9.37007
35	1.37795	86	3.38582	137	5.39370	188	7.40157	239	9.40944
36	1.41732	87	3.42519	138	5.43307	189	7.44094	240	9.44881
37	1.45669	88	3.46456	139	5.47244	190	7.48031	241	9.48818
38	1.49606	89	3.50393	140	5.51181	191	7.51968	242	9.52755
39	1.53543	90	3.54330	141	5.55118	192	7.55905	243	9.56692
40	1.57480	91	3.58267	142	5.59055	193	7.59842	244	9.60629
41	1.61417	92	3.62204	143	5.62992	194	7.63779	245	9.64566
42	1.65354	93	3.66141	144	5.66929	195	7.67716	246	9.68503
43	1.69291	94	3.70078	145	5.70866	196	7.71653	247	9.72440
44	1.73228	95	3.74015	146	5.74803	197	7.75590	248	9.76378
45	1.77165	96	3.77952	147	5.78740	198	7.79527	249	9.80315
46	1.81102	97	3.81889	148	5.82677	199	7.83464	250	9.84252
47	1.85039	98	3.85826	149	5.86614	200	7.87401	251	9.88189
48	1.88976	99	3.89763	150	5.90551	201	7.91338	252	9.92126
49	1.92913	100	3.93700	151	5.94488	202	7.95275	253	9.96063
50	1.96850	101	3.97637	152	5.98425	203	7.99212	254	10.00000
51	2.00787	102	4.01574	153	6.02362	204	8.03149	...	....

VOLUMETRIC FLOW RATES

gallons per minute, US (gpm)	liters per second (l per sec)	0.008434
	cubic feet per minute (cfm)	0.1337
	cubic feet per hour (cu ft per hr)	8.022
gallons per minute, UK or Canadian (gpm)	liters per second (l per sec)	0.0101
	cubic feet per minute (cfm)	0.1606
	cubic feet per hour (cu ft per hr)	9.634
cubic feet per second (cfs)	gpm (UK or Canadian)	373.77
	gpm (US)	448.86
	liters per second (l per sec)	1699.2
liters per second (l per sec)	cubic feet per minute (cfm)	2.119
	gpm (UK or Canadian)	13.20
	gpm (US)	15.85
millions of gallons per day, US (MGD)	liters per second (l per sec)	43.81
	cubic feet per minute (cfm)	92.85
	gallons per minute, US (gpm)	694.44

PRESSURE

Pascals (Pa)	pounds per square inch (psi)	0.0001450
	pounds per square foot (lb per ft^2)	0.02089
	newtons per square meter	1
pounds per square inch (psi)	atmospheres, std. (atm)	0.0680
	pounds per square foot (lb per ft^2)	144
	Pascals (Pa)	6894.8
	foot of water (ft of H_2O) 60F	2.301
atmospheres (atm), standard	psi	14.70
	lb per ft^2	2116.8
	Pa	101325
inch of water, 60F (in of H_2O)	psi	0.03609
	lb per ft^2	5.197
	Pa	248.84
foot of water, 60F (ft of H_2O)	psi	0.4331
	lb per ft^2	62.36
	Pa	2985.9

WEIGHT, MASS, INERTIA

pounds (lb)*	kilograms (kg)	0.4536
	ounces (oz)	16
kilograms (kg)	pounds (lb)	2.205
	ounces (oz)	35.27

WEIGHT, MASS, INERTIA, continued

COLUMN A

To Convert From...	To	Multiply Col. A by
tons (short)	metric tons	0.9072
	kilograms (kg)	907.2
	pounds (lb)	2000
metric tons	tons (short)	1.102
	kilograms	1000
	pounds	2205
pounds, weight (lb)	slugs, mass (lb-sec^2 per ft)	0.03106
pound-foot^2 (lb-ft^2)	kilogram-meters^2 (kg-m^2)	0.04214

*pounds and ounces are avoirdupois

FORCE AND TORQUE

pounds (lb)	newtons(N)	4.448
newtons (N)	pounds (lb)	0.2248
newton-meters (N-m)	pound-feet (lb-ft)	0.7376
	pound-inches (lb-in)	8.851
	ounce-inches (oz-in)	141.60

ounce-inches (oz-in)	lb-ft	0.005208
	N-m	0.007062
	lb-in	0.0625
pound-inches (lb-in)	lb-ft	0.0833
	N-m	0.11298
	oz-ln	16
pound-feet (lb-ft)	N-m	1.356
	lb -in	12
	oz-ln	192

POWER

horsepower (hp)	kilowatts (kW)	0.7457
	foot-pounds per second (ft-lb per sec)	550
	foot-pounds per minute (ft-lb per min.)	33000
kilowatts (kW)	horsepower (hp)	1.341

FORMULAS AND CONSTANTS†

1 HP = 33,000 Foot-pounds of work per minute.
1 HP = .746 K.W. = K.W.÷ 1.341.
1 HP = 2547 B.T.U. per hour.
1 B.T.U. = Heat required to raise 1 lb. water 1°F.
1 B.T.U. = 777.6 Foot-pounds work.
1 Kilowatt Hour = 3415 B.T.U.
Heat Value of Carbon = 14,600 B.T.U. per pound.
Latent Heat of Fusion of Ice = 143.15 B.T.U. per pound.
Latent Heat of Evaporation of Water at 212°F. =
    970.4 B.T.U. per pound.
Total Heat of Saturated Steam at atmospheric pressure =
    1,150.4 B.T.U. per pound.
1 Ton of Refrigeration = 288,000 B.T.U. per 24 hours.
g = Acceleration of Gravity (commonly taken as 32.16
    feet per second per second).
1 Radian = 57.296 degrees.

† Also Look In General Index Under Weights, Measures, Or The Subject Material Required

TEMPERATURE

		Use This Relationship
degrees Fahrenheit (F)	degrees Celsius (C)	C =5/9 (F-32)
degrees Celsius (C)	degrees Fahrenheit (F)	F=9/5C+32
degrees Fahrenheit (F)	degrees Rankine (R)	R =F+459.69
degrees Celsius (C)	degrees Kelvin (K)	K=C+273.16

Examples: 1. Convert 12F to C.   C = 5/9 (F-32) = 5/9 (12—32) = 5/9 (-20)
                      Answer = -11.1C
                   2. Convert 40C to F.   F = 9/5C + 32 = 9/5 (40) + 32 = 72 + 32
                      Answer = 104F

V-Belt Drive Formulas

V-belt tensioning In cases where tensioning of a drive effects belt pull and bearing loads, the following formulas may be used.

$$T_1\text{-}T_2 = 33,000\left(\frac{HP}{V}\right)$$

where: T_1 = tight side tension, pounds
T_2 = slack side tension, pounds
HP = design horsepower
V = belt speed, feet per minute

$$T_1 + T_2 = 33,000\,(2.5\text{–}G)\left(\frac{HP}{GV}\right)$$

where: T_1 = tight side tension, pounds
T_2 = slack side tension, pounds
HP = design horsepower
V = belt speed, feet per minute
G = arc of contact correction factor*

$$T_1/T_2 = \frac{1}{1\text{–}0.8G} \ (\text{Also } T_1/T_2 = eK\varnothing)$$

where: T_1 = tight side tension, pounds
T_2 = slack side tension, pounds
G = arc of contact correction factor*
e = base of natural logarithms
K = .51230, a constant for V-belt drive design
$\varnothing$ = arc of contact in radians

$$T_1 = 41,250\left(\frac{HP}{GV}\right)$$

where: T_1 = tight side tension, pounds
HP = design horsepower
V = belt speed, feet per minute
G = arc of contact correction factor*

$$T_2 = 33,000\,(1.25\text{–}G)\left(\frac{HP}{GV}\right)$$

where: T_2 = slack side tension, pounds
HP = design horsepower
V = belt speed, feet per minute
G = arc of contact correction factor*

Belt Speed

$$V = \frac{(PD)\,(rpm)}{3.82} = (PD\,(rpm)\,(.262))$$

where: V = belt speed, feet per minute
PD = pitch diameter of sheave or pulley
rpm = revolutions per minute of the same sheave or pulley

*See Table 1, at left.

Table 1—Arc of Contact Correction Factors G and R

$\frac{D\text{-}d}{C}$	Small Sheave Arc of Contact	Factor G	Factor R	$\frac{D\text{-}d}{C}$	Small Sheave Arc of Contact	Factor G	Factor R
.00	180°	1.00	1.000	.80	133°	.87	.917
.10	174°	.99	.999	.90	127°	.85	.893
.20	169°	.97	.995	1.00	120°	.82	.866
.30	163°	.96	.989	1.10	113°	.80	.835
.40	157°	.94	.980	1.20	106°	.77	.800
.50	151°	.93	.968	1.30	99°	.73	.760
.60	145°	.91	.954	1.40	91°	.70	.714
.70	139°	.89	.937	1.50	83°	.65	.661

D = Diam. of large sheave. d = Diam. of small sheave.
C = Center distance.

Table 2—Allowable Sheave Rim Speed

Sheave Material	Rim Speed in Feet per Minute
Cast Iron	6,500
Ductile Iron	8,000
Steel	10,000

Note: Above rim speed values are maximum for normal considerations. In some cases these values may be exceeded. Consult factory and include complete details of proposed application.

Bearing Load Calculations

To find actual bearing loads it is necessary to know machine component weights and values of all other forces contributing to the load. Sometimes it becomes desirable to know the bearing load imposed by the V-belt drive alone. This can be done if you know bearing spacing with respect to the sheave center and shaft load and apply it to the following formulas:

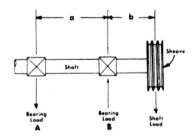

Overhung Sheave

Load at B, lbs. $= \dfrac{\text{Shaft Load X }(a+b)}{a}$

Load at A, lbs. $= \text{Shaft Load X }\dfrac{b}{a}$

Where : a and b = Spacing, inches

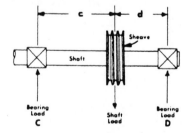

Sheave Between Bearings

Load at D, lbs. $= \dfrac{\text{Shaft Load X }c}{c+d}$

Load at C, lbs. $= \dfrac{\text{Shaft Load X }d}{c+d}$

Where : c and d = Spacing, inches